Selected Titles in This Series

(*Continued in the back of this publication*)

Nonstandard Methods in Commutative Harmonic Analysis

Translations of
MATHEMATICAL
MONOGRAPHS

Volume 164

Nonstandard Methods in Commutative Harmonic Analysis

E. I. Gordon

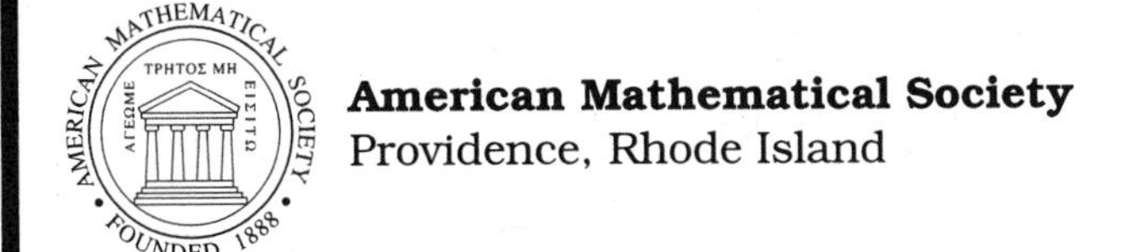

American Mathematical Society
Providence, Rhode Island

Е. И. Гордон

НЕСТАНДАРТНЫЕ МЕТОДЫ В НЕКОММУТАТИВНОМ
ГАРМОНИЧЕСКОМ АНАЛИЗЕ

Translated from an original Russian manuscript by H. H. McFaden.

The investigations in this book were supported in part by the
Russian Foundation for Fundamental Research, grant no. 95-01-00673.

1991 *Mathematics Subject Classification.* Primary 43–02, 03–01, 26E35, 43A25, 03H05;
Secondary 28E05, 47S20, 42A38, 22B05, 22E35, 54J05, 03E70, 46S10, 46F99.

ABSTRACT. This monograph contains some investigations by the author in the field of nonstandard analysis (NSA) and its applications to harmonic analysis on locally compact Abelian (LCA) groups. A new notion of approximation of topological groups by finite groups is introduced and investigated. On the basis of this notion some new results are obtained about convergence of finite Fourier transformations (FT) to the FT on an LCA group. These results are proved by means of NSA. They are formulated in standard terms in the Introduction. Some new results about nonstandard methods are also included in the book: the theory of relatively standard elements, and extensions of results about S-integrable liftings in Loeb measure spaces to the case of σ-finite Loeb measures. The basic concepts of NSA are given in Chapter 0.

Library of Congress Cataloging-in-Publication Data

Gordon, E. I. (Evgenii Izrailevich), 1949–
Nonstandard methods in commutative harmonic analysis / E. I. Gordon.
p. cm. — (Translations of mathematical monographs, ISSN 0065-9282 ; v. 164)
Translated from the original Russian manuscript.
Includes bibliographical references (p. –).
ISBN 0-8218-0419-7 (alk. paper)
1. Harmonic analysis. 2. Nonstandard mathematical analysis. I. Title. II. Series.
QA403.G67 1997
515′.785—dc21 97-7187
CIP

∞ The paper used in this book is acid-free and falls within the guidelines
established to ensure permanence and durability.
Visit the AMS homepage at URL: `http://www.ams.org/`

10 9 8 7 6 5 4 3 2 1 02 01 00 99 98 97

This book is dedicated to the fond memory of my parents, the mathematicians I. I. Gordon and N. A. Gubar′.

Contents

Notation

Preface

Nonstandard analysis, which was discovered by Robinson in the 1960's, has found diverse applications in many areas of mathematics in the course of the past few decades: for example, the theory of Banach spaces, stochastic analysis, singular perturbations of differential equations, and other areas. The known applications of nonstandard analysis are reflected most completely in the 1986 monograph *Nonstandard methods in stochastic analysis and mathematical physics* by Albeverio and his co-authors ([**1**]).

Nonstandard analysis has also stimulated the development of new investigations in the foundations of mathematics. Axiomatic systems for nonstandard set theory have been constructed, among which Nelson's internal set theory (IST) is foremost and best known. Somewhat by itself is the alternative set theory (AST) of Vopěnka, in which a completely new way of looking at infinity is developed. In this theory all sets are finite, though they can be very large, and this manifests itself in the violation of the induction principle: phenomena of the 'pile paradox' type were first formalized in AST.

An approach analogous to AST can also be formalized in a natural way in the framework of nonstandard analysis, with 'very large' sets being sets whose cardinality is an infinite natural number in the sense of nonstandard analysis. Such sets (they are called hyperfinite sets) can contain subclasses that are not sets, a fact connected with the violation of the induction principle. These subclasses are called semisets in AST, and external sets in nonstandard analysis. In this approach continuous objects are obtained by factorization of hyperfinite sets with respect to certain external equivalence relations that are understood intuitively as indiscernibility relations (as they are called in AST).

Many important applications of nonstandard analysis to the investigation of continuous and other infinite objects are based on precisely such a 'finite' modeling of them. Here we should mention first and foremost the construction of Loeb measures, which permits the modeling of countably additive measures by measures on hyperfinite sets, thus making it possible to use the intuition of elementary probability theory more directly in the study of complicated random processes.

An analogous approach to the construction of a theory of Fourier series on the basis of nonstandard analysis was developed in the 1972 paper *A nonstandard analysis approach to Fourier analysis* by one of the founders of nonstandard analysis, W. A. J. Luxemburg [**45**]. The basic idea of this approach is to approximate the unit circle by the group of Nth roots of unity, where N is an infinite natural number, and to approximate the Fourier transformation (FT) on the unit circle by the Fourier transformation on this group.

In the author's papers the approach of Luxemburg was extended to arbitrary locally compact Abelian (LCA) groups. The results obtained there are presented in this book.

The construction of a commutative harmonic analysis using nonstandard analysis is based on a treatment of hyperfinite Abelian groups. Two special subgroups are singled out in such a group: the subgroup of bounded elements and the subgroup of infinitesimal elements. Then a (standard) LCA group is obtained as the factor group of the first by the second, and the Fourier transformation is constructed on the LCA group from the Fourier transformation on the hyperfinite group. The latter has all the basic properties of the Fourier transformation on the usual finite groups.

Some development of the techniques of nonstandard analysis was required, and that is also reflected in the monograph. This includes an extension of the theory of integration with respect to the Loeb measure to the case when the internal measure of the whole space is infinite, and the theory of relative standardness, which generalizes the Benninghofen–Richter–Stroyan theory of superinfinitesimals. The concept of relative standardness is formulated in the framework of Nelson's internal set theory. This also has a certain interest from the point of view of axiomatic set theory, since it leads to a proof that one of the basic postulates of IST, the standardization principle, is independent of the rest of its axioms.

This construction can easily be formalized in the framework of AST, and it is in agreement with the constructions of topological spaces in that theory on the basis of the bi-equivalences studied by Zlatoš and his colleagues. However, such a construction of commutative harmonic analysis apparently provides the first example of a substantive mathematical theory that has been constructed in the framework of the alternative set theory and turns out to be completely adequate for the corresponding classical theory. We do not present the alternative set theory in the book, but we do give an axiomatics of nonstandard set theory in whose framework it can easily be interpreted. The axiomatics presented is new and stands in the same relation to IST as the familiar von Neumann–Bernays–Gödel axiomatics to the Zermelo–Fraenkel axiomatics.

The results obtained have a natural interpretation in terms of classical mathematics. In essence, the question investigated here is the approximability of the Fourier transformation on an LCA group by the Fourier transformation on a finite group. For this we introduce the new concept of an approximation of a topological group by finite groups in which the imbeddings of these finite groups in the topological group being approximated are not homomorphisms, but converge to homomorphisms in some natural sense. This concept enables us to also encompass LCA groups that, in contrast to the unit circle, do not contain arbitrarily dense finite subgroups. In fact, any LCA group turns out to be approximable in the indicated sense. This can be easily understood by starting from the fact that both inductive and projective limits are particular cases of the approximation introduced. In the book's Introduction this definition of approximation is formulated along with the basic results in classical terms without any use of the language of nonstandard analysis.

It is of interest to consider the question of extending the constructions described to the noncommutative case. Here the situation is considerably more complicated, because far from all locally compact noncommutative groups can be approximated by finite groups in the sense described above. At the present time it is known only that approximable groups must be unimodular, but this condition is not sufficient. Of course, inductive and projective limits of finite groups can be approximated in our sense in view of the above remarks, but this is unknown for the group SO(3), for example. We remark that, by analogy with the Fourier transformation on LCA groups, the irreducible unitary representations of approximable compact groups are approximated by irreducible unitary representations of the approximating finite groups. Therefore, the question of approximability of topological groups is of interest not only from the point of view of the 'nonstandard' foundations of mathematics, but also for concrete problems connected with the approximation of operators acting in spaces of functions on topological groups and homogeneous spaces.

The reader is not assumed to have a preliminary knowledge of nonstandard analysis, nor even of the elements of mathematical logic. In this connection we included a Chapter 0, which contains the basic concepts and results of nonstandard analysis. In the process of working on the book this chapter grew considerably in length, and is a short introductory course in nonstandard analysis intended for mathematicians, including students of advanced university mathematics courses. The difference between this course and other textbooks is due to the basic content of the book, which, as is clear from the preceding, is oriented not only toward applications of nonstandard methods, but also toward questions of the foundations. On the other hand, the author has tried to take into account the interests of the reader who is not attracted to the formalized exposition that is typical for courses in mathematical logic.

Nonstandard analysis has one feature that causes a certain amount of difficulty in its perception and even a certain repulsion for mathematicians working in analysis, geometry, and other areas far from mathematical logic. This feature is that the expression of substantive mathematical propositions in some formal logical language is here one of the basic instruments for carrying out proofs and obtaining new results. Although it is clear at present that an exposition of nonstandard analysis is possible without such a formalization yet with the level of rigor commonly acceptable in mathematics (see H. J. Keisler's book, *Elementary calculus: an approach using infinitesimals*, Prindle, Weber, and Schmidt, 1976), we had to forgo this method of presentation here, in part because of the above questions relating to the foundations of mathematics.

Either the language of superstructures or the language of formal set theory is usually taken as a formal logical language in courses in nonstandard analysis. To more easily overcome the difficulties mentioned in the preceding paragraph we have chosen another language here; namely, the language of elementary analysis described in Manin's book *The provable and the unprovable* [**59**]. Although this language is less universal than the other languages mentioned, it suffices for the formalization of all the necessary propositions. At the same time the very formalization in it is, in the author's opinion, essentially more simple and natural than that in more universal languages. The first section of Chapter 0 is devoted to examples of the expression of substantive mathematical propositions in the formal language of elementary analysis. It seems that the contents of this section may be useful irrespective of nonstandard analysis in view of the ever more widespread use of formalization in mathematics, a fact connected with programming languages and with typesetting systems like TeX.

The sixth section of the first chapter deals with standard and nonstandard axiomatic systems in set theory. The readers interested only in approximation questions, that is, applications of nonstandard analysis, can skip this section with scarcely any detriment to the understanding of what follows. The basic techniques of nonstandard analysis are developed in the third and fourth sections of Chapter 0. There we postulate the existence of nonstandard universes possessing all the necessary properties such as concurrence, saturation, and so on. The construction of such universes is carried out in the fifth section, which can also be skipped by the same category of readers. On the other hand, although the concrete form of nonstandard universes is not used in studying their properties, the knowledge of it can facilitate such study. In this case we recommend reading through the fifth section directly after the third.

Except for the aforementioned axiomatics of nonstandard set theory, Chapter 0 does not contain any new results and bears a textbook character. On the other hand, the results making up the remaining two chapters are mainly due to the author. The nonstandard analysis needed for the techniques to follow is developed in Chapter I. Chapter II is devoted to commutative harmonic analysis proper. The whole exposition there is in terms of nonstandard analysis, but in the last subsection all the assertions obtained with the help of the well-known techniques of Nelson are carried over to the standard language. The basic approximation results formulated in the Introduction are thereby proved.

Acknowledgements

First of all, I would like to recall with gratitude my late teachers, Yu. V. Glebskiĭ and D. A. Gudkov. Under the guidance of Yu. V. Glebskiĭ I began my scientific work and studied mathematical logic. I worked about twenty years in the department headed by D. A. Gudkov, constantly under his protection and always aware of his concern for me.

V. A. Lyubetskiĭ taught me axiomatic set theory. I learned nonstandard analysis from the lectures and seminars of A. G. Dragalin at Moscow University. Much of the content of this book was developed from ideas expounded by him at that time.

My many years of collaboration with S. S. Kutateladze and A. G. Kusraev has had great significance for the whole of my research work, including that reflected in this book. The numerous discussions of various aspects of my book with S. Albeverio, M. A. Antonets, A. M. Vershik, M. Wolff, P. Zlatoš, V. G. Kanoveĭ, V. M. Tikhomirov, S. N. Samborskiĭ, I. A. Shereshevskiĭ, and M. A. Shubin have been very useful. I express my deep gratitude to them all.

I am also very grateful to my wife, I. N. Gordon, for her great help in the preparation of the manuscript.

The investigations in this book were supported in part by the Russian Foundation for Fundamental Research, grant no. 95-01-00673.

Introduction

1°. The history of nonstandard analysis goes back all the way to Leibnitz; it is covered in [**20**]. Here we recall only that in contrast to Newton, who regarded infinitesimals as variable quantities tending to zero in the process of their variation, Leibnitz understood them as certain ideal constant quantities "less than any quantity that can be given". Newton's approach was taken as a rigorous basis for analysis, and by the end of the nineteenth century there was the firm conviction that the constant infinitesimals of Leibnitz could not be introduced in analysis on the level of mathematical rigor then accepted, or, as Luzin said, "in the calculus". As a result of the development of mathematical logic, this idea was refuted by Robinson, who constructed an ordered extension of the field $\mathbb{R}$ that in a certain sense preserves the basic properties of $\mathbb{R}$, but that at the same time contains elements greater than all real numbers, infinite numbers, as well as their opposites, infinitesimal numbers. The fact that the extension ${}^*\mathbb{R}$ constructed (this is the customary notation) preserves the basic properties of the field $\mathbb{R}$ permits analysis to be developed in it; in particular, it can be used to prove the theorems of ordinary real analysis in a way similar to the way theorems on integers are proved in analytic number theory by passing from the ring $\mathbb{Z}$ to its extension the field $\mathbb{C}$. Analysis in ${}^*\mathbb{R}$ has received the name (unfortunate, in my opinion) nonstandard analysis.

It is not difficult to construct an ordered extension of the field $\mathbb{R}$: for example, we can consider the field $\mathbb{R}(x)$ of rational functions of a variable x and order it by setting $0 < x < a$ $\forall a \in \mathbb{R}$, $a > 0$. It is well known that this ordering can be extended in a unique way to the whole field $\mathbb{R}(x)$, and the element $x \in \mathbb{R}(x)$ is infinitesimal, since it is positive and less than all positive real numbers in the ordering. However, it does not seem to be possible to develop analysis in this field—not even the elementary functions can be extended to it.

In attempting to construct an ordered extension of $\mathbb{R}$ in which analysis can be developed, one should reckon with the fact that all proper extensions of $\mathbb{R}$ lack a basic property of it: the least upper bound axiom. Indeed, the set of all infinitesimals cannot have a least upper bound, since it is trivial to show that both the assumption that this least upper bound is an infinitesimal and the assumption that it is not an infinitesimal lead directly to a contradiction. On the other hand, the set is obviously bounded above by any positive real number.

This obstacle is overcome in nonstandard analysis as follows. In the field ${}^*\mathbb{R}$ one singles out a sufficiently broad class of subsets called internal sets that enjoys the least upper bound property, that is, any internal subset of ${}^*\mathbb{R}$ that is bounded above has a least upper bound. What is more, any true proposition about $\mathbb{R}$ carries over to ${}^*\mathbb{R}$ if in it any expression of the form "for any set" or "there exists a set" is replaced by "for any internal set" or "there exists an internal set", respectively. This last assertion is called the transfer principle. In order to formulate it precisely we must describe a formal language in which the propositions of analysis are written, and give a precise definition of the truth of propositions in the language. This will be done in §§1–3 of Chapter 0, but now we describe one of the possible ways of constructing the field $^*\mathbb{R}$.

Let μ be a two-valued finitely additive measure on the algebra $\mathcal{P}(\mathbb{N})$ of all subsets of $\mathbb{N}$, that is, it takes only the values 0 and 1. On the partially ordered ring $\mathbb{R}^{\mathbb{N}}$ we introduce the relation $\sim$ of coincidence almost everywhere with respect to the measure μ, that is, $x_n \sim y_n \iff \mu(\{n \mid x_n = y_n\}) = 1$. The quotient ring of the ring $\mathbb{R}^{\mathbb{N}}$ by the ideal of elements equivalent to zero will now be a linearly ordered field $^*\mathbb{R}$. It is easy to see that $^*\mathbb{R}$ really is a linearly ordered field. For if it is false that $x_n \sim 0$, then the measure of the set $A = \{n \mid x_n = 0\}$ is not equal to 1, that is, it is equal to 0 (since μ is two-valued), which means that an element y_n equal to x_n^{-1} for $n \notin A$ and arbitrary for $n \in A$ has the property that $x_n \cdot y_n = 1$ almost everywhere. This shows that a nonzero element of $^*\mathbb{R}$ is invertible. It can be established similarly that $^*\mathbb{R}$ is linearly ordered.

The field $\mathbb{R}$ is imbedded in $^*\mathbb{R}$ in the obvious way: associated with each number is the class in $^*\mathbb{R}$ of the sequence identically equal to this number. The images in $^*\mathbb{R}$ of the real numbers are denoted by the numbers themselves. If the measure μ has the additional property that the measure of each finite set is zero, then $^*\mathbb{R}$ is actually a proper extension of $\mathbb{R}$. In particular, the classes in $^*\mathbb{R}$ of sequences converging to zero are infinitesimal elements, while the classes of sequences converging to infinity are infinite elements. But if μ does not have this property, then its two-valuedness implies the existence of an $n \in \mathbb{N}$ with $\mu(\{n\}) = 1$, and then $^*\mathbb{R}$ coincides with $\mathbb{R}$.

Each sequence of subsets A_n of $\mathbb{R}$ determines a subset $\widetilde{A} \subseteq {}^*\mathbb{R}$ such that $\{x_n\} \in \widetilde{A} \iff x_n \in A_n$ for almost all $n \in \mathbb{N}$. Here $\{x_n\}$ is the image in $^*\mathbb{R}$ of the sequence x_n. Such sets $\widetilde{A}$ are called internal sets. The fact that each internal set that is bounded above has a least upper bound can be established as follows. It is easy to see that the boundedness of $\widetilde{A}$ is equivalent to the boundedness of A_n in $\mathbb{R}$ for almost all n. The element $\{x_n\} \in {}^*\mathbb{R}$ with $x_n = \sup A_n$ for almost all n is now the least upper bound of the set $\widetilde{A}$ in $^*\mathbb{R}$. An internal function $\widetilde{f}\colon {}^*\mathbb{R} \to {}^*\mathbb{R}$ is defined similarly in terms of sequences of functions $f_n\colon \mathbb{R} \to \mathbb{R}$.

If $A_n = A$ for all or almost all $n \in \mathbb{N}$, then the set $\widetilde{A}$ is called the nonstandard extension of the set A and is denoted by $^*A$. The nonstandard extension $^*f\colon {}^*\mathbb{R} \to {}^*\mathbb{R}$ of a function $f\colon \mathbb{R} \to \mathbb{R}$ is defined similarly. It is clear that $t \in {}^*A \iff t \in A$ for any real number t, and that $f(x) = y \longrightarrow {}^*f(x) = y$ for any real numbers x and y. Thus, even all real functions, and not just elementary ones, extend to $^*\mathbb{R}$.

If all the sets A_n are finite, then $\widetilde{A}$ is said to be hyperfinite. It is easy to show that if the cardinalities of the sets A_n are not uniformly bounded (at least almost everywhere), then $\widetilde{A}$ is infinite, and even uncountable. Nevertheless, in view of the transfer principle mentioned above, it has many properties of finite sets. This lies at the basis of diverse applications of nonstandard analysis. As an example we describe the construction of Loeb spaces [**44**].

Let $\widetilde{A}$ be a hyperfinite set, and consider the algebra $\mathfrak{A}$ of hyperfinite subsets of $\widetilde{A}$. A finitely additive measure ν can be defined on this algebra as follows. Suppose that the hyperfinite subset $\widetilde{B}$ of $\widetilde{A}$ is generated by the sequence B_n, and consider the numerical sequence $\nu_n = |B_n|/|A_n|$. Denote by $\mathcal{F}$ the ultrafilter of subsets $D \subseteq \mathbb{N}$ with $\mu(D) = 1$. Then $\nu(\widetilde{B}) = \lim_{\mathcal{F}} \nu_n$. It turns out that this measure can be extended to a countably additive measure on the σ-algebra $\sigma(\mathfrak{A})$ generated by

$\mathfrak{A}$. This σ-algebra does not consist solely of internal sets, of course, but it can be shown that each of its elements coincides up to a set of measure zero with some hyperfinite set in $\mathfrak{A}$. The space constructed is called a Loeb space.

If now $\widetilde{\nu}(\widetilde{B})$ is the class in ${}^*\mathbb{R}$ of the sequence ν_n, then $(\widetilde{A}, \mathfrak{A}, \widetilde{\nu})$ is an internal measure space with a hyperfinite finitely additive probability measure. In view of the transfer principle such spaces enjoy many of the properties of ordinary finite probability spaces. By the foregoing, the 'genuine' probability space $(\widetilde{A}, \sigma(\mathfrak{A}), \nu)$ is well approximated by the space $(\widetilde{A}, \mathfrak{A}, \widetilde{\nu})$. On the other hand, the spaces of the form $(\widetilde{A}, \sigma(\mathfrak{A}), \nu)$ are sufficiently complicated. It is known that many complicated random processes can be realized in them; for example, Brownian motion. It is clear from the preceding that such a realization enables us more directly to employ the intuition of elementary probability theory in the study of these processes. Numerous applications of Loeb spaces based on this can be found in the book [**1**]. Here (Chapter 0, §4) we use Loeb spaces as an example to realize Fürstenberg's well-known passage from subsets of $\mathbb{N}$ of positive upper density to measure spaces for an ergodic proof of the famous van der Waerden–Szemerédi theorem about arithmetic progressions (see, for example, the book [**75**] or the survey [**70**]). Loeb spaces allow this to be done intuitively in a clear and direct manner, and in a more general situation, for arbitrary amenable groups.

We consider one more type of internal set that is often used in applications of nonstandard analysis. If all the elements of a sequence A_n determining an internal set $\widetilde{A}$ are finite-dimensional spaces over, say, $\mathbb{R}$, then $\widetilde{A}$ itself is a linear space over ${}^*\mathbb{R}$, and hence also over $\mathbb{R}$. Such spaces are called hyperfinite-dimensional spaces. Again, if the dimensions of the spaces A_n are not uniformly bounded, at least up to sets of μ-measure zero, then the space $\widetilde{A}$ is infinite-dimensional, but it has by the transfer principle many properties of ordinary finite-dimensional spaces. Further, any infinite-dimensional space can be imbedded in a hyperfinite-dimensional space, a fact that also has been the basis for various applications. In particular, it is this path that led to the first well-known application of nonstandard analysis—the Bernstein–Robinson proof of Halmos's conjecture that if some polynomial in an operator is compact, then the operator has an invariant subspace.

The theory of Banach spaces often makes use of another construction that is similar to the construction of Loeb measures. We now assume that all the linear spaces A_n are normed, but no longer necessarily finite-dimensional. Then for each element $\{a_n\} \in \widetilde{A}$ we can define an element $\{\|a_n\|\} \in {}^*\mathbb{R}$ which it is natural to call the internal norm of $\{a_n\}$. Two subspaces (no longer internal) can now be singled out in $\widetilde{A}$: the subspace $\widetilde{A}_b$ of elements with bounded norm (that is, not infinite), and the subspace $\widetilde{A}_0$ of elements with infinitesimal norm. On the space $\widetilde{A}^\# = \widetilde{A}_b/\widetilde{A}_0$ it is then natural to define the norm of an element to be the limit of the sequence $\{\|a_n\|\}$ with respect to the ultrafilter $\mathfrak{F}$ defined above. It can be established that $\widetilde{A}^\#$ is a Banach space in this norm, and that it is nonseparable in the case when $\widetilde{A}$ is infinite-dimensional (see the preceding paragraph). If A is an ordinary Banach space, then ${}^*A^\#$ is called the nonstandard hull of A. Numerous examples of applications of nonstandard hulls in the theory of Banach spaces can be found in the survey [**39**]. In this book Banach spaces of the form $\widetilde{A}^\#$ with $\widetilde{A}$ a hyperfinite-dimensional space play a large role. For lack of a better term we also call them nonstandard hulls of the spaces $\widetilde{A}$.

A completely analogous treatment can be given for sets of the type $\widetilde{A}$ constructed with the help of a two-valued measure on the set of all subsets of an arbitrary set I, not necessarily $\mathbb{N}$. Moreover, it need not be required that all the sets A_n necessarily be subsets of $\mathbb{R}$. Constructions of this kind are called ultraproducts, and sets of the form $^*A$ are called ultrapowers. They arose in mathematical logic and first found applications in the general theory of algebraic systems [**76**], then also in analysis, and not even in the context of nonstandard analysis. Here we can point to, for instance, the survey [**40**], and also [**77**], where they were used to construct new examples of nonhyperfinite II_1 factors. However, these constructions did not become widespread in analysis except in connection with nonstandard analysis, apparently because of their complexity and extreme abstractness. In fact, the existence of finitely additive two-valued measures on $\mathcal{P}(\mathbb{N})$ that are equal to zero on all finite sets can be proved only with the help of the axiom of choice. Consequently, it is impossible to construct a single concrete example of such a measure!

Thus, from the point of view of applications one can regard nonstandard analysis as a system of concepts that develop intuition in working with ultraproducts. But certainly this is a very narrow way of looking at the subject. In fact, a model of the field $^*\mathbb{R}$ satisfying the requirements formulated at the beginning of this section, for example, the transfer principle, need not be realized in the form of an ultrapower of the field $\mathbb{R}$ and can even be given axiomatically in general. At the present time an axiomatic system has been worked out for nonstandard set theory. It will be considered briefly in §6 of Chapter 0. Of course, the other models of nonstandard analysis are equally nonconstructible. In particular, it is impossible to give an unambiguous description of a single concrete infinitesimal quantity. But this is in complete agreement with the idea of Leibnitz (mentioned at the very beginning), according to which standard numbers are quantities that can be well described, while infinitesimal quantities are ideal elements smaller than all standard elements, and hence they cannot be well described.

Therefore, from the point of view of the foundations of mathematics, nonstandard analysis represents a completely different approach to the idealizations lying at the basis of ideas about infinity. In this connection it is not without meaning to consider the following. At the present time we can regard as established that it would not have been impossible in principle to have based mathematical analysis from the very beginning 'on Leibnitz' and not 'on Newton', although the historical reasons why this was not done are perfectly understandable, of course, and we shall not discuss them here. One can imagine that if mathematics had developed in that way, then many constructions, especially in infinite-dimensional analysis, would have a different appearance. In particular, it is entirely possible that instead of Lebesgue measure a large role would have been played by measures on hyperfinite spaces, and that separable Hilbert spaces would have been replaced in prominence by nonstandard hulls of hyperfinite-dimensional Euclidean spaces, that is, again spaces of functions on hyperfinite sets.

The question is the following. To what degree can mathematics be constructed on the basis of hyperfinite sets, and how? These sets are an idealization of finite but very large sets. In nonstandard analysis the role of such very large natural numbers is played by elements of the set $^*\mathbb{N} \setminus \mathbb{N}$. If Ω is an element of this set, then it is easy to show that it is greater than all the standard natural numbers, and hence also larger than the real numbers. The set $\{n \in {}^*\mathbb{N} \mid n < \Omega\}$ is a typical example of a hyperfinite set. A characteristic feature here is the violation of the principle of mathematical induction—this set cannot be exhausted by successive enumeration from a single element. Indeed, the set of standard natural numbers satisfies all the premises of the induction principle yet is a proper subset of the indicated set. Here the reason is the same as in the violation of the least upper bound axiom in $^*\mathbb{R}$—the induction principle is valid for internal sets, but the set of all standard natural numbers is not an internal subset of $^*\mathbb{N}$. It can now be seen that in this approach to infinity a phenomenon is formalized that falls completely outside the realm of classical mathematics and that can be expressed, for example, in the facetious 'pile paradox'. Let us recall what the latter is.

A single grain of sand does not make up a pile of sand, of course. It is also clear that if n grains of sand do not make up a pile of sand, then neither do $n+1$ grains of sand. But when does a pile of sand appear? The answer is that a pile of sand contains such a large number of grains that it cannot be well described; that is, from our point of view the infinite natural number Ω is an element of the set $^*\mathbb{N} \setminus \mathbb{N}$.

On the level of set theory an attempt to answer this question is contained in the work of Vopěnka and his students on the alternative set theory (see, for example, [**78**]), which is essentially a theory of hyperfinite sets. The concluding §6 of Chapter 0 contains a rather different approach to the theory of hyperfinite sets that is more explicitly based on nonstandard analysis. But the main content of the book is a construction of commutative harmonic analysis on the basis of hyperfinite sets. It turns out that here it is possible to construct a theory that is completely adequate for the classical theory. However, some new concepts and results arise in classical commutative harmonic analysis. These results relate to the tabular approximation of operators acting in spaces of functions on LCA groups, mainly of operators connected with the Fourier transformation. In the second section of the Introduction the main concepts and the results obtained are formulated in purely classical terms, without any mention of nonstandard analysis.

From the point of view of the theory of hyperfinite sets our approach consists in the following. In a hyperfinite Abelian group G we single out two subgroups: the subgroup G_b of bounded elements, which is the union of a certain countable sequence of internal subsets of G (intuitively, the nth internal subset is the set of elements not more than n units away from zero, where n is a standard natural number), and G_0 is the subgroup of infinitesimal elements, which is the intersection of a countable sequence of internal sets (here the nth internal subset is intuitively the set of elements not more than n^{-1} units away from zero. A topology is defined on the factor group $G^\# = G_b/G_0$ in a certain natural way (the images of internal subsets of G_b form a base of closed sets). Under definite conditions on G_b and G_0 the group $G^\#$ is compact, and it turns out that any LCA group can be realized in this way. The Fourier transformation on an LCA group can be obtained from the Fourier transformation on a hyperfinite group also by means of certain formalizations. All this is described in detail in Chapter 2.

It is interesting that this very abstract construction leads to the perfectly concrete concept of approximation of topological groups by finite groups, which can be formulated in classical terms. Let us now proceed to its description.

2°. In this book we consider only approximation of locally compact groups, and, moreover, in this section it is assumed that the groups being approximated are separable.

The numbering of definitions and theorems in the Introduction is independent of the numbering in the other parts of the book.

DEFINITION 1. Let $\mathcal{G}$ be a locally compact group, and ρ a left-invariant metric on $\mathcal{G}$. We say that $\mathcal{G}$ is approximable by finite groups if for any compact set $K \subseteq \mathcal{G}$ and any $\varepsilon > 0$ there exist a finite group G and an injective mapping $j: G \to \mathcal{G}$ satisfying the following three conditions:

(1) $j(G)$ is an ε-net for K with respect to the metric ρ;

(2) $\rho(j(g_1, \cdot g_2^{\pm 1}), j(g_1) \cdot j(g_2)^{\pm 1}) < \varepsilon \ \forall g_1, g_2 \in j^{-1}(K)$;

(3) $j(e_G) = e$, where e_G and e are the identities of G and $\mathcal{G}$, respectively.

A pair (G, j) satisfying (1)–(3) will be called a (K, ε)-subgroup of the group $\mathcal{G}$. If $\mathcal{G}$ is compact, then a $(\mathcal{G}, \varepsilon)$-subgroup of $\mathcal{G}$ will simply be called an ε-subgroup. Compact groups are approximated only by ε-subgroups below.

If an increasing sequence of compact sets K_n is such that $\bigcup_{n=1}^{\infty} K_n = \mathcal{G}$, and the sequence $\varepsilon_n \to 0$, then a sequence of (K_n, ε_n)-subgroups of $\mathcal{G}$ is called an approximating sequence (AS).

We remark that (K, ε)-subgroups are not in general actual subgroups of $\mathcal{G}$, because the mapping j is not necessarily a homomorphism, but only close to a homomorphism in the sense of the condition (2) of Definition 1.

Obviously, the definition of approximability of a locally compact group and of an approximating sequence do not depend on the choice of the metric ρ.

For approximable groups the integral with respect to Haar measure can be represented as a limit of Riemann integral sums.

THEOREM 1. *Suppose that $\mathcal{G}$ is an approximable group, μ is left Haar measure on $\mathcal{G}$, U is a relatively compact neighborhood of the identity, $\mathcal{K}$ is the family of all compact subsets of $\mathcal{G}$, and (G_n, j_n) is an AS of $\mathcal{G}$. In this case if the function $f: \mathcal{G} \to \mathbb{C}$ is bounded, is continuous almost everywhere with respect to the measure μ, and satisfies the condition*

$$\begin{gathered} \forall \varepsilon > 0 \ \exists K \in \mathcal{K} \ \exists n_0 \ \forall n > n_0 \ \forall B \subseteq j_n^{-1}(\mathcal{G} \setminus K) \\ \Delta_n \cdot \sum_{g \in B} |f(j_n(g))| < \varepsilon, \end{gathered} \tag{1}$$

where $\Delta_n = \frac{\mu(U)}{|j_n^{-1}(U)|}$, then $f \in L_1(\mu)$, and

$$\int_{\mathcal{G}} f \, d\mu = \lim_{n \to \infty} \Delta_n \cdot \sum_{g \in G_n} f(j_n(g)). \tag{2}$$

Obviously, if $\mathcal{G}$ is a compact group, then the condition (1) holds automatically. A sequence Δ_n satisfying the equalities (1) and (2) is called a normalizing multiplier of the AS (G_n, j_n). It is clear that any sequence Δ'_n equivalent to the sequence Δ_n defined in the theorem can be chosen as a normalizing multiplier.

For compact groups $|G_n|^{-1}$ can always be chosen as a normalizing multiplier, and for discrete groups $\Delta_n \equiv 1$ can be chosen.

It is natural to call functions satisfying the conditions of Theorem 1 Riemann-integrable. The class of Riemann-integrable functions will be denoted by $\mathcal{R}(\mathcal{G})$. Although the definition of this class depends on the choice of the AS, for each concrete group $\mathcal{G}$ it is possible to get a criterion for Riemann integrability independent of the AS. Moreover, the class of Riemann-integrable functions on $\mathcal{G}$ always contains the class of compactly supported functions that are continuous and bounded almost everywhere with respect to Haar measure, and it simply coincides with this class in the case when $\mathcal{G}$ is compact.

It is clear from Theorem 1 that approximable locally compact groups are unimodular. However, this condition does not suffice for approximability, since there are even discrete groups that are not approximable. The class of discrete groups approximable by finite groups in the sense of Definition 1 was studied by Vershik and the author in **[79]**, where such groups are said to be locally imbeddable in the class of finite groups. It is easy to understand that a discrete group $\mathcal{G}$ is locally imbeddable in the class of finite groups if and only if each finite square cut out from its Cayley table can be supplemented to form the Cayley table of some finite group. This class contains all the locally residually finite groups, in particular, all the matrix groups. Thus, the classical example of a nonunimodular group—the semidirect product of the additive group $\mathbb{R}$ and the multiplicative group $\mathbb{R}_+$—is also an example of a group that is approximable as a discrete group but not approximable as a topological group. On the other hand, groups with finitely many generators and defining relations (finitely presented groups) belong to this class if and only if they are residually finite. This fact permits one to give examples of discrete groups that are not approximable. Such groups include, in particular, finitely presented groups for which the word problem is not solvable.

The question of a description of the class of approximable locally compact groups is open. It is clear that inductive and projective limits of finite groups are approximable. However, it is not known whether, for example, the group SO(3) is approximable. On the other hand, no examples of nonapproximable compact groups are known. For compact groups the property of approximability can turn out to be very useful in view of the fact that irreducible representations of an approximable compact group $\mathcal{G}$ are approximated by irreducible representations of finite groups approximating $\mathcal{G}$. To formulate the last assertion precisely we first give a definition.

DEFINITION 2. Suppose that $\mathcal{G}$ is a compact group, (G_n, j_n) is an AS of it, and Γ is a base of neighborhoods of the identity of $\mathcal{G}$. A sequence φ_n of finite-dimensional unitary representations of the groups G_n is said to be equicontinuous if

$$\forall \varepsilon > 0\ \exists U \in \Gamma\ \exists n_0\ \forall n > n_0\ \forall g, h \in G_n$$
$$(j_n(g \cdot h^{-1}) \in U \longrightarrow \|\varphi_n(g) - \varphi_n(h)\| < \varepsilon).$$

PROPOSITION 1. *If φ_n is an equicontinuous sequence of representations, then the set $\{\dim \varphi_n \mid n \in \mathbb{N}\}$ is finite.*

This proposition implies that an equicontinuous sequence of representations always has a subsequence for which all the representations have the same dimension. Since a subsequence of an AS is itself an AS, we shall always assume that the representations of an equicontinuous sequence of representations act in one and the same space.

Below we denote by $\mathbb{U}_n$ the group of unitary $m \times m$ matrices.

THEOREM 2. i) *Let $\{\varphi_n \colon G_n \to \mathbb{U}_m\}$ be an equicontinuous sequence of unitary representations of fixed dimension m. Then there exists a unitary representation*

$\{\varphi\colon \mathcal{G} \to \mathbb{U}_m\}$ of the group $\mathcal{G}$, denoted below by $\lim_{n\to\infty} \varphi_n$, such that if $\xi \in \mathcal{G}$ and $\lim_{n\to\infty} j_n(g_n) = \xi$ for a sequence $g_n \in G_n$, then $\varphi(\xi) = \lim \varphi_n(g_n)$.

If, furthermore, all the φ_n are irreducible, then so is φ.

ii) *If $\{\varphi\colon \mathcal{G} \to \mathbb{U}_m\}$ is an irreducible unitary representation of an approximable compact group $\mathcal{G}$, then for any AS (G_n, j_n) of $\mathcal{G}$ there exist a subsequence (G_{n_k}, j_{n_k}) and an equicontinuous sequence $\{\varphi_{n_k}\colon G_{n_k} \to \mathbb{U}_m\}$ of irreducible unitary representations that converges to φ.*

The main content of the book relates to LCA groups. Here basic questions about approximation are answered; in particular, any separable LCA group $\mathcal{G}$ is approximable. It turns out that the Fourier transformation on $\mathcal{G}$ is approximated by the Fourier transformation on the groups G_n. Before presenting the precise formulations we explain what we have in mind.

If a function f is given on $\mathcal{G}$, then we can form its table at the points of the set $j_n(G_n)$. We can apply the Fourier transformation on G_n to the function h_n obtained on G_n. As a result we obtain a function $\widehat{h}_n$ on the group $\widehat{G}_n$ dual to G_n. It turns out that there is an AS $(\widehat{G}_n, \widehat{j}_n)$ for the group $\widehat{\mathcal{G}}$ dual to $\mathcal{G}$ such that the table of the Fourier transform $\widehat{f}$ of f, computed at the points of the set $\widehat{j}_n(\widehat{G}_n)$, is approximated by the function $\widehat{h}_n$.

In approximating the Fourier transformation on $\mathbb{R}$ by a finite Fourier transformation it is common in practice to form the table of the Fourier transform of the function with step equal to l^{-1}, where l is the length of the interval on which the table of the function itself is formed. Here we refer to Kotel′nikov's theorem, which, first, relates only to functions with compact support, and, second, does not say anything about the approximation described above. Below it is shown that this rule itself is a particular case of the general approach studied in the present book.

Recall that the Fourier transformation $\mathcal{F}\colon L_2(\mathcal{G}) \to L_2(\widehat{\mathcal{G}})$ is defined by

$$\mathcal{F}(f)(\chi) = \int_{\mathcal{G}} f(g)\overline{\chi(g)}\, d\mu(g).$$

Here $\chi \in \widehat{G}$ is a character of $\mathcal{G}$, and μ is Haar measure on $\mathcal{G}$. We recall that $\widehat{\mathcal{G}}$ has a Haar measure $\widehat{\mu}$ such that the Fourier transformation preserves the inner product.

Suppose again that (G_n, j_n) is an AS of the group $\mathcal{G}$. Since $\mathcal{G}$ is commutative, all the G_n can also be assumed to be commutative, and since they are finite, $\widehat{G}_n$ is isomorphic to G_n. It turns out that $\widehat{\mathcal{G}}$ can be approximated by the groups $\widehat{G}_n$. Below we consider only such approximations of dual groups.

Accordingly, let $(\widehat{G}_n, \widehat{j}_n)$ be an AS of the group $\widehat{\mathcal{G}}$. We denote by $\widehat{\Gamma}$ the base of neighborhoods of the identity in $\widehat{\mathcal{G}}$ made up of neighborhoods of the form $\mathcal{U}(K, \varepsilon)$, where $K \in \mathcal{K}$ is a compact subset of $\mathcal{G}$, $\varepsilon > 0$, and $\mathcal{U}(K, \varepsilon) = \{\chi \in \widehat{\mathcal{G}} \mid |\chi(g) - 1| < \varepsilon\ \forall g \in K\}$. If $\gamma \in \widehat{G}_n$, then we say that γ occurs strongly in $\mathcal{U}(K, \varepsilon)$ ($\gamma \in_s \mathcal{U}(K, \varepsilon)$) if $\widehat{j}_n(\gamma) \in \mathcal{U}(K, \varepsilon)$, and that γ occurs weakly in $\mathcal{U}(K, \varepsilon)$ ($\gamma \in_w \mathcal{U}(K, \varepsilon)$) if $|\gamma(g)-1| < \varepsilon$ $\forall g \in j_n^{-1}(K)$.

We denote by $\widehat{\mathcal{K}}$ the family of all compact subsets of $\widehat{\mathcal{G}}$.

DEFINITION 3. An AS $(\widehat{G}_n, \widehat{j}_n)$ of the group $\widehat{\mathcal{G}}$ is said to be dual or conjugate to an AS (G_n, j_n) if the following two conditions hold:

1. $\forall \mathcal{U} \in \widehat{\Gamma}\ \exists \mathcal{U}' \in \widehat{\Gamma}\ \exists n_0 \in \mathbb{N}\ \forall n > n_0\ \forall \gamma \in \widehat{G}_n\ (\gamma \in_w \mathcal{U}' \longrightarrow \gamma \in_s \mathcal{U})$;

2. $\forall \varepsilon > 0\ \forall K \in \mathcal{K}\ \forall \widehat{K} \in \widehat{\mathcal{K}}\ \exists n_0 \in \mathbb{N}\ \forall n > n_0\ \forall g \in j_n^{-1}(K)\ \forall \gamma \in \widehat{j}_n^{-1}(\widehat{K})$

$$(|\widehat{j}_n(\gamma)(j_n(g)) - \gamma(g)| < \varepsilon).$$

THEOREM 3. *For any separable LCA group there are pairs of conjugate AS's.*

Suppose now that Δ_n is the number defined in Theorem 1. Let

$$\widehat{\Delta}_n = (|G_n| \cdot \Delta_n)^{-1}.$$

PROPOSITION 2. *For the AS $(\widehat{G}_n, \widehat{j}_n)$ the sequence $\widehat{\Delta}_n$ is a normalizing multiplier (see Theorem 1).*

Let $L_2(G_n)$ be the space of all complex-valued functions on G_n, with the norm $\|\varphi\|_n = \left(\Delta_n \cdot \sum_{g \in G_n} |\varphi(g)|^2\right)^{1/2}$, and let $L_2(\widehat{G}_n)$ be the space of all complex-valued functions on $\widehat{G}_n$, with the norm $\|\psi\|_{\hat{n}} = \left(\widehat{\Delta}_n \cdot \sum_{\gamma \in \widehat{G}_n} |\psi(\gamma)|^2\right)^{1/2}$.

Then the Fourier transformation on G_n is the operator $F_n\colon L_2(G_n) \to L_2(\widehat{G}_n)$ given by the formula

$$F_n(\varphi)(\gamma) = \Delta_n \cdot \sum_{g \in G_n} \varphi(g) \cdot \overline{\gamma(g)}.$$

The inverse operator is given by the formula

$$F_n^{-1}(\psi)(g) = \widehat{\Delta}_n \cdot \sum_{\gamma \in \widehat{G}_n} \psi(\gamma)\gamma(g).$$

Finally, if f is a complex-valued function on $\mathcal{G}$, then $T_n(f)$ denotes its table at the points of the set $j_n(G_n)$, that is, $T_n(f) = f \circ j_n$. Similarly, $\widehat{T}_n(\varphi)$ denotes the table at the points of $\widehat{j}_n(\widehat{G}_n)$ for a complex-valued function φ on $\widehat{\mathcal{G}}$, that is, $\widehat{T}_n(\varphi) = \varphi \circ \widehat{j}_n$.

We can now formulate the main approximation theorem.

THEOREM 4. *Suppose that (G_n, j_n) is an AS for the LCA group $\mathcal{G}$, $(\widehat{G}_n, \widehat{j}_n)$ is an AS of $\widehat{\mathcal{G}}$ dual to it, and f is a function on $\mathcal{G}$ that is almost everywhere continuous with respect to Haar measure μ and such that $|f|^2 \in \mathcal{R}(\mathcal{G})$ and $|\mathcal{F}(f)|^2 \in \mathcal{R}(\widehat{\mathcal{G}})$. Then*

$$\lim_{n \to \infty} \|\widehat{T}_n(\mathcal{F}(f)) - F_n(T_n(f))\|_{\hat{n}} = 0. \tag{3}$$

We now give examples of approximation of concrete LCA groups.

a) Let $\mathcal{G} = \mathbb{S}^1$ be the unit circle. It is convenient to represent it in the form of the half-open interval $(-1/2, 1/2]$ with addition modulo 1. Then $\widehat{\mathcal{G}}$ is isomorphic to $\mathbb{Z}$. Here it is convenient to take the mapping $k \to \chi_k$ as the isomorphism, where $\chi_k \in \widehat{\mathcal{G}}$ is such that $\forall \xi \in \mathcal{G}$

$$\chi_k(\xi) = \exp 2\pi i k \xi. \tag{4}$$

The additive groups of residues $\bmod n$, $n \in \mathbb{N}$, are taken as the approximating finite groups G_n, and it is convenient to confine ourselves to odd $n = 2m + 1$ and to represent G_n in the form $\{-m, \dots, m\}$. Then the isomorphism between G_n and $\widehat{G}_n$ is given by the mapping $k \to \gamma_k$, where $\gamma_k \in \widehat{G}_n$ is such that $\forall l \in G_n$

$$\gamma_k(l) = \exp \frac{2\pi i k l}{n}. \tag{5}$$

We now define $j_n\colon G_n \to \mathcal{G}$ by the formula

$$j_n(l) = \frac{l}{m}, \quad l \in \{-m, \dots, m\}.$$

Identifying $\widehat{\mathcal{G}}$ with $\mathbb{Z}$ according to the formula (4), and $\widehat{G}_n$ with G_n according to the formula (5), we define $\widehat{j}_n\colon G_n \to \mathbb{Z}$ by setting

$$\widehat{j}_n(k) = k, \quad k \in \{-m, \dots, m\}.$$

Then (G_n, j_n) is an AS of the group $\mathcal{G}$, and $(\widehat{G}_n, \widehat{j}_n)$ is an AS of $\widehat{\mathcal{G}}$ conjugate to it. In this case the equality (3) takes the form

$$\lim_{m\to\infty} \sum_{k=-m}^{m} \left| \int_{-1/2}^{1/2} f(x)\exp(-2\pi ikx)\,dx - \frac{1}{n}\cdot \sum_{l=-m}^{m} f\left(\frac{l}{n}\right)\exp\left(-\frac{2\pi ikl}{n}\right) \right|^2 = 0,$$

where f is Riemann-integrable, and $n = 2m+1$.

This equality is in essence contained in Luxemburg's paper [**45**].

b) Now suppose that $\mathcal{G}$ is the additive group of the field $\mathbb{R}$. Then $\widehat{\mathcal{G}}$ is also isomorphic to $\mathbb{R}$. We give the isomorphism between them by the mapping $\xi \to \chi_\xi$, where $\xi \in \mathbb{R}$, and $\chi_\xi \in \widehat{\mathcal{G}}$ is such that $\forall \eta \in \mathbb{R}$

$$\chi_\xi(\eta) = \exp 2\pi i\xi\eta. \tag{6}$$

Then the Fourier transformation $\mathcal{F}\colon L_2(\mathbb{R}) \to L_2(\mathbb{R})$ is given by

$$\mathcal{F}(f)(\xi) = \int_{-\infty}^{+\infty} f(\eta)\exp(-2\pi i\xi\eta)\,d\eta.$$

We recall that in this case Haar measure μ is $d\eta$, and the Haar measure dual to it is $\frac{1}{2\pi}d\xi$.

The same groups as in the preceding example are taken as approximating groups G_n. The isomorphism between G_n and $\widehat{G}_n$ is again given by the formula (5). To specify the mappings j_n we consider a sequence $\Delta_n \to 0$ such that $n\cdot\Delta_n \to \infty$ and let

$$j_n(k) = k\Delta_n, \quad k \in \{-m, \dots, m\}.$$

According to the general theory we now define $\widehat{\Delta}_n = (n\Delta_n)^{-1}$, and, using the isomorphisms (5) and (6), we can regard $\widehat{j}_n$ as a mapping from G_n to $\mathbb{R}$. We specify it by the formula

$$\widehat{j}_n(l) = l\widehat{\Delta}_n, \quad l \in \{-m, \dots, m\}.$$

Then again (G_n, j_n) is an AS of the group $\mathcal{G}$ and $(\widehat{G}_n, \widehat{j}_n)$ an AS of the group $\widehat{\mathcal{G}}$, and we can rewrite the limit relation (3) for this concrete case:

$$\frac{1}{n\Delta_n}\sum_{l=-m}^{m} \left| \int_{-\infty}^{+\infty} f(\eta)\exp\left(-\frac{2\pi il\eta}{n\Delta_n}\right)d\eta - \Delta_n \sum_{k=-m}^{m} f(k\Delta_n)\exp\left(-\frac{2\pi ikl}{n}\right)\right|^2 \to 0$$

as $n \to \infty$.

We note that $n\Delta_n$ is exactly equal to the length of the interval on which the table of the function f is formed, that is, the relation between the steps of the tables of the function and of its Fourier transform coincides with the relation taken from Kotel′nikov's theorem.

In this case the condition (1) of Theorem 1 is equivalent to the condition

$$\int_{-\infty}^{+\infty} f(x)\,dx = \lim_{h\to 0} h\cdot \sum_{n=-\infty}^{+\infty} f(nh).$$

This condition is satisfied by a very broad class of absolutely integrable functions on $\mathbb{R}$ (see, for example, [**80**]).

c) We fix some sequence $\tau = \{a_n \mid n \in \mathbb{Z}\}$ of integers greater than 1 and take $\mathcal{G}$ to be the additive group $\mathbb{Q}_\tau$ of τ-adic numbers. Let us recall briefly how it is defined. Details about the groups considered below in this section can be found in [**29**].

The elements of $\mathbb{Q}_\tau$ are all possible sequences $\xi = \{x_i \mid i \in \mathbb{Z}\}$ satisfying the following two conditions: 1) $0 \le x_i < a_i\ \forall i$; 2) $\exists m\ \forall i < m\ (x_i = 0)$ (here m depends on ξ, of course).

The operation of addition is defined as follows. Below, $\mathrm{qt}(a, b)$ and $\mathrm{rem}(a, b)$ denote the respective quotient and remainder when a is divided by b.

Let

$$\{x_i\} + \{y_i\} = \{z_i\}.$$

If $x_i = y_i = 0\ \forall i \in \mathbb{Z}$, then $z_i \equiv 0$. Otherwise, we define the sequence $\{z_i\}$ by induction. To do this we define another auxiliary sequence $\{t_i\}$ of positive integers. Let r be the largest integer such that $x_i = y_i = 0\ \forall i < r$. Then $z_i = t_i = 0\ \forall i < r$, $t_r = \mathrm{qt}(x_r + y_r, a_r)$, and $z_r = \mathrm{rem}(x_r + y_r, a_r)$. Further, if t_{k-1} has been defined, then $t_k = \mathrm{qt}(x_k + y_k + t_{k-1}, a_k)$ and $z_k = \mathrm{rem}(x_k + y_k + t_{k-1}, a_k)$.

It is not hard to verify that the operation defined in this way turns $\mathbb{Q}_\tau$ into an Abelian group, which is called the group of τ-adic numbers. In fact, it is possible to specify in addition a multiplication that turns $\mathbb{Q}_\tau$ into a commutative associative ring. We shall not describe it here, but mention only that if $\tau(i) = p\ \forall i$, where p is some prime number, then $\mathbb{Q}_\tau = \mathbb{Q}_p$, the field of p-adic numbers.

The topology in $\mathbb{Q}_\tau$ is given by the system $\{\Lambda_k \mid k \in \mathbb{Z}\}$ of neighborhoods of zero, where Λ_k consists of the $\xi = \{x_i\}$ such that $x_i = 0\ \forall i < k$. This topology can also be given by the metric $\rho(\xi, \eta) = 2^{-m}$, where $m = \min\{i \mid \xi(i) \ne \eta(i)\}$, and $\rho(\xi, \xi) = 0$.

To construct an approximation of the group $\mathbb{Q}_\tau$ we first make a remark.

If $s_0, \dots, s_r$ is a sequence of integers greater than 1, and $S = \prod_{i=0}^{r} s_i$, then every number $m \in \{0, \dots, S-1\}$ is uniquely representable in the form

$$m = x_0 + x_1 s_0 + x_2 s_0 s_1 + \dots + x_r s_0 s_1 \cdots s_{r-1}, \quad 0 \le x_i < s_i,\ 0 \le i \le r. \tag{7}$$

Suppose now that G_n is the additive group of the ring of residues mod S represented in the form $\{0, \dots, S-1\}$, where $S = a_{-n} a_{-n+1} \cdots a_0 \cdots a_{n-1} a_n$.

We let $r = 2n$ and $s_i = a_{i-n}$, $0 \le i \le r$, and we define the mapping $j_n\colon G_n \to \mathbb{Q}_\tau$ by setting $j_n(m) = \xi\ \forall m \in G_n$, where

$$\xi(i) = \begin{cases} 0, & |i| > n, \\ x_{i+n}, & |i| \le n, \end{cases}$$

and the sequence $x_0, \dots, x_r$ is defined by the formula (7).

It is now not hard to show that (G_n, j_n) is an AS of the group $\mathbb{Q}_\tau$. Further, if the Haar measure μ on $\mathbb{Q}_\tau$ is normalized so that $\mu(\Lambda_0) = 1$, then the normalizing multiplier Δ_n is equal to $a_0 a_1 \cdots a_n$.

To construct a conjugate AS we first describe the dual group $\widehat{\mathbb{Q}}_\tau$. We define the sequence $\hat{\tau}$ by setting $\hat{\tau}(i) = \tau(-i)$. It turns out that $\widehat{\mathbb{Q}}_\tau$ is isomorphic to $\mathbb{Q}_{\hat{\tau}}$. As an isomorphism we can take the mapping $\eta \to \chi_\eta$, $\eta = \{y_i\} \in \mathbb{Q}_{\hat{\tau}}$, where $\forall \xi = \{x_i\} \in \mathbb{Q}_\tau$

$$\chi_\eta(\xi) = \exp\left[2\pi i\left(\sum_{n=-\infty}^{\infty} x_n\left(\sum_{s=n}^{\infty} \frac{y_{-s}}{a_n a_{n+1}\cdots a_s}\right)\right)\right]. \tag{8}$$

It is easy to see that in each of the sums here only finitely many terms are nonzero (a sum in which the lower limit of the summation index is greater than the upper limit is assumed to be equal to zero).

As above, we identify G_n and $\widehat{G}_n$, defining the character γ_k for each $k \in G_n$ by the formula

$$\gamma_k(l) = \exp\frac{2\pi i k l}{S}. \tag{9}$$

The mapping $\widehat{j}_n \colon G_n \to \mathbb{Q}_{\hat{\tau}}$ is now defined from $\hat{\tau}$ just as j_n is defined from τ.

After identification of $\widehat{\mathbb{Q}}_\tau$ with $\mathbb{Q}_{\hat{\tau}}$ and $\widehat{G}_n$ with G_n, according to the respective formulas (8) and (9), it turns out that $(\widehat{G}_n, \widehat{j}_n)$ is an AS of $\widehat{\mathbb{Q}}_\tau$ conjugate to the AS (G_n, j_n). It is now easy to write a variant of the formula (3) for this case, and we leave it to the reader to do so.

d) The neighborhood Λ_0 of zero forms a subgroup, and even a subring, of $\mathbb{Q}_\tau$ called the ring of τ-adic integers and denoted by $\mathbb{Z}_\tau$. In considering the additive group $\mathbb{Z}_\tau$ we do not use the terms of the sequence τ with negative indices, and thus in this example we assume that τ has the form $\{a_n \mid n \in \mathbb{N}\}$. The group G_n is again defined as the group of residues $\bmod S$, but in the given case $S = a_0 a_1 \cdots a_n$.

Here the mapping $j_n \colon G_n \to \mathbb{Z}_\tau$ is again defined with the help of the formula (7) as follows: $j_n(m) = \{x_i \mid i \in \mathbb{N}\}$ $\forall m \in G_n$, where x_i is defined for $0 \le i \le n$ by the formula (7) with $s_i = a_i$, and $x_i = 0$ for $i > n$.

Now (G_n, j_n) is an AS of the group $\mathbb{Z}_\tau$. Further, since $\mathbb{Z}_\tau$ is commutative, the normalizing multiplier Δ_n is equal to S^{-1}, according to the conventions adopted earlier.

To construct the dual group we denote by $\mathbb{Q}^{(\tau)}$ the subgroup of the additive group $\mathbb{Q}$ consisting of all the rational numbers whose denominators have the form $a_0 a_1 \cdots a_m$ for $m \in \mathbb{N}$, and we set $\mathbb{Z}^{(\tau)} = \mathbb{Q}^{(\tau)}/\mathbb{Z}$. Then $\widehat{\mathbb{Z}}_\tau$ is isomorphic to $\mathbb{Z}^{(\tau)}$. Indeed, each element $\zeta \in \mathbb{Z}^{(\tau)}$ can be uniquely represented in the form $\zeta = y/a_0 a_1 \cdots a_r$, where $r \in \mathbb{N}$ and $0 \le y < a_0 a_1 \cdots a_r$. As an isomorphism we can now take the mapping $\zeta \to \chi_\zeta$, where $\chi_\zeta \in \widehat{\mathbb{Z}}_\tau$ is such that $\forall \xi = \{x_i\} \in \mathbb{Z}_\tau$

$$\chi_\zeta(\xi) = \exp 2\pi i m \zeta, \tag{10}$$

where m is found from the formula (7) with $s_i = a_i$ for $0 \le i \le r$.

We again identify G_n and $\widehat{G}_n$ with the help of the formula (9) (but with the S defined in this example), and we define $\widehat{j}_n \colon G_n \to \mathbb{Z}^{(\tau)}$ by setting $\widehat{j}_n(k) = k/S$ $\forall k \in G_n$. After the identifications (9) and (10) it turns out that $(\widehat{G}_n, \widehat{j}_n)$ is an AS of $\widehat{\mathbb{Z}}_\tau$ conjugate to the AS (G_n, j_n) of $\mathbb{Z}_\tau$. As in the preceding example, it is left to the reader to write out the equality (3) for this case.

e) We now take $\mathcal{G}$ to be the compact group Σ_τ, the τ-adic solenoid. Here τ is the same sequence as in the preceding example. As a set, Σ_τ is $[0,1) \times \mathbb{Z}_\tau$. The operation of addition is defined by the formula

$$\langle x, \xi\rangle + \langle y, \eta\rangle = \langle \{x+y\}, \xi+\eta+[x+y]\rangle,$$

where $\{t\}$ and $[t]$ are the fractional and integer parts of a number t.

The topology in Σ_τ is given by the system $\{U_n \mid n \in \mathbb{N}\}$ of neighborhoods of zero, where

$$U_n = [0, 1/n) \times \Lambda_n \cup (1-1/n, 1) \times (1+\Lambda_n).$$

Here Λ_n is the same as in the two preceding examples. It is also not hard to write out a metric determining this topology, but we do not need it.

Suppose now that S is the same as in the preceding example. Since we need the AS's defined there, we keep the notation (G_n, j_n) and $(\widehat{G}_n, \widehat{j}_n)$ for them, and we let (H_n, ι_n) and $(\widehat{H}_n, \widehat{\iota}_n)$ denote an AS of Σ_τ and an AS of $\widehat{\Sigma}_\tau$ conjugate to it, respectively.

Then H_n is the group of residues $\bmod S^2$, represented in the form $\{0, \dots, S^2-1\}$. The mapping $\iota_n \colon H_n \to \Sigma_\tau$ is defined as follows: $\forall k \in H_n$

$$\iota_n(k) = \langle \{k/S\}, j_n([k/S])\rangle.$$

Here j_n is the same as in the preceding example. Note that $[k/S] < S$, that is, it lies in G_n.

To construct the conjugate approximation we recall that $\widehat{\Sigma}_\tau$ is isomorphic to $\mathbb{Q}^{(\tau)}$. This isomorphism is defined as follows.

Let $\xi = \{x_i\} \in \mathbb{Z}_\tau$. We denote by $\psi_r(\xi)$ the number m obtained from ξ according to the formula (7) with $s_i = a_i$. In this case if $u = m/a_0 \cdots a_r$, $m \in \mathbb{Z}$, then

$$\chi_u(\xi) = \exp \frac{2\pi i (x + \psi_r(\xi))m}{a_0 \cdots a_r}. \tag{11}$$

We represent the group $\widehat{H}_n$ as the group of absolutely least residues $\bmod S^2$, that is, $\widehat{H}_n = \{n \in \mathbb{Z} \mid |n| \le \frac{S^2}{2}\}$. The mapping $\widehat{\iota}_n \colon \widehat{H}_n \to \mathbb{Q}^{(\tau)}$ is determined by the formula $\widehat{\iota}_n(k) = k/S^2$ $\forall k \in \widehat{H}_n$. By identifying $\mathbb{Q}^{(\tau)}$ with $\widehat{\Sigma}_\tau$ according to (11), it is easy to verify that $(\widehat{H}_n, \widehat{\iota}_n)$ is an AS of $\widehat{\Sigma}_\tau$ conjugate to the AS (H_n, ι_n).

If the sequence τ is such that $\tau(n) = n+1$ $\forall n \in \mathbb{N}$, then it is easy to see that $\mathbb{Q}^{(\tau)} = \mathbb{Q}$. An AS for the group $\mathbb{Q}$ in the discrete topology has thereby been constructed.

Of course, the LCA groups considered in this section also have other AS's. The question of comparing the different AS's of one and the same group has so far not been treated.

3°. In the concluding section of the Introduction we dwell on the issue of why nonstandard analysis proves to be useful in the investigation of the problems described above, and we briefly describe the content of the rest of the book.

Two basic approaches can be singled out in problems of approximation of operators. In the first approach a finite-dimensional subspace H of an infinite-dimensional space E is fixed, a projection $P \colon E \to H$ is constructed, and the operator PAP is regarded as an approximation of the operator A. Then the convergence of the approximations is investigated as the finite-dimensional subspaces become larger. Most of the literature in classical approximation theory is devoted to just such an approach. This approach was also used in the first applications of nonstandard analysis to operator theory. In it a nonstandard extension ${}^*E$ of a standard Hilbert space E is considered along with a hyperfinite-dimensional subspace H of it satisfying the condition $E \subset H \subset {}^*E$. The operator $P \cdot {}^*A \cdot P$ is regarded as a hyperfinite-dimensional approximation of the operator A, where $P\colon {}^*E \to H$ is an internal orthogonal projection. By the transfer principle, this operator has many properties of finite-dimensional operators, while its values on standard elements coincide with the values of A. These facts have been responsible for the success of the use of nonstandard analysis in problems where it was required to carry over some properties of finite-dimensional operators to infinite-dimensional operators. The Bernstein–Robinson result about invariant subspaces mentioned above was obtained in this way.

The other approach often used in practical computations is tabular approximation (grid methods). In this approach the infinite-dimensional space E is a function space. For functions in E tables of their values at a finite set of points are formed, and an approximation of the operator $A\colon E \to E$ is taken to be an operator carrying the table of an $f \in E$ into a vector that is close in some sense to the table of the function $A(f)$. The investigation of questions involving the construction and convergence of such approximations often encounters difficulties of a fundamental character.

First of all, the question arises as to which points it is most natural to use in forming the tables of functions.

In grid methods, when the functions are given on $\mathbb{R}^n$ or on some domain in $\mathbb{R}^n$, one usually considers a uniform grid with step Δ and investigates the behavior of the approximations as $\Delta \to 0$. Even in this case a certain difficulty in the investigation is that the finite-dimensional spaces obtained (the spaces of tables of functions) are not imbedded in each other in any canonical way, and this means the absence of an imbedding of them not only in the original space E, but also in some infinite-dimensional space associated with them in a natural way. The fact that there are no canonical imbeddings is due in part to the fact that in these finite-dimensional spaces the norms or metrics approximating the norm or metric in E depend on Δ as a rule.

Here nonstandard analysis has an advantage in that when a table is formed with a constant infinitesimal step Δ, one has to work with a single hyperfinite-dimensional space H whose norm depends on Δ. Furthermore, as a rule there is a natural imbedding $T\colon E \to H^\#$, where $H^\#$ is the nonstandard hull of H. The meaning of this imbedding is that $T(f)$ is the class of the table of the function $f \in E$, computed with step Δ. It is clear that every hyperfinite-dimensional operator $B\colon H \to H$ with bounded (that is, not infinite) norm induces a bounded operator $B^\#\colon H^\# \to H^\#$. Further, the hyperfinite operator B approximates the operator $A\colon E \to E$ if $T \circ A = B^\# \circ T$, because in this case the B-image of the table of a function f is infinitely close to the table of $A(f)$. The last equality is easy to verify in many concrete cases. Further, there are well-known procedures (for example, the Nelson algorithm) for transforming nonstandard propositions into standard ones. Applying them to a proposition on approximability of an operator A by a hyperfinite-dimensional operator, we can get an assertion about convergence of the corresponding standard finite-dimensional approximations.

Here this method is used to investigate Hilbert–Schmidt operators in addition to the Fourier transformation. A table of the kernel $K(x, y)$ of the original operator is taken as the matrix of an approximating operator. There is an essential difference between the results in the two cases. In the case of Hilbert–Schmidt operators the steps along the x- and y-axes can be chosen arbitrarily, with a single natural restriction connected with the fact that the number of nodes in the table must be the same with respect to each of the axes, the steps tend to zero, and the lengths of the intervals on which the tables are formed tend to infinity. But in the case of the Fourier transformation (see example b) the relation between the steps must be such that the table $K(m\Delta, k\widehat{\Delta})$ turns out to be the matrix of the finite Fourier transformation. There are examples when convergence fails for a different choice of steps.

The question of choosing the nodes of the table becomes more complicated in the investigation of tabular approximations in the case when the functions are given on sets different from domains in $\mathbb{R}^n$. Since it is mainly integral operators and operators associated with them that are considered here, we dwell first on approximation of integrals. One of the most widespread methods for computing integrals (especially multiple integrals) is the Monte Carlo method, in which the value of the integral is approximated by the arithmetic mean of the values of the function at random points. There are many different ways of generating random sequences of points, more precisely, sequences whose behavior nicely models that of a random sequence. However, the precise definition (going back to Kolmogorov and Martin-Löf) of what it means for an individual sequence of points to be random is based on the theory of algorithms, and it is not clear at present how to use it in analytical investigations of convergence of approximations, not to mention the case of arbitrary measure spaces.

In nonstandard analysis it is possible to formulate a natural definition of randomness of an individual point of an arbitrary measure space. Namely, a random point does not lie in the nonstandard extension of any standard set of measure zero. As a rule, such a point is not standard. Using this definition, we can in arbitrary finite (and even countably finite) measure spaces single out hyperfinite sets of points at which to form tables of functions for approximation of integrals and integral operators, and then reduce the problems on these spaces to problems on the Loeb spaces whose advantages were discussed in the first section.

Questions connected with random elements are expounded in §2 of Chapter 1. Section 3 of Chapter 1 is devoted to general questions of hyperfinite tabular approximations of arbitrary operators and Hilbert–Schmidt operators.

If the measure space on which the functions are defined is also equipped with some structures closely connected with the operator being approximated, then the choice of the nodes of the table must in some sense approximate these structures. This is the idea at the basis of the theory of approximation of topological groups by finite groups discussed in the preceding section. Passage to nonstandard analysis in this theory enables us, in particular, to approximate Haar measures by Loeb measures on hyperfinite groups. True, there is a problem here involving the fact that the theory of integration of Loeb measures was previously developed only for finite measures. But if the LCA group is not compact, then the Haar measure of the whole group is not finite. In this connection it was necessary to extend the theory of integration of Loeb measures to the case when the whole space is a countable union of internal sets having finite internal measure. This material is also contained in §2 of Chapter 1.

Finally, of no small importance is the essential simplification of the definitions upon passage to hyperfinite-dimensional approximations. It is a well-known fact that nonstandard analysis makes the formulations of many definitions in classical analysis considerably simpler and more natural. For example, $\lim_{x\to a} f(x) = b$ means that when the number x is infinitely close to a, then the number $f(x)$ is infinitely close to b. In our case approximability by a hyperfinite group means that the imbedding j of the hyperfinite group G in the nonstandard extension of the locally compact group $\mathcal{G}$ is such that, first, each standard point in $\mathcal{G}$ is infinitely close to the image of some element in G, and, second, the image of the product of two elements is infinitely close to the product of their images. This is the nonstandard version of Definition 1.

Here also there is a difficulty that should be considered. The fact is that significant simplifications are achieved for various formulations only when nonstandard analysis is applied to standard objects. But if we have a standard function of two variables, and a nonstandard parameter Δ is substituted for one of them, then the resulting function of the other variable is no longer standard, and this limits the possibilities for using the traditional transfer technique of nonstandard analysis. In the problems taken up in this book we shall constantly have to deal with such difficulties in the investigation of noncompact groups, where an essential role is played by the minimum of two independent or weakly dependent nonstandard parameters; for example, an infinite length of the interval on which the table of a function is formed, and an infinitesimal step in this table. To overcome these problems a theory of relative standardness is worked out here that enables us in certain cases to apply the transfer technique also to essentially nonstandard objects. This theory is expounded in §1 of Chapter 1.

CHAPTER 0

Basic Concepts of Nonstandard Analysis

§1. The language $\mathcal{LR}$ of real analysis

1°. In this section we define the basic formal language with which we shall work, and we show how substantive mathematical propositions are written in this language.

DEFINITION 0.1.1. The language $\mathcal{LR}$ includes the following objects.

(1) For each $n \in \mathbb{N}$ a countable set $\mathcal{A}_n$ whose elements are called variables of type n. The variables of type 0 are also called object variables. The type of a variable is indicated in its notation in parentheses: $x^{(0)}$, $f^{(1)}$, $h^{(2)}$, $\varphi^{(n)}$, and so on. The type is often clear from the context, and then the type is omitted in the notation for the variable.

(2) Constants $\mathbf{0}$ and $\mathbf{1}$ of type 0, and constants $+$ and $\cdot$ of type 2.

(3) The relation symbols $=$ and $<$.

(4) The logical connectives $\wedge$ (conjunction), $\vee$ (disjunction), $\longrightarrow$ (implication), $\neg$ (negation), and $\longleftrightarrow$ (equivalence), along with the quantifiers $\forall$ (universality) and $\exists$ (existence).

(5) Left and right parentheses, commas (square brackets and curly brackets are sometimes used for clarity).

The formulas of the language $\mathcal{LR}$ are finite sequences of its elements and are formed according to definite rules formulated below. They denote certain statements about real numbers and functions of a real variable. We show by examples that the language $\mathcal{LR}$ is sufficiently rich to express almost any proposition of analysis.

DEFINITION 0.1.2 (the terms of the language $\mathcal{LR}$).

(1) Variables and constants of type 0 are terms of $\mathcal{LR}$.

(2) If $t_1, \dots, t_n$ are terms of $\mathcal{LR}$, and $\varphi^{(n)}$ is a variable of type n, then the quantity $\varphi^{(n)}(t_1, \dots, t_n)$ is a term of $\mathcal{LR}$.

(3) If t_1 and t_2 are terms, then so are $t_1 + t_2$ and $t_1 \cdot t_2$.

(4) Every term of $\mathcal{LR}$ is obtained according to the rules 1–3.

DEFINITION 0.1.3 (the formulas of the language $\mathcal{LR}$).

(1) If t_1 and t_2 are terms, then $t_1 < t_2$ and $t_1 = t_2$ are formulas of $\mathcal{LR}$.

(2) If $\mathcal{A}$ and $\mathcal{B}$ are formulas, then so are $(\neg\mathcal{A})$, $(\mathcal{A} \wedge \mathcal{B})$, $(\mathcal{A} \vee \mathcal{B})$, $(\mathcal{A} \longrightarrow \mathcal{B})$, $(\mathcal{A} \longleftrightarrow \mathcal{B})$, $\exists x\mathcal{A}$, and $\forall x\mathcal{A}$. Here x is a variable of any type.

(3) Every formula of $\mathcal{LR}$ is obtained according to the rules 1–2.

In the writing of formulas, use is made of natural abbreviations that are, as a rule, clear from the context: superfluous parentheses are dropped, $\forall x, y, z$ is written instead of $\forall x \forall y \forall z$, and so on. Moreover, we use the convention that the operation

$\wedge$ is implemented before the operations $\vee$, $\longrightarrow$, or $\longleftrightarrow$, which allows us to write $A \wedge B \vee C$ instead of $(A \wedge B) \vee C$, and so on.

If we consider elements of the language $\mathcal{LR}$ as symbols of an alphabet, then the formulas are words in this alphabet that satisfy definite rules. It can be assumed that the elements of each set $\mathcal{A}_n$ in Definition 0.1.1 are effectively enumerated. Then it is easy to see that there is an algorithm that recognizes, for any word in the alphabet of $\mathcal{LR}$, whether or not this word is a formula. A subword of a formula that is itself a formula is called a subformula.

An occurrence of a variable α in a formula $\mathcal{A}$ is said to be bound if it is an occurrence in a subformula of the form $\exists\alpha\mathcal{B}$ or $\forall\alpha\mathcal{B}$. Otherwise the occurrence of this variable in $\mathcal{A}$ is said to be free. The expression $\mathcal{A}(\alpha_1, \dots, \alpha_n)$ will mean that all the variables having free occurrences in $\mathcal{A}$ are in the list $\alpha_1, \dots, \alpha_n$.

As already mentioned, each formula in the language $\mathcal{LR}$ denotes a proposition about real numbers and functions. Further, $\neg\mathcal{A}$ denotes the proposition "it is false that $\mathcal{A}$", $\mathcal{A}\wedge\mathcal{B}$ the proposition "$\mathcal{A}$ and $\mathcal{B}$", $\mathcal{A}\vee\mathcal{B}$ the proposition "$\mathcal{A}$ or $\mathcal{B}$", $\mathcal{A} \longrightarrow \mathcal{B}$ the proposition "if $\mathcal{A}$, then $\mathcal{B}$", $\mathcal{A} \longleftrightarrow \mathcal{B}$ the proposition "$\mathcal{A}$ if and only if $\mathcal{B}$", $\exists f\mathcal{A}$ the proposition "there exists an f satisfying $\mathcal{A}$", and $\forall f\mathcal{A}$ the proposition "$\mathcal{A}$ holds for any f".

Thus, $\mathcal{A} \longleftrightarrow \mathcal{B}$ is equivalent to $(\mathcal{A} \longrightarrow \mathcal{B}) \wedge (\mathcal{B} \longrightarrow \mathcal{A})$. Instead of the word "equivalent" we shall usually employ the symbol $\equiv$, that is, $\mathcal{A} \longleftrightarrow \mathcal{B} \equiv (\mathcal{A} \longrightarrow \mathcal{B}) \wedge (\mathcal{B} \longrightarrow \mathcal{A})$. In contrast to $\longleftrightarrow$, the symbol $\equiv$ is not a logical connective (it does not occur in the language $\mathcal{LR}$). The symbol $\Longleftrightarrow$ is used to mean the same as $\equiv$.

The formula $\mathcal{A}(\alpha_1, \dots, \alpha_n)$ denotes a variable proposition that becomes either true or false when real numbers are substituted for the free variables of type 0 and functions from $\mathbb{R}^n$ to $\mathbb{R}$ are substituted for the variables of type $n > 0$. The truth or falsity of the proposition obtained depends on just which numbers and functions are substituted for the free variables. Such variable propositions are usually called *predicates* in logic.

Formulas without free variables are called *sentences.* Sentences are true or false independently of what variables appear in them. True sentences are mathematical theorems.

2°. We shall consider some examples illustrating the expressive possibilities of the language $\mathcal{LR}$.

In this subsection the object variables are denoted by the letters a, b, c, x, y, z, possibly with subscripts, and the function variables are denoted by the letters f, g, h, φ, ψ, also sometimes with subscripts. It is usually clear from the context what type $n > 0$ a variable has. Otherwise this type is indicated according to the conventions adopted above.

$$\text{a)} \qquad \exists!\alpha\mathcal{A}(\alpha) \rightleftharpoons \exists\alpha\mathcal{A}(\alpha) \ \wedge\ \forall\beta,\gamma\ (\mathcal{A}(\beta) \wedge \mathcal{A}(\gamma) \longrightarrow \beta = \gamma).$$

Here α is a variable of arbitrary kind. If for some formula $\mathcal{A}(\alpha)$ not having other free variables the formula $\exists!\alpha\mathcal{A}(\alpha)$ is true, then there exists a unique object of the same type as α (that is, a real number or a function from $\mathbb{R}^n$ to $\mathbb{R}$) that satisfies $\mathcal{A}(\alpha)$. In this case one says that the corresponding object is definable in $\mathcal{LR}$. It can be assumed that for each object definable in $\mathcal{LR}$ there is a constant of corresponding type in the language. A natural number $m \in \mathbb{N}$ is defined in $\mathcal{LR}$ by the formula

$$x = \underbrace{(\dots(1+1)+\dots+1)}_{m \text{ times}}.$$

Consequently, it can be assumed that corresponding to each $m \in \mathbb{N}$ the language $\mathcal{LR}$ has a constant, which we also denote by m.

A positive rational number $r = \frac{m}{n}$ is defined in $\mathcal{LR}$ by the formula $n \cdot x = m$. For any positive $r \in \mathbb{Q}$ the formula $r + x = 0$ now defines the number $(-r)$. It is easy to show that any algebraic numbers are definable in $\mathcal{LR}$. For example, $\sqrt{2}$ is defined by the formula $x \cdot x = 2$. We show below how to define certain transcendental numbers in $\mathcal{LR}$; for example, e and π.

As in the case of natural numbers, for arbitrary objects definable in $\mathcal{LR}$ and having commonly accepted notations we use these notations also for the constants in $\mathcal{LR}$ corresponding to the given objects. Obviously, the set of objects definable in $\mathcal{LR}$ is countable, because the set of formulas in $\mathcal{LR}$ is countable.

b) $$\mathrm{Set}^{(n)}(f) \rightleftharpoons \forall x_1, \dots, x_n \ (f(x_1, \dots, x_n) = 0 \ \vee \ f(x_1, \dots, x_n) = 1),$$
$$\langle x_1, \dots, x_n \rangle \in f \rightleftharpoons \mathrm{Set}^{(n)}(f) \ \wedge \ f(x_1, \dots, x_n) = 1.$$

Identifying the sets with these characteristic functions and using the formulas written above, we can assume that for each $n > 0$ there are variables in $\mathcal{LR}$ whose values are subsets of $\mathbb{R}^n$. By definition these variables also have type n. Below they are denoted by capital letters of the Latin alphabet with the type of variable indicated, if needed, by a superscript in parentheses, and they are called set variables. Moreover, it can be assumed that there is a predicate $\langle x_1, \dots, x_n \rangle \in X^{(n)}$ in $\mathcal{LR}$ for each $n > 0$ (see the second of the formulas indicated above):

$$X^{(n)} \subseteq Y^{(n)} \rightleftharpoons \forall x_1, \dots, x_n \ (\langle x_1, \dots, x_n \rangle \in X^{(n)} \longrightarrow \langle x_1, \dots, x_n \rangle \in Y^{(n)}).$$

For any formula $\mathcal{A}(z_1, \dots, z_m, \alpha_1^{(n_1)}, \dots, \alpha_k^{(n_k)})$ the following formula is true:

$$\forall \alpha_1^{(n_1)}, \dots, \alpha_k^{(n_k)} \exists! X^{(m)} \forall z_1, \dots, z_m \ (\langle z_1, \dots, z_m \rangle \in X^{(m)} \longleftrightarrow \mathcal{A}(z_1, \dots, z_m, \alpha_1^{(n_1)}, \dots, \alpha_k^{(n_k)})).$$

This defines an operation in $\mathcal{LR}$ that associates with each k-tuple of objects of types $n_1, \dots, n_k$ a certain subset of $\mathbb{R}^m$. The result of this operation is denoted by

$$\{\langle z_1, \dots, z_m \rangle \mid \mathcal{A}(z_1, \dots, z_m, \alpha_1^{(n_1)}, \dots, \alpha_k^{(n_k)})\}.$$

Such configurations can be used in the formulas of $\mathcal{LR}$ in place of variables of type m. If the corresponding operation has some special notation, then it is carried over also to $\mathcal{LR}$. For example,

$$(a, b) = \{t \mid a < t \ \wedge \ t < b\}$$
$$\mathbb{R}_+ = \{t \mid 0 \le t\}$$
$$X^{(n)} \times Y^{(m)} = \{(x_1, \dots, x_n, y_1, \dots, y_m) \mid (x_1, \dots, x_n) \in X^{(n)} \ \wedge \ (y_1, \dots, y_m) \in Y^{(m)}\}.$$

It is also easy to write the definitions for the Boolean operations on sets.

We use the following abbreviations below:

$$\forall x \in X \ \mathcal{A} \rightleftharpoons \forall x \ (x \in X \longrightarrow \mathcal{A}),$$
$$\exists x \in X \ \mathcal{A} \rightleftharpoons \exists x \ (x \in X \ \wedge \ \mathcal{A}).$$

In particular, $Q\,x \leq t \ \mathcal{A} \rightleftharpoons Q\,x \in (-\infty, t] \ \mathcal{A}$. Here Q stands for either $\exists$ or $\forall$. If X is a set variable of type $n > 0$, then the set of type k

$$\operatorname{dom}^{(k)} X = \{\langle x_1, \dots, x_k\rangle \mid \exists \langle x_{k+1}, \dots, x_n\rangle \ (\langle x_1, \dots, x_n\rangle \in X)\}$$

and the set of type $n-k$

$$\operatorname{rng}^{(k)} X = \{\langle x_{k+1}, \dots, x_n\rangle \mid \exists \langle x_1, \dots, x_k\rangle \ (\langle x_1, \dots, x_n\rangle \in X)\}$$

are defined for any k with $1 \leq k < n$.

Now the formula $\operatorname{Fnc}^{(k)}(X)$ given by

$$\forall \langle x_1, \dots, x_k\rangle \in \operatorname{dom}^{(k)} X \ \exists! x_{k+1}, \dots, x_n \ (\langle x_1, \dots, x_n\rangle \in X)$$

expresses the fact that $X \subseteq \mathbb{R}^k \times \mathbb{R}^{n-k}$ is a function relation.

The formula $\operatorname{Map}(F^{n+m}, A^{(n)}, B^{(m)})$ given by

$$\operatorname{Fnc}^{(n)}(F^{n+m}) \ \wedge \ \operatorname{dom}^{(n)} F^{n+m} \subseteq A^{(n)} \ \wedge \ \operatorname{rng}^{(n)} F^{n+m} \subseteq B^{(m)}$$

means that F^{n+m} is a mapping from $A^{(n)}$ to $B^{(m)}$. Below we write $F\colon A \longrightarrow B$ instead of $\operatorname{Map}(F, A, B)$.

Ordered tuples of real numbers will be denoted by $\overline{a}$, $\overline{b}$, $\overline{x}$, $\overline{y}$, etc. In the expression $\overline{x} \in X$ the length of the tuple $\overline{x}$ is equal to the type of the variable X. We set

$$\overline{y} = F(\overline{x}) \rightleftharpoons \operatorname{Fnc}(F) \ \wedge \ \langle \overline{x}, \overline{y}\rangle \in F.$$

Here and below, the type of F is equal to the sum of the lengths of the tuples $\overline{x}$ and $\overline{y}$; that is, a countable family of definitions is written in the preceding expression. The formulas

$$\operatorname{IMap}(F, A, B) \rightleftharpoons \operatorname{Map}(F, A, B) \ \wedge \ \forall \overline{y} \in \operatorname{rng}(F) \exists! \overline{x} \ (\overline{y} = F(\overline{x})),$$
$$\operatorname{SMap}(F, A, B) \rightleftharpoons \operatorname{Map}(F, A, B) \ \wedge \ \operatorname{rng}(F) = B,$$
$$\operatorname{BMap}(F, A, B) \rightleftharpoons \operatorname{IMap}(F, A, B) \ \wedge \ \operatorname{SMap}(F, A, B)$$

mean that the mapping $F\colon A \longrightarrow B$ is injective, surjective, or bijective, respectively. We shall also use the following abbreviations:

$$F\colon A \longmapsto B \rightleftharpoons \operatorname{IMap}(F, A, B),$$
$$F\colon A \Longrightarrow B \rightleftharpoons \operatorname{SMap}(F, A, B),$$
$$F\colon A \Longleftrightarrow B \rightleftharpoons \operatorname{BMap}(F, A, B).$$

c) An important fragment of the language $\mathcal{LR}$ is the language $\mathcal{ER}$—the elementary theory of the field $\mathbb{R}$—obtained if only variables of type 0 are considered in the language $\mathcal{LR}$ while preserving all the remaining objects in it (see Definition 0.1.1). It is clear from Definition 0.1.2 that the terms of the language $\mathcal{ER}$ can be interpreted in $\mathbb{R}$, and also in an arbitrary commutative associative ring, as polynomials in several variables with natural number coefficients. Consequently, each elementary formula (see Definition 0.1.3) of the language $\mathcal{ER}$ can be written in the form $p(x_1, \dots, x_n) > 0$ or $p(x_1, \dots, x_n) = 0$, where $p(x_1, \dots, x_n)$ is a polynomial with *integer* coefficients. The axioms of a linearly ordered field are written in the language $\mathcal{ER}$ as the following sentences:

1. $\forall x, y, z\ ((x+y)+z = x+(y+z)\ \wedge\ (x \cdot y)\cdot z = x\cdot(y\cdot z))$;
2. $\forall x, y\ (x\cdot y = y \cdot x)$;
3. $\forall x, y, z\ ((x+y)\cdot z = x\cdot z + y\cdot z)$;
4. $\forall x\ (x+0 = x\ \wedge\ x\cdot 1 = x)$;
5. $\forall x\ (\exists y(x+y=0)\ \wedge\ (\neg\, x = 0 \longrightarrow \exists z(x\cdot z = 1)))$;
6. $\forall x\ (\neg\, x < x)$;
7. $\forall x, y\ (x < y\ \vee\ y < x\ \vee\ y = x)$;
8. $\forall x, y, z\ (x < y\ \wedge\ y < z \longrightarrow x < z)$;
9. $\forall x, y, z\ (x < y \longrightarrow x + z < y + z)$;
10. $\forall x, y, z\ (x < y\ \wedge\ z > 0 \longrightarrow x\cdot z < y \cdot z)$.

We denote the conjunction of sentences 1–10 by LOF (linearly ordered field). If binary operations $+$ and $\cdot$ are given in some set along with a binary relation $<$, and the sentence LOF of the language $\mathcal{ER}$ is true in the algebraic system obtained, then this system is a linearly ordered field.

If the sentence

$$\forall a_0, \ldots, a_{2n} \exists x\ (x^{2n+1} + a_{2n}x^{2n} + \cdots + a_1 x + a_0 = 0)$$

is true for each $n \in \mathbb{N}$ in some linearly ordered field P, then P is called a real closed field.

TARSKI'S THEOREM. *For any formula $\mathcal{A}(x_1, \ldots, x_n)$ in the language $\mathcal{ER}$ it is possible to construct a quantifier-free formula $\mathcal{B}(x_1, \ldots, x_n)$ in this language such that in any real closed field P the formula $\mathcal{A}(x_1, \ldots, x_n) \longleftrightarrow \mathcal{B}(x_1, \ldots, x_n)$ is true.*

A comparatively short proof of Tarski's theorem is given in [**56**] (see also [**57**]).

In logic it is proved that every quantifier-free formula is equivalent to a formula of the form

$$\bigvee_{i=1}^{k} \bigwedge_{j=1}^{n_i} \mathcal{A}_{ij},$$

where each $\mathcal{A}_{ij}$ is either an elementary formula or the negation of an elementary formula. By using the form of elementary formulas it is now not hard to show that for each formula $\mathcal{A}(x_1, \ldots, x_n)$ the set

$$\{\langle t_1, \ldots, t_n\rangle \in P^n \mid \mathcal{A}(t_1, \ldots, t_n) \text{ is true}\}$$

is a finite union of sets, each of which is the collection of all solutions of some finite system of polynomial inequalities of the form

$$\begin{cases} p_1(x_1, \ldots, x_n) > 0 \\ \qquad\vdots \\ p_k(x_1, \ldots, x_n) > 0 \\ q_1(x_1, \ldots, x_n) \geq 0 \\ \qquad\vdots \\ q_l(x_1, \ldots, x_n) \geq 0, \end{cases}$$

where $k, l \geq 0$, and p_i and q_j are polynomials with integer coefficients. The nonstrict inequalities arise because there can be negations of elementary formulas among the formulas $\mathcal{A}_{ij}$.

If $\mathcal{A}$ is a sentence in the language $\mathcal{ER}$, then each $\mathcal{A}_{ij}$ in the quantifier-free sentence equivalent to it by Tarski's theorem is $n > 0$, or $n = 0$, or the negation of an elementary formula of this form, since each term without variables in any ordered field is equal to some natural constant (every linearly ordered field has characteristic zero). Consequently, the truth or falsity of a quantifier-free sentence is independent of just which linearly ordered field the given sentence is being considered in. Thus, the following statement is a consequence of Tarski's theorem.

Every sentence in the language $\mathcal{ER}$ is either true in all real closed fields or false in all real closed fields. In particular, no set of sentences in the language $\mathcal{ER}$ describes the field $\mathbb{R}$ uniquely up to an isomorphism.

d) If we add to $\mathcal{ER}$ even just variables of type 1, then the situation changes radically (in fact, since $\mathbb{R}^n$ has the same cardinality as $\mathbb{R}$, any assertion about $\mathbb{R}$ that can be expressed in the language $\mathcal{ER}$ can be expressed as a sentence containing only variables of types 0 and 1). Indeed, let

$$B(X) \rightleftharpoons \exists t \forall x \in X \ (x \leq t),$$

that is, $B(X)$ is a formula expressing the boundedness of a set $X \subseteq \mathbb{R}$ from above. It is also easy to write the relation $y = \sup X$ in $\mathcal{LR}$ (exercise). The sentence

$$\text{LUB} \rightleftharpoons \forall X \ (B(X) \longrightarrow \exists y (y = \sup X))$$

is now none other than the least upper bound (LUB) axiom. The next result is well known.

THEOREM 0.1.4. *If the sentence* LUB *is true in the ordered field P, then P is isomorphic to $\mathbb{R}$.*

In fact, it is possible to interpret the concept of truth of sentences in $\mathcal{LR}$ in a somewhat more general sense so that the above theorem ceases to be true. It is this circumstance that lies at the basis of all the constructions of nonstandard analysis, and we shall investigate it in detail in the next section.

EXERCISE. Express the following assertions in the language $\mathcal{LR}$:
(1) $a = \lim_{x \to b} f(x)$;
(2) the function f is (uniformly) continuous on $[a, b]$;
(3) the function f is differentiable, and $g = f'$.

We can now define all the elementary functions in $\mathcal{LR}$. For example, the formula $f' = f \ \wedge \ f(0) = 1$ defines $\exp(x)$, the formula $f'' = -f \ \wedge \ f(0) = 0 \ \wedge \ f'(0) = 1$ defines the function $\sin x$, and so on. It is left to the reader as an exercise to define the remaining elementary functions in $\mathcal{LR}$. According to the discussion in part a) it can now be assumed that there are constants in $\mathcal{LR}$ for all the elementary functions, and the notation for them coincides with the commonly accepted notation for these functions.

The number e can now be defined in $\mathcal{LR}$ as $\exp(1)$, and the number π can be defined by, for example, the formula

$$\pi(x) \rightleftharpoons \forall y \ (\sin(y + 2x) = \sin y) \ \wedge \ 0 < x < 4.$$

We remark that, in fact, all the special functions can be defined in $\mathcal{LR}$.

e) We now define in $\mathfrak{LR}$ the set $\mathbb{N}$ of natural numbers. It is defined by the formula

$$\text{Nat}(X) \rightleftharpoons$$
$$0 \in X \ \wedge \ \forall y \in X(y+1 \in X) \ \wedge \ \forall Y[0 \in Y \ \wedge \ \forall y \in Y(y+1 \in Y) \longrightarrow X \subseteq Y].$$

The induction principle for $\mathbb{N}$ follows immediately from this definition:

$$\forall A \subseteq \mathbb{N} \ (0 \in A \ \wedge \ \forall x \in A(x+1 \in A) \longrightarrow A = \mathbb{N}).$$

It is now easy to define the sets $\mathbb{Z}$ and $\mathbb{Q}$ of integers and rational numbers.

We do not touch here on questions of a systematic construction of the field $\mathbb{R}$, that is, of a linearly ordered field satisfying LUB, for which, conversely, the set of natural numbers with the succession operation $(+1)$ is defined from the start as a system satisfying the Peano axioms. With an approach that is not completely formal the existence of such a system is postulated, while with a strictly formal approach it is proved in axiomatic set theory—see the concluding section of this chapter.

The formula $\text{Seq}(F, A) \rightleftharpoons \text{Map}(F, \mathbb{N}, A)$ expresses the fact that F is a sequence of elements of the set A. Instead of $\text{Seq}(F, \mathbb{R})$ we simply write $\text{Seq}(F)$. The next sentence of the language $\mathfrak{LR}$ expresses the well-known theorem on the existence and uniqueness of functions defined by induction (or, what is the same, by recursion):

$$\begin{aligned} \forall f^{(n+2)} \forall g^{(n)} \exists! F \ [\text{Seq}(F) \ \wedge \ F(0) = g^{(n)}(a_1, \dots, a_n) \\ \wedge \ \forall k \in \mathbb{N} \ (F(k+1) = f^{(n+2)}(a_1, \dots, a_n, k, F(k)))]. \end{aligned}$$

This theorem asserts the existence and uniqueness (for any values of the parameters $a_1, \dots, a_n$) of a sequence $F_{a_1,\dots,a_n} \colon \mathbb{N} \longrightarrow \mathbb{R}$ such that

$$\begin{cases} F_{a_1,\dots,a_n}(0) = g(a_1, \dots, a_n), \\ F_{a_1,\dots,a_n}(k+1) = f(a_1, \dots, a_n, k, F(k)) \end{cases}$$

for an arbitrary n-place function g and an arbitrary $(n+2)$-place function f.

In ordinary mathematics courses, sequences defined by recursion are used without any reference to this theorem, which is assumed to be obvious. Its formulation and proof is usually contained in the courses in which more thorough attention is paid to questions of foundations (see, for example, [**58**]). In particular, $g(n) = \sum_{i=0}^{n} f(i)$ is none other than a sequence defined by recursion. Its definition is written in $\mathfrak{LR}$ as the formula

$$\text{Seq}(f) \wedge \text{Seq}(g) \wedge f(0) = g(0) \ \wedge \ \forall k \in \mathbb{N} \ (g(k+1) = g(k) + f(k+1)).$$

An analogous definition applies to $\prod_{i=0}^{n} f(i)$. For those familiar with the theory of algorithms we mention that all partial recursive functions are definable in the language $\mathfrak{LR}$. We note also that in $\mathfrak{LR}$ we can speak also of sequences of functions. Indeed, the formula $\text{Map}(F, \mathbb{N} \times A, B)$ expresses the fact that F is a mapping from $\mathbb{N} \times A$ to B, or, what is the same, a sequence of mappings from A to B. For $k \in \mathbb{N}$ we denote by $[k]$ the set $\{0, 1, \dots, k-1\}$, which is defined in $\mathfrak{LR}$ by the formula $\{x \mid x \in \mathbb{N} \ \wedge \ x < k\}$. Then the formula

$$\text{Fin}(X) \rightleftharpoons \exists k, F \ (F \colon [k] \Longleftrightarrow X)$$

means that X is a finite set. We denote by $\mathrm{Card}(k, X)$ the formula

$$k \in \mathbb{N} \ \wedge \ \mathrm{Fin}(X) \ \wedge \ \exists F \ (F\colon [k] \Longleftrightarrow X).$$

It is clear that the sentence

$$\forall X(\mathrm{Fin}(X) \longrightarrow \exists! k \ \ \mathrm{Card}(k, X))$$

is true. Instead of $\mathrm{Card}(k, X)$ we write $|X| = k$, as is customary (the natural number k is the cardinality of the finite set X). If the finite set X is $\subseteq \mathbb{R}$, then we can define the number $s = \sum_{x \in X} x$. Actually, this is done on the basis of the following theorem (the proof of which is left to the reader):

$$\mathrm{Fin}(X) \longrightarrow \forall F, G$$
$$\left[(F\colon [|X|] \Longleftrightarrow X) \wedge (G\colon [|X|] \Longleftrightarrow X) \longrightarrow \sum_{i=0}^{|X|-1} F(i) = \sum_{i=0}^{|X|-1} G(i)\right].$$

To bring this in correspondence with the definition $g(n) = \sum_{i=0}^{n} f(i)$ formulated above we can assume that F and G are defined on $\mathbb{N}$ and are equal to zero outside $[|X|]$.

Now

$$s = \sum_{x \in X} x \rightleftharpoons \forall F\left[(F\colon [|X|] \Longleftrightarrow X) \longrightarrow s = \sum_{i=0}^{|X|-1} F(i)\right].$$

For arbitrary sets A and B the formula $F\colon A \Longleftrightarrow B$ says that the sets have the same cardinality, and, as usual, we write $|A| = |B|$ in place of it. Note that, in contrast to the case of finite sets, for infinite sets no objects are defined here that represent their cardinalities (this is done in axiomatic set theory), but only the relation of being of equal cardinality is defined.

It is now clear that certain sentences independent of the axioms of set theory can also be formulated in the language $\mathcal{LR}$. For example, the continuum hypothesis can be formulated as

$$\forall X^{(1)}\big(\mathrm{Fin}(X^{(1)}) \ \vee \ |X^{(1)}| = |\mathbb{N}| \ \vee \ |X^{(1)}| = |\mathbb{R}|\big).$$

Questions of independence are not treated in this book. It is assumed that all arguments are carried out in some fixed model of set theory in which the continuum hypothesis is either true or false. Just which of these two possibilities is realized is not essential for us.

f) Exercises. Express the following assertions in the language $\mathcal{LR}$:

(1) $s = \int_a^b f(x)\,dx$ (Riemann integral);
(2) the series $\sum_{i=0}^{\infty} a_i$ converges (absolutely, conditionally);
(3) the sequence of functions $f_n(x)$ converges pointwise (uniformly) to a function $f(x)$;
(4) the series $\sum_{i=0}^{\infty} f_i(x)$ converges (absolutely, uniformly);
(5) uniform convergence theorems;
(6) $t = \mu(X)$, where $X \subseteq \mathbb{R}$ is open (closed), and μ is Lebesgue measure;
(7) the set X is Lebesgue-measurable, and $t = \mu(X)$;
(8) the function f is Lebesgue-measurable;
(9) $s = (L)\int_{\mathbb{R}} f\,dx$ (Lebesgue integral);

(10) the sequence of measurable functions $f_n(x)$ converges almost everywhere (in measure, in the mean);
(11) the function f is absolutely continuous (has bounded variation on $[a, b]$).

g) The examples above show that practically any sentence of elementary analysis can be formulated in the language $\mathcal{LR}$. We now show that a significant portion of functional analysis and general topology can also be formalized in $\mathcal{LR}$. It is clear that for this it suffices to learn how to write in $\mathcal{LR}$ propositions about families of subsets of a set.

Let X be a set variable of type $k > 0$, and let $\mathcal{Y}$ be a set variable of type $k+1$. Then $\mathcal{Y}$ determines a family of subsets of X indexed by the elements of $\operatorname{dom}^{(1)} \mathcal{Y}$:

$$\mathcal{Y} \subseteq \mathcal{P}(X) \rightleftharpoons \forall t_1, t_2 \in \operatorname{dom}^{(1)} \mathcal{Y}\ \exists x_1, x_2 \in X \\ (\langle t_1, x_1\rangle \in \mathcal{Y}\ \wedge\ \langle t_2, x_2\rangle \in \mathcal{Y}\ \wedge\ (\langle t_1, x_2\rangle \notin \mathcal{Y}\ \vee\ \langle t_2, x_1\rangle \notin \mathcal{Y})).$$

Here it is also indicated that different subsets correspond to different indices. It is easy to see that in $\mathcal{LR}$ we can only work with families of subsets with cardinality not exceeding the continuum. If $\mathcal{Y} \subseteq \mathcal{P}(X)$, then

$$Y \in \mathcal{Y} \rightleftharpoons \exists t \in \operatorname{dom}^{(1)} \mathcal{Y}\ \forall x \in X (x \in Y \longleftrightarrow \langle t, x\rangle \in \mathcal{Y}),$$

$$\mathcal{Y}_1 \subseteq \mathcal{Y}_2 \rightleftharpoons \forall t \in \operatorname{dom}^{(1)} \mathcal{Y}_1\ \exists s \in \operatorname{dom}^{(1)} \mathcal{Y}_2\ \forall x \in X \\ (\langle t, x\rangle \in \mathcal{Y}_1 \longleftrightarrow \langle s, x\rangle \in \mathcal{Y}_2).$$

We can now formulate the assertion that $\mathcal{Y}$ is a base for a topology on X:

$$\operatorname{Top}(X, \mathcal{Y}) \rightleftharpoons \forall x \in X \exists Y \in \mathcal{Y}\ (x \in Y) \\ \wedge\ \forall Y_1, Y_2 \in \mathcal{Y}\ \forall x \in Y_1 \cap Y_2\ \exists Z \in \mathcal{Y}\ (Z \subseteq Y_1 \cap Y_2\ \wedge\ x \in Z).$$

h) EXERCISES. Write the following assertions in $\mathcal{LR}$ assuming the sentence $\operatorname{Top}(X, \mathcal{Y})$:
(1) A is an open (closed, dense, nowhere dense, first-category) subset of the space X;
(2) A is the closure (interior) of a set B in the topological space X;
(3) $\mathcal{K}$ is an open covering of X;
(4) A is a compact (locally compact) subset of X;
(5) $f\colon X_1 \longrightarrow X_2$ is a continuous mapping of the topological space X_1 into the topological space X_2;
(6) A is a connected (arcwise-connected) subset of X.

Formulate the following in $\mathcal{LR}$:
(7) all the separation axioms for topological spaces;
(8) that $\mathcal{S}$ is a σ-algebra of subsets of the set X (generated by the family $\mathcal{Y}$);
(9) that $(X, \mathcal{Y}, \mu)$ is a measure space;
(10) that $\langle G, \cdot, \mathcal{Y}\rangle$ is a topological group, and μ is Haar measure on G;
(11) that $(X, +, \cdot)$ is a linear space over the field K (normed, countably normed, Banach, pre-Hilbert, Hilbert, finite-dimensional, infinite-dimensional);
(12) that Y is the space $C(X)$ of continuous functions on the topological space X;
(13) the Arzelà and Stone-Weierstrass theorems.

In an exposition of results on independence of the continuum hypothesis, the book [**59**] used a language $\mathrm{L_2Real}$ differing from $\mathcal{LR}$ only in the absence of variables of types $n > 1$ in it. The expressive possibilities of $\mathrm{L_2Real}$ are the same as for $\mathcal{LR}$. Here the language $\mathcal{LR}$ is used just for convenience.

§2. Interpretations of the language $\mathcal{LR}$

1°. In the preceding section it was shown how to write substantive mathematical assertions as sentences in the formal language $\mathcal{LR}$. The content of that section is in a certain sense analogous to the content of textbooks on programming languages and even on stenography.

In nonstandard analysis the proofs of mathematical theorems make essential use of their formal expressions in $\mathcal{LR}$. In this connection it is necessary to give a rigorous definition of a model of $\mathcal{LR}$, as well as of the truth or falsity of sentences of the language in this model.

DEFINITION 0.2.1. A model $\mathcal{M}$ of the language $\mathcal{LR}$ is defined to be a set M in which elements $m_0, m_1 \in M$ are distinguished, binary operations $f_+, f_\cdot \colon M^2 \longrightarrow M$ are defined, and a binary relation $P_< \subseteq M^2$ is given. Moreover, for each $n > 0$ a class $\mathcal{C}_n$ of n-place operations on M is defined, with $\mathcal{C}_0 = M$. For this class, $\mathcal{C} = \bigcup_{n\in\mathbb{N}} \mathcal{C}_n$ is closed with respect to superposition, and $f_+, f_\cdot, \chi_< \in \mathcal{C}_2$, where $\chi_<$ is the characteristic function of the relation $P_<$. The elements of $\mathcal{C}_n$ with $n > 0$ are called *internal operations.* Correspondingly, the subsets of $\mathbb{R}^k$ whose characteristic functions are internal functions are called *internal sets.* If for each $n > 0$ the class $\mathcal{C}_n$ is the class of *all* n-place operations on M, then the model is said to be *standard.* Otherwise the model is said to be *nonstandard.* We say that $\mathcal{M}$ is a model of the language $\mathcal{LR}$ on the basis of the set M, the operations f_+ and $f_\cdot$, and the relation $P_<$. The elements of $\mathcal{C}_n$ are called objects of type n.

Below we write 0, 1, $+$, $\cdot$, and $<$ instead of m_0, m_1, f_+, $f_\cdot$, and $P_<$, respectively (possibly with some indices), and we write the model itself in the form

$$\mathcal{M} = \langle M; \{0, 1, +, \cdot, <\}; \{\mathcal{C}_n \mid n \in \mathbb{N}\}\rangle.$$

The standard model on the basis of the ordered field $\mathbb{R}$ is also denoted by $\mathbb{R}$. Beginning with the next section, we consider only models constructed on the basis of the field $\mathbb{R}$ and certain of its extensions of special form. In this section all arguments are carried out in the general case of an arbitrary set M and binary operations and a binary relation on it.

DEFINITION 0.2.2. Suppose that along with the model $\mathcal{M}$ we are also given a model

$$\mathcal{M}' = \langle M'; \{0', 1', +', \cdot', <'\}; \{\mathcal{C}'_n \mid n \in \mathbb{N}\}\rangle.$$

A homomorphism of the model $\mathcal{M}$ into the model $\mathcal{M}'$ is defined to be a sequence $\Phi = \{\varphi_n \colon \mathcal{C}_n \longrightarrow \mathcal{C}'_n\}$ such that $\varphi_0(0) = 0'$, $\varphi_0(1) = 1'$, $\varphi_2(+) = +'$, $\varphi_2(\cdot) = \cdot'$, $\varphi_2(\chi_<) = \chi_{<'}$, and

$$\varphi_n(f)(\varphi_0(m_1), \dots, \varphi_0(m_n)) = \varphi_0(f(m_1, \dots, m_n)) \quad \forall f \in \mathcal{C}_n\ \forall m_1, \dots, m_n \in M. \tag{0.2.1}$$

Notation: $\Phi \colon \mathcal{M} \longrightarrow \mathcal{M}'$.

If all the φ_n are injective, then the homomorphism is called an *imbedding,* and $\mathcal{M}'$ an *extension* of $\mathcal{M}$. Here it is usually assumed that $M \subseteq M'$ and that φ_0 is the identity imbedding. The equality (0.2.1) shows that $\varphi_n(f)|M^n = f$ ($\varphi_n(f) \colon M'^n \longrightarrow M'$) in this case. If φ_0 is a bijection, then the equalities (0.2.1) uniquely determine mappings $\varphi_n \colon \mathcal{C}_n \longrightarrow \mathcal{C}'_n$ which are also bijections. In this case the homomorphism Φ is called an *isomorphism*. It is obvious that a model isomorphic to a standard model is itself standard.

2°. If $\mathcal{A}(\alpha_1, \dots, \alpha_n)$ is a formula in the language $\mathcal{LR}$, then upon substitution of objects $a_1, \dots, a_n$ in M of corresponding types in place of the free variables $\alpha_1, \dots, \alpha_n$ the formula $\mathcal{A}(a_1, \dots, a_n)$ becomes either true (takes the value $\mathbf{T}$) or false (takes the value $\mathbf{F}$). We now give a formal definition of the truth or falsity of $\mathcal{A}$ in the model $\mathcal{M}$. It is based on the following definition of truth and falsity for the logical connectives.

DEFINITION 0.2.3. If $\mathbb{B} = \{\mathbf{T}, \mathbf{F}\}$, then the logical connectives $\vee$, $\wedge$, $\longrightarrow$, $\longleftrightarrow$ are binary operations on $\mathbb{B}$, and $\neg$ is a unary operation on $\mathbb{B}$. They are given by the following truth tables.

P	Q	$\neg P$	$P \wedge Q$	$P \vee Q$	$P \longrightarrow Q$	$P \longleftrightarrow Q$
T	**T**	**F**	**T**	**T**	**T**	**T**
T	**F**	**F**	**F**	**T**	**F**	**F**
F	**T**	**T**	**F**	**T**	**T**	**F**
F	**F**	**T**	**F**	**F**	**T**	**T**

There is no doubt that this table adequately reflects the natural interpretation of the logical connectives (see §1, 1°). However, the definition of the value of $P \longrightarrow Q$ in the last two rows of the table is perhaps in need of some clarification. The meaning of the implication $P \longrightarrow Q$ in connection with a mathematical assertion is that it is possible to obtain Q from the assertion P by arguing correctly. It is clear that both a true and a false statement can be obtained from a false statement by arguing correctly.

PROPOSITION 0.2.4. *The logical connectives $\wedge$ and $\vee$, understood as operations on $\mathbb{B} = \{\mathbf{T}, \mathbf{F}\}$, are commutative, associative, and distributive each with respect to the other. Moreover, the following identities hold:*

(1) $P \longrightarrow Q \equiv \neg P \vee Q$;
(2) $\neg(P \vee Q) \equiv \neg P \wedge \neg Q$;
(3) $\neg(P \wedge Q) \equiv \neg P \vee \neg Q$.

The identities (2) and (3) are called the De Morgan laws. The identities (1)–(3) show that in fact it is possible to manage with either of the two pairs $\{\neg, \wedge\}$ and $\{\neg, \vee\}$ of logical connectives.

DEFINITION 0.2.5. Let

$$\mathcal{M} = \langle M; \{0, 1, +, \cdot, <\}; \{\mathcal{C}_n \mid n \in \mathbb{N}\}\rangle$$

be a model of the language $\mathcal{LR}$.

(1) An interpretation of the variables of $\mathcal{LR}$ is defined as a map $i\colon \bigcup_{n\in\mathbb{N}} \mathcal{A}_n \longrightarrow \bigcup_{n\in\mathbb{N}} \mathcal{C}_n$ (see Definition 0.2.1) such that $i(\mathcal{A}_n) \subseteq \mathcal{C}_n$, $i(0) = 0$, $i(1) = 1$, $i(+) = +$, and $i(\cdot) = \cdot$.

(2) Each interpretation of the variables determines an interpretation of the terms (also denoted by i) according to the following rules:
(a) if t is a variable or constant of type 0, then $i(t)$ is defined in item (1);
(b) if f is a variable of type $n > 0$, $t_1, \dots, t_n$ are terms, $i(t_1), \dots, i(t_n)$ are already defined, and $t = f(t_1, \dots, t_n)$, then $i(t) = i(f)(i(t_1), \dots, i(t_n))$.

(3) Any interpretation of the variables i gives a mapping (also denoted by i) of the set of all formulas in $\mathcal{LR}$ into the set $\mathbb{B} = \{\mathbf{T}, \mathbf{F}\}$ according to the following rules:
(a) $i(t_1 < t_2) \Longleftrightarrow i(t_1) < i(t_2)$, and $i(t_1 = t_2) \Longleftrightarrow i(t_1) = i(t_2)$;
(b) if $i(\mathcal{A})$ and $i(\mathcal{B})$ are already defined, then

$$i(\neg\mathcal{A}),\ i(\mathcal{A} \vee \mathcal{B}),\ i(\mathcal{A} \wedge \mathcal{B}),\ i(\mathcal{A} \longrightarrow \mathcal{B}),\ i(\mathcal{A} \longleftrightarrow \mathcal{B})$$

are computed according to the table in Definition 0.2.3;
(c) if $j(\mathcal{A})$ is defined for any interpretation j of the variables, and α is a variable (occurring freely in $\mathcal{A}$ as a rule, although formally this is not necessary), then $i(\exists\alpha\mathcal{A}) = \mathbf{T}$ if and only if there exists an interpretation i' of the variables such that $i(\beta) = i'(\beta)$ for $\beta \neq \alpha$ and $i'(\mathcal{A}) = \mathbf{T}$, and $i(\forall\alpha\mathcal{A}) = \mathbf{T}$ if and only if $i'(\mathcal{A}) = \mathbf{T}$ for any interpretation i' having this property (that is, differing from i at most on α).

If $i(\mathcal{A}) = \mathbf{T}$, then one says that $\mathcal{A}$ is true in $\mathcal{M}$ for the given interpretation of the variables. In the opposite case $\mathcal{A}$ is false in $\mathcal{M}$.

It is clear from the given definition that the value of $i(\mathcal{A})$ depends only on the values of $i(\alpha_1), \dots, i(\alpha_n)$, where $\alpha_1, \dots, \alpha_n$ is a complete list of the free variables occurring in $\mathcal{A}$. In particular, if $\mathcal{A}$ is a sentence, then the value of $i(\mathcal{A})$ does not depend on i. In this connection we use the following notation for sentences: $\mathcal{M} \models \mathcal{A} \rightleftharpoons i(\mathcal{A}) = \mathbf{T}$ for any interpretation i. In precisely the same way, if $\mathcal{A} = \mathcal{A}(\alpha_1, \dots, \alpha_n)$ and $a_1, \dots, a_n \in \mathcal{C}$, then $\mathcal{M} \models \mathcal{A}(a_1, \dots, a_n)$ means that $i(\mathcal{A}) = \mathbf{T}$ for any interpretation i such that $i(\alpha_k) = a_k$.

Thus, $\mathcal{A}(\alpha_1, \dots, \alpha_n)$ can be regarded as a predicate on M. It is often necessary to also consider predicates obtained from $\mathcal{A}(\alpha_1, \dots, \alpha_n)$ by substituting elements of $\mathcal{C}$ in place of part of the variables.

REMARK. Similarly, if t is a term, and $\alpha_1, \dots, \alpha_n$ are all the variables occurring in it (as in the case of formulas, the notation $t = t(\alpha_1, \dots, \alpha_n)$ is used in this case), then $i(t)$ depends only on the values of $i(\alpha_1), \dots, i(\alpha_n)$, that is, each term determines a mapping

$$\mathcal{C}_{k_1} \times \cdots \times \mathcal{C}_{k_n} \longrightarrow \mathcal{C}_0,$$

where k_j is the type of the variable α_j. The value of this mapping for $i(\alpha_l) = a_l$ is denoted by $t(a_1, \dots, a_n)$.

3°. Formulas $\mathcal{A}(\alpha_1, \dots, \alpha_n)$ and $\mathcal{B}(\alpha_1, \dots, \alpha_n)$ are said to be equivalent in the model $\mathcal{M}$ (notation: $\mathcal{A} \equiv_{\mathcal{M}} \mathcal{B}$) if

$$\forall a_1, \dots, a_n \in M\ (\mathcal{M} \models \mathcal{A}(a_1, \dots, a_n) \Longleftrightarrow \mathcal{M} \models \mathcal{B}(a_1, \dots, a_n)).$$

If this holds for any model $\mathcal{M}$, then the formulas $\mathcal{A}$ and $\mathcal{B}$ are said to be logically equivalent (notation: $\mathcal{A} \equiv \mathcal{B}$). The next proposition follows easily from Definition 0.2.5.

PROPOSITION 0.2.6. *The following logical equivalences are valid:*
1) $\neg\forall\alpha\mathcal{A} \equiv \exists\alpha\neg\mathcal{A}$;
2) $\neg\exists\alpha\mathcal{A} \equiv \forall\alpha\neg\mathcal{A}$;
3) $\forall\alpha(\mathcal{A}\wedge\mathcal{B}) \equiv \forall\alpha\mathcal{A} \;\wedge\; \forall\alpha\mathcal{B}$;
4) $\exists\alpha(\mathcal{A}\vee\mathcal{B}) \equiv \exists\alpha\mathcal{A} \;\vee\; \exists\alpha\mathcal{B}$;
5) $\forall\alpha(\mathcal{A}(\alpha)\vee\mathcal{B}) \equiv \forall\alpha\mathcal{A}\vee\mathcal{B}$ ($\mathcal{B}$ *does not have free occurrences of* α);
6) $\exists\alpha(\mathcal{A}(\alpha)\wedge\mathcal{B}) \equiv \exists\alpha\mathcal{A}\wedge\mathcal{B}$ ($\mathcal{B}$ *does not have free occurrences of* α);
7) $\forall\alpha\forall\beta\mathcal{A} \equiv \forall\beta\forall\alpha\mathcal{A}$;
8) $\exists\alpha\exists\beta\mathcal{A} \equiv \exists\beta\exists\alpha\mathcal{A}$;
9) $\exists\alpha\forall\beta\mathcal{A} \longrightarrow \forall\beta\exists\alpha\mathcal{A} \equiv \mathbf{T}$;
10) *if* $\mathcal{A}(\alpha)$ *does not have occurrences of a variable* β *(neither free nor bound) of the same type as* α*, then* $\exists\alpha\mathcal{A}(\alpha)\equiv\exists\beta\mathcal{A}(\beta)$ *and* $\forall\alpha\mathcal{A}(\alpha)\equiv\forall\beta\mathcal{A}(\beta)$;
11) *if* $\mathcal{M}$ *is a standard model, then*

$$\forall x\in X\exists y\in Y\ \mathcal{A}(x,y)\equiv_{\mathcal{M}}\exists f\in Y^X\forall x\ \mathcal{A}(x,f(x)),$$

where $\mathcal{A}$ *may possibly contain free variables besides* x *and* y.

We do not present the simple proof of this proposition here, noting only that 11) is equivalent to the axiom of choice.

By using the equivalences 1)–6) and 10) it is easy to prove the next result.

PROPOSITION 0.2.7. *For any formula* $\mathcal{A}(\alpha_1,\dots,\alpha_n)$ *it is possible to construct a formula* $\mathcal{B}(\alpha_1,\dots,\alpha_n)$ *logically equivalent to it and having the form*

$$\mathcal{B}(\alpha_1,\dots,\alpha_n)\equiv Q_1\beta_1\dots Q_m\beta_m C(\alpha_1,\dots,\alpha_n,\beta_1,\dots,\beta_m), \tag{0.2.2}$$

where $Q_i=\forall$ *or* $\exists$ *for* $i=1,\dots,m$*, and the formula* C *does not contain quantifiers.*

A formula of the form (0.2.2) is said to have *prenex normal form.*

4°. Models $\mathcal{M}_1$ and $\mathcal{M}_2$ are said to be *elementarily equivalent* if for any sentence $\mathcal{A}$ in the language $\mathcal{LR}$

$$\mathcal{M}_1\models\mathcal{A}\iff\mathcal{M}_2\models\mathcal{B}.$$

PROPOSITION 0.2.8. *Isomorphic models are elementarily equivalent.*

PROOF. Let $\Phi=\{\varphi_n\}$ be an isomorphism between $\mathcal{M}$ and $\mathcal{M}'$ (see Definition 0.2.2). Then, as mentioned, $\varphi_0\colon M\longrightarrow M'$ is a bijection, and the φ_n are uniquely determined by the equalities (0.2.1). We show that

$$\mathcal{M}\models\mathcal{A}(a_1,\dots,a_n)\iff\mathcal{M}'\models\mathcal{A}(\varphi_{k_1}(a_1),\dots,\varphi_{k_n}(a_n)) \tag{0.2.3}$$

for any formula $\mathcal{A}(\alpha_1,\dots,\alpha_n)$ with α_j a variable of type k_j and for any n-tuple $\langle a_1,\dots,a_n\rangle\in\mathcal{C}_{k_1}\times\dots\times\mathcal{C}_{k_n}$. Our proposition is obtained from (0.2.3) for $n=0$. Let i be an arbitrary interpretation of variables in $\mathcal{M}$. We denote by $\Phi\circ i$ the interpretation i' of the variables in $\mathcal{M}'$ defined as follows. If x is a variable of type 0, then $i'(x)=\varphi_0\circ i(x)$, and if f is a variable of type $n>0$, then $i'(f)$ is determined from the equality (0.2.1). Since all the φ_n are bijective, each interpretation i' in $\mathcal{M}'$ has the form $\Phi\circ i$ for some interpretation i in $\mathcal{M}$. Further,

$$i'(\alpha_j)=\varphi_{k_j}(a_j)\iff i(\alpha_j)=a_j. \tag{0.2.4}$$

We show that for each term $t(\alpha_1,\dots,\alpha_n)$

$$i'(t)(\varphi_{k_1}(a_1),\dots,\varphi_{k_n}(a_n))=\varphi_0(i(t)(a_1,\dots,a_n)) \tag{0.2.5}$$

(see Definition 0.2.5 (2)). For terms that are variables or constants of type 0 this is obvious from the definition of the interpretation $\Phi \circ i$. Let $t = f(s_1, \dots, s_m)$ and suppose that the required equality has already been proved for the terms $s_j = s_j(\alpha_1, \dots, \alpha_n)$. Below we write $\overline{\xi}$ instead of $\xi_1, \dots, \xi_l$, and $\overline{\eta}(\overline{\xi})$ instead of $\eta_1(\xi_1), \dots, \eta_l(\xi_l)$. Then $t = t(\overline{\alpha}, f)$. If $i(f) = b$, then $i'(f) = \varphi_m(b)$, and it is required to prove that $i'(t)(\overline{\varphi}(\overline{a}), \varphi_m(b)) = \varphi_0(i(t)(\overline{a}, b))$.

In view of Definition 0.2.5 (2), the equality (0.2.1), and the induction hypothesis we have that

$$
\begin{aligned}
i'(t)(\overline{\varphi}(\overline{a}), \varphi_m(b)) &= \varphi_m(b)(i'(\overline{s})(\overline{\varphi}(\overline{a}))) \\
&= \varphi_m(b)(i(\overline{s})(\overline{a})) = \varphi_0(b(i(\overline{s})(\overline{a}))) = \varphi_0(i(t)(\overline{a}, b)),
\end{aligned}
$$

which is what was required. We first prove (0.2.3) for a formula $\mathcal{A}$ having the form $t_1 = t_2$. Let i be an arbitrary interpretation such that $i(\alpha_j) = a_j$, and let $i' = \Phi \circ i$. Then $\mathcal{M} \models t_1 = t_2 \iff i(t_1) = i(t_2)$ and $\mathcal{M}' \models t_1 = t_2 \iff i'(t_1) = i'(t_2)$, but $i'(t_l) = \varphi_0(i(t_l))$ in view of (0.2.5), and (0.2.3) follows from the fact that φ_0 is a bijection.

If $\mathcal{A}$ has the form $t_1 < t_2$, then

$$\mathcal{M} \models t_1 < t_2 \iff i(t_1) < i(t_2), \quad \mathcal{M}' \models t_1 < t_2 \iff i'(t_1) < i'(t_2).$$

In view of (0.2.5) the equivalence (0.2.3) now follows from the equivalence $m_1 < m_2 \iff \varphi_0(m_1) < \varphi_0(m_2)$, which, in turn, follows from the fact that $\chi_{<'} = \varphi_2(\chi_<)$ (see Definition 0.2.2). If (0.2.3) has been proved for formulas $\mathcal{A}$ and $\mathcal{B}$, then in view of Definition 0.2.5 (3b) it holds also for any formulas obtained from $\mathcal{A}$ and $\mathcal{B}$ with the help of propositional connectives. We show that it holds also for formulas obtained from $\mathcal{A}$ and $\mathcal{B}$ by quantification.

Suppose that $\mathcal{M} \models \exists \alpha_1 \mathcal{A}(\alpha_1, a_2, \dots, a_n)$. Then there exists an interpretation j differing from i at most at α_1 and such that $\mathcal{M} \models \mathcal{A}(b, a_2, \dots, a_n)$, where $b = j(\alpha_1)$. If $j' = \Phi \circ j$, then, by (0.2.4), $j'(\alpha_1) = \varphi_{k_1}(b)$ and $j'(\alpha_l) = \varphi_{k_l}(a_l)$ for $l = 2, \dots, n$. By the induction hypothesis, $\mathcal{M}' \models \mathcal{A}(\varphi_{k_1}(b), \varphi_{k_2}(a_2), \dots, \varphi_{k_n}(a_n))$, and hence $\mathcal{M}' \models \exists \alpha_1 \mathcal{A}(\alpha_1, \varphi_{k_2}(a_2), \dots, \varphi_{k_n}(a_n))$. The proof is analogous in the reverse direction, with account taken of the fact that every interpretation j' in $\mathcal{M}'$ has the form $j' = \Phi \circ j$ (see above). For the universal quantifier the assertion now follows from Proposition 0.2.6 (1).

DEFINITION 0.2.9. An imbedding $\Phi\colon \mathcal{M} \longrightarrow \mathcal{M}'$ (see Definition 0.2.2) is said to be *elementary* if

$$\mathcal{M} \models \mathcal{A}(a_1, \dots, a_n) \iff \mathcal{M}' \models \mathcal{A}(\varphi_{k_1}(a_1), \dots, \varphi_{k_n}(a_n)) \tag{0.2.6}$$

for an arbitrary formula $\mathcal{A}(\alpha_1, \dots, \alpha_n)$ and arbitrary objects $a_1, \dots, a_n \in \mathcal{C}$ of corresponding types, where k_i is the type of the variable α_i. We recall that $\Phi = \{\varphi_n\colon \mathcal{C}_n \longrightarrow \mathcal{C}'_n \mid n \in \mathbb{N}\}$ and $\varphi_0(m) = m \ \forall m \in M$, that is, $M \subseteq M'$. Below, if the type of an object $a \in \mathcal{C}$ is not indicated, then $\Phi(a)$ denotes $\varphi_k(a)$, where a is of type k.

If $\Phi\colon \mathcal{M} \longrightarrow \mathcal{M}'$ is an elementary imbedding, then $\mathcal{M}'$ is called an *elementary extension* of $\mathcal{M}$, and, correspondingly, $\mathcal{M}$ is called an elementary submodel of $\mathcal{M}'$. If $a \in \mathcal{C}$, then $\Phi(a)$ is called an $\mathcal{M}$-element of the model $\mathcal{M}'$. If $\mathcal{M}$ is a standard model, then the $\mathcal{M}$-elements of $\mathcal{M}'$ are also called standard elements.

It will be shown below that any model has proper elementary extensions. Identifying the internal sets with their characteristic functions, we call an object $A \in \mathcal{C}$ an internal set if $\mathcal{M} \models \mathrm{Set}^{(n)}(A)$ for some $n \in \mathbb{N}$. The formula $\mathrm{Set}^{(n)}$ was defined in §1, 2°, part b. Internal objects satisfying the other formulas in the same subsection are defined similarly: relations, mappings, and so on. Moreover, by the equivalence (0.2.6), if an object $A \in \mathcal{C}$ satisfies one of these formulas, then the object $\Phi(A) \in \mathcal{C}'$ also satisfies the same formula.

PROPOSITION 0.2.10.

1. *The sets* $A, B \subseteq M^k$ *are internal in* $\mathcal{M}$ *if and only if* $\varphi_k(A), \varphi_k(B) \subseteq M'^k$ *are internal in* $\mathcal{M}'$. *Furthermore:*
   a) $\varphi_k(A) \cap M^k = A$;
   b) $A \subseteq B \Longleftrightarrow \varphi_k(A) \subseteq \varphi_k(B)$ *and* $A = B \Longleftrightarrow \varphi_k(A) = \varphi_k(B)$;
   c) $\Phi(A \delta B) = \Phi(A) \delta \Phi(B)$, *where* $\delta = \cup, \cap, \setminus, \times$.
2. *If* $\mathrm{dom}^{(n)} A$ *is an internal set in* $\mathcal{M}$, *then* $\Phi(\mathrm{dom}^{(n)} A) = \mathrm{dom}^{(n)} \Phi(A)$, *and similarly for* $\mathrm{rng}^{(n)} A$.
3. $\mathcal{M} \models F\colon A \longrightarrow B \Longleftrightarrow \mathcal{M}' \models \Phi(F)\colon \Phi(A) \longrightarrow \Phi(B)$ *(recall that* $\mathcal{M} \models F\colon A \longrightarrow B$ *if and only if* F, A, *and* B *are internal objects, and* F *is a mapping from* A *to* B*). Furthermore:*
   a) F *is injective (surjective, bijective) if and only if* $\Phi(F)$ *is injective (surjective, bijective);*
   b) $\Phi(F)|A = F$;
   c) *if* $\mathcal{M} \models F\colon A \longrightarrow B$, *then* $F = G \Longleftrightarrow \Phi(F) = \Phi(G)$.

PROOF. In part 1 we confine ourselves to the case when $k = 1$. Let $m \in M$. In view of (0.2.6) and the fact that $\varphi_0(m) = m$,

$$m \in \varphi(A) \Longleftrightarrow \mathcal{M}' \models m \in \varphi_1(A) \Longleftrightarrow \mathcal{M}' \models \varphi_0(m) \in \varphi_1(A) \Longleftrightarrow \mathcal{M} \models m \in A \Longleftrightarrow m \in A.$$

This implies that $\varphi_1(A) \cap M = A$. The remaining assertions of part 1 follow directly from (0.2.6). Let $\mathcal{D} = \mathrm{dom}^{(1)} A \in \mathcal{C}$. Then

$$\mathcal{M} \models \forall \overline{x}\ (\overline{x} \in \mathcal{D} \longleftrightarrow \exists \overline{y}\ \langle \overline{x}, \overline{y} \rangle \in A).$$

By (0.2.6),

$$\mathcal{M}' \models \forall \overline{x}\ (\overline{x} \in \Phi(\mathcal{D}) \longleftrightarrow \exists \overline{y}\ \langle \overline{x}, \overline{y} \rangle \in \Phi(A)). \tag{0.2.7}$$

On the other hand, an internal set $E \in \mathcal{C}$ is $\mathrm{dom}^{(n)} \Phi(A)$ if and only if

$$\mathcal{M}' \models \forall \overline{x}\ (\overline{x} \in E \longleftrightarrow \forall \overline{y}\ \langle \overline{x}, \overline{y} \rangle \in \Phi(A)). \tag{0.2.8}$$

Comparing (0.2.7) and (0.2.8), we get that $\Phi(\mathcal{D}) = \mathrm{dom}^{(n)} \Phi(A)$, which is what was required.

The assertions in part 3 follow at once from (0.2.6) and the assertions in part 2.

REMARK. If $\mathcal{M}$ is a standard model, then arbitrary sets are internal, that is, the class of internal sets is closed under Boolean operations, Cartesian products, and projection. It follows from (0.2.6) and Proposition 0.2.10 that in an arbitrary elementary extension of a standard model the class of internal sets is closed under these operations.

The preceding considerations can be carried over also to certain classes of internal sets.

DEFINITION 0.2.11. i). Suppose that $\mathfrak{M} \subseteq \mathcal{C}_{k_1} \times \cdots \times \mathcal{C}_{k_n}$. $\mathfrak{M}$ is called an *internal subclass* of the model $\mathcal{M}$ if there exist a formula $\mathcal{A}(\alpha_1, \ldots, \alpha_n, \beta_1, \ldots, \beta_l)$ and elements $b_1, \ldots, b_l \in \mathcal{C}$ such that

$$\mathfrak{M} = \{\langle a_1, \ldots, a_n \rangle \mid \mathcal{M} \models \mathcal{A}(a_1, \ldots, a_n, b_1, \ldots, b_l)\}.$$

In this case it is also said that the class $\mathfrak{M}$ is definable (by the formula $\mathcal{A}$) in terms of the elements $b_1, \ldots, b_l$. If $\mathcal{M}$ is a standard model, then the internal class $\mathfrak{M}$ is said to be standard in $\mathcal{M}$.

ii). If $\mathcal{M}'$ is an elementary extension of $\mathcal{M}$, then an internal subclass $\mathfrak{M}'$ of $\mathcal{M}$ will be called an $\mathcal{M}$-class (a standard class if $\mathcal{M}$ is a standard model) if it is definable in terms of the elements $\Phi(b_1), \ldots, \Phi(b_l)$, where $b_1, \ldots, b_l \in \mathcal{C}$. If, furthermore, $\mathfrak{M}$ is a subclass of the model $\mathcal{M}$ that is definable by the same formula in terms of the elements $b_1, \ldots, b_l$, then the notation $\mathfrak{M}' = \Phi(\mathfrak{M})$ is used.

REMARKS. 1. Each internal set $A \subseteq M^k$ is an internal subclass of $\mathcal{M}$:

$$A = \{\langle m_1, \ldots, m_k \rangle \mid \mathcal{M} \models \langle m_1, \ldots, m_k \rangle \in A\}.$$

The converse does not hold in general, that is, there can be internal subclasses $\mathfrak{M} \subseteq M^k$ that are not internal sets in the sense of Definition 0.2.1. In particular, $\mathrm{dom}^{(n)} A$ is always an internal class if A is an internal set.

2. If $\mathcal{M}'$ is an elementary extension of $\mathcal{M}$ and A is an internal set in $\mathcal{M}$, then $\varphi_k(A) = \Phi(A)$, where $\Phi(A)$ is the internal class in $\mathcal{M}'$ introduced in Definition 0.2.11 ii). Thus, the notation $\Phi(\mathfrak{M})$ is consistent with the notation $\Phi(a)$ introduced in Definition 0.2.9.

PROPOSITION 0.2.12. *Let $\mathcal{M}'$ be an elementary extension of the model $\mathcal{M}$, and let $\mathfrak{M}_1$ and $\mathfrak{M}_2$ be internal subclasses of $\mathcal{M}$. Then:*
(1) $\forall \overline{a} \in \mathcal{C}^k \ (\Phi(\overline{a}) \in \Phi(\mathfrak{M}_1) \iff \overline{a} \in \mathfrak{M}_1)$;
(2) $\Phi(\mathfrak{M}_1) \subseteq \Phi(\mathfrak{M}_2) \iff \mathfrak{M}_1 \subseteq \mathfrak{M}_2$;
(3) $\Phi(\mathfrak{M}_1 \delta \mathfrak{M}_2) = \Phi(\mathfrak{M}_1) \delta \Phi(\mathfrak{M}_2)$, *where* $\delta = \cup, \cap, \setminus, \times$;
(4) *if* $\mathfrak{M}_2 = \mathrm{dom}^{(n)} \mathfrak{M}_1$, *then* $\Phi(\mathfrak{M}_2) = \mathrm{dom}^{(n)} \Phi(\mathfrak{M}_1)$.

PROOF. This repeats the proof of the corresponding assertions in Proposition 0.2.10.

It is easy to see that the families of internal subsets considered in §1, 2°, part g are internal subclasses of the model $\mathcal{M}$ in the sense of Definition 0.2.11.

REMARK. To make possible a uniform treatment of internal sets, families of internal sets, families of families, and so on, it is common in books on nonstandard analysis to consider the superstructure $V(\mathbb{R}) = \bigcup_{n \in \mathbb{N}} V_n(\mathbb{R})$, where $V_0(\mathbb{R}) = \mathbb{R}$ and $V_{n+1}(\mathbb{R}) = \mathcal{P}(V_n(\mathbb{R}))$, along with the language L for formalization of sentences about this superstructure, and elementary extensions of it. However, a large part of the substantive mathematics is 'distributed' in the first two or three 'storeys' of this superstructure, and, as shown above, the language $\mathcal{LR}$ is quite sufficient for 'servicing' these storeys. $\mathcal{LR}$ seems to us to be essentially simpler and more natural than the language L for the superstructure. In §6 of this chapter we consider briefly the language of set theory, which is universal for the whole of mathematics, along with modifications of it that are adapted for the treatment of nonstandard analysis in the most general form.

§3. Elementary extensions of the model $\mathbb{R}$

1°. We consider the standard model of $\mathcal{LR}$ constructed on the basis of the field of real numbers, that is, the model $\langle \mathbb{R}, \{0, 1, +, \cdot, <\}, \{\mathcal{C}_n \mid n \in \mathbb{N}\}\rangle$, where $\langle \mathbb{R}, \{0, 1, +, \cdot, <\}\rangle$ is the ordered field of real numbers, $\mathcal{C}_0 = \mathbb{R}$, and $\mathcal{C}_n$ is the set of all mappings from $\mathbb{R}^n$ to $\mathbb{R}$ for $n > 0$. The model itself is also denoted by $\mathbb{R}$. Since all complete (that is, satisfying the least upper bound axiom) ordered fields are isomorphic to $\mathbb{R}$, any two standard models constructed on the basis of complete ordered fields are also isomorphic (see Definition 0.2.2), that is, the following result holds.

PROPOSITION 0.3.1. *If $\mathcal{M}$ is a standard model of $\mathcal{LR}$ and $\mathcal{M} \models$ LOF $\wedge$ LUB (see §1, 2°, parts* c, d), *then $\mathcal{M}$ is isomorphic to the model $\mathbb{R}$.*

The basic objects of nonstandard analysis are the elementary extensions of the model $\mathbb{R}$. Such an elementary extension is usually denoted by

$$\langle {}^*\mathbb{R}, \{{}^*0, {}^*1, {}^*+, {}^*\cdot, {}^*<\}, \{{}^*\mathcal{C}_n \mid n \in \mathbb{N}\}\rangle. \tag{0.3.1}$$

Here it is assumed that $\mathbb{R} \subseteq {}^*\mathbb{R}$, that is, ${}^*0 = 0$ and ${}^*1 = 1$. Moreover, in place of ${}^*+$, ${}^*\cdot$, and ${}^*<$ we simply write $+$, $\cdot$, and $<$. For each $n > 1$ the imbedding of $\mathcal{C}_n$ in ${}^*\mathcal{C}_n$ is also commonly denoted by *, that is, if $f \in \mathcal{C}_n$, then its image in ${}^*\mathcal{C}_n$ is denoted by ${}^*f$ with the exception of the case $n = 0$ (since in this case ${}^*t = t$ for $t \in \mathbb{R}$) and, as already mentioned, of the images of $+$, $\cdot$, and $<$, for which the same notation is kept. Similarly, if $\mathfrak{M}$ is a standard class in $\mathbb{R}$, then its image in ${}^*\mathbb{R}$ (see Definition 0.2.11) is also denoted by ${}^*\mathfrak{M}$. The model (0.3.1) itself is also denoted by ${}^*\mathbb{R}$. If $a \in \mathcal{C}$ or a is a standard class of $\mathfrak{M}$, then ${}^*a$ is called the nonstandard extension of a.

The equality (0.2.1) for this case is written in the form

$$\forall t_1, \ldots, t_n \in \mathbb{R}\ \forall f \colon \mathbb{R}^n \longrightarrow \mathbb{R}\ [{}^*f(t_1, \ldots, t_n) = f(t_1, \ldots, t_n)],$$

and the property of elementary equivalence in this case is one of the most important principles of nonstandard analysis. Here it bears the name

TRANSFER PRINCIPLE. *If $\mathcal{A}(\alpha_1, \ldots, \alpha_n)$ is a formula in $\mathcal{LR}$, and $a_1, \ldots, a_n \in \mathcal{C}$ are objects of the same types as the variables $\alpha_1, \ldots, \alpha_n$, then*

$$\mathbb{R} \models \mathcal{A}(a_1, \ldots, a_n) \Longleftrightarrow {}^*\mathbb{R} \models \mathcal{A}({}^*a_1, \ldots, {}^*a_n).$$

This implies, in particular, that ${}^*\mathbb{R}$ is an ordered field.

In §1 it was demonstrated that most theorems of mathematical analysis can be written as sentences in the language $\mathcal{LR}$. We are now in a position to give this statement a more precise meaning. The meaning is that associated with a mathematical sentence $\mathcal{A}$ is a formal sentence A of $\mathcal{LR}$ such that $\mathcal{A}$ is a true mathematical theorem if and only if $\mathbb{R} \models A$. By virtue of the transfer principle this holds if and only if ${}^*\mathbb{R} \models A$. Informally, the latter means the truth in ${}^*\mathbb{R}$ of the sentence obtained when each occurrence in $\mathcal{A}$ of the words "for all ... " is replaced by "for all internal ... ", and each occurrence of "there exists ... " is replaced by "there exists an internal ... ".

EXAMPLE 0.3.2. It follows from the transfer principle that

$$^*\mathbb{R} \models \forall B \subseteq {}^*\mathbb{N}\ (0 \in B\ \wedge\ \forall n(n \in B \longrightarrow (n+1) \in B) \longrightarrow B = {}^*\mathbb{N}).$$

This means that any subset $B \subseteq {}^*\mathbb{N}$ that is internal in $^*\mathbb{R}$ (that is, $B \in \mathcal{C}_1$) and has the property that

$$0 \in B\ \wedge\ \forall n(n \in B \longrightarrow (n+1) \in B) \tag{0.3.2}$$

coincides with $^*\mathbb{N}$. It follows from Proposition 0.2.10 that $\mathbb{N} \subseteq {}^*\mathbb{N}$. In the next subsection it will be shown that if $^*\mathbb{R}$ is a proper extension of $\mathbb{R}$, then also $^*\mathbb{N} \neq \mathbb{N}$. On the other hand, $^*\mathbb{N}$ obviously satisfies (0.3.2), that is, in any proper elementary extension $^*\mathbb{R}$ of the model $\mathbb{R}$ the set $\mathbb{N}$ is not internal.

EXAMPLE 0.3.3. In Remark 1 after Definition 0.2.11 it was noted that an internal subclass of M^k in the model $\mathcal{M}$ need not in general be an internal set. We show, however, that this is the case when $\mathcal{M} = {}^*\mathbb{R}$ is an elementary extension of the model $\mathbb{R}$.

Obviously, for any formula $\mathcal{A}(x_1, \dots, x_n, \alpha_1, \dots, \alpha_k)$ with variables $x_1, \dots, x_n$ of type 0 (see §1, 2°, part b)

$$\mathbb{R} \models \forall \alpha_1, \dots, \alpha_k \exists X^{(n)} \forall x_1, \dots, x_n$$
$$(\langle x_1, \dots, x_n \rangle \in X^{(n)} \longleftrightarrow \mathcal{A}(x_1, \dots, x_n \alpha_1, \dots, \alpha_k)).$$

By the transfer principle, this sentence is true also in $^*\mathbb{R}$. The latter means that for any internal objects $a_1, \dots, a_k \in {}^*\mathcal{C}$ ($^*\mathcal{C} = \bigcup_{m \in \mathbb{N}} {}^*\mathcal{C}_m$) there exists an *internal set* $A \in {}^*\mathcal{C}_n$ such that

$$A = \{\langle t_1, \dots, t_n \rangle \in {}^*\mathbb{R}^n \mid {}^*\mathbb{R} \models \mathcal{A}(t_1, \dots, t_n, a_1, \dots, a_k)\}.$$

Of course, this argument also works for elementary extensions of other standard models.

EXAMPLE 0.3.4. By the transfer principle, $^*\mathbb{R} \models$ "Any set $A \subseteq {}^*\mathbb{R}$ that is bounded above has a least upper bound." (In fact, here we should write the sentence LUB, which is a formalization of the assertion in quotes; instead of the formal analogues of mathematical statements we shall in what follows write the statements themselves in quotes, with a view to clarity.) This means that any *internal* subset of $^*\mathbb{R}$ that is bounded above has a least upper bound. In the next subsection we consider examples of subsets of $^*\mathbb{R}$ not having a least upper bound even though they are bounded above, that is, subsets that are not internal.

2°. We first study an ordered field $^*\mathbb{R}$ that is a proper extension of the field $\mathbb{R}$. In this (and only this) subsection it is not required that $^*\mathbb{R}$ be an elementary extension of $\mathbb{R}$.

DEFINITION 0.3.5.
i) An element $\alpha \in {}^*\mathbb{R}$ is said to be infinitesimal if $\forall t \in \mathbb{R}\ (t > 0 \longrightarrow |\alpha| < t)$.
ii) An element $\Omega \in {}^*\mathbb{R}$ is said to be infinite if $\forall t \in \mathbb{R}\ (t > 0 \longrightarrow |\Omega| > t)$.
iii) An element $\xi \in \mathbb{R}$ is said to be bounded if it is not infinite.

Below we use the following notation and terminology:
$\alpha \approx 0 \longleftrightarrow \alpha$ is infinitesimal, and $\alpha \approx \beta \longleftrightarrow \alpha - \beta \approx 0$;
$\Omega \sim \infty \longleftrightarrow \Omega$ is infinite;

$\mu(0) = \{\alpha \mid \alpha \approx 0\}$ is the monad or halo of zero, and, similarly, $\mu(t) = t+\mu(0) = \{\xi \in {}^*\mathbb{R} \mid \xi \approx t\}$ is the monad or halo of t $\forall t \in \mathbb{R}$;

${}^*\mathbb{R}_b$ is the set of bounded elements of ${}^*\mathbb{R}$ (it is obvious from the definition that $\mu(0) \subseteq {}^*\mathbb{R}_b$);

$|t| \ll +\infty \longleftrightarrow t \in {}^*\mathbb{R}_b$

The proof of the following lemma is trivial and is left to the reader.

LEMMA 0.3.6.
i) *If $\alpha \in \mathbb{R}$ and $\alpha \neq 0$, then $\alpha \approx 0 \Longleftrightarrow \alpha^{-1} \sim \infty$.*
ii) *A sum of infinitesimal elements is infinitesimal.*
iii) *A sum of bounded elements is bounded, as is their product.*
iv) *The product of an infinitesimal element and a bounded element is infinitesimal.*

THEOREM 0.3.7. *In any proper extension of the ordered field $\mathbb{R}$ there exist infinitesimal and infinite elements, and for every bounded element $\xi \in {}^*\mathbb{R}$ there exists a unique $t \in \mathbb{R}$ such that $t \approx \xi$. This t is called the standard part or shadow of the number ξ and is denoted by ${}^\circ\xi$ or $\mathrm{st}(\xi)$.*

PROOF. Let ξ be a bounded element of ${}^*\mathbb{R}$, and suppose that $\xi \notin \mathbb{R}$. Then $\exists s \in \mathbb{R}$ $(|\xi| < s)$ (see Definition 0.3.5 iii). We consider the set $M = \{u \in \mathbb{R} \mid u < \xi\}$. Then $M \neq \emptyset$ $(-s \in M)$ and is bounded above by the number s, that is, $\exists t = \sup M \in \mathbb{R}$. We show that $\xi - t \approx 0$. Let us consider two cases.

1) $t - \xi > 0$. Suppose that $\exists w \in \mathbb{R}$ $(t - \xi > w > 0)$. Then $\xi < t - w$ and $\forall u \in M$ $(u < t-w)$, that is, $\sup M \leq t-w$. This contradiction shows that $t-\xi \approx 0$.

2) $\xi - t > 0$. Again suppose that $\exists w \in \mathbb{R}$ $(\xi - t > w > 0)$. Then $\xi > t + w$, that is, $t + w \in M$, and hence $\sup M \geq t + w$. We again get a contradiction, so $\xi - t \approx 0$.

Thus, $0 \neq \xi - t \approx 0$, and $(\xi - t)^{-1} \sim +\infty$.

THEOREM 0.3.8. i) *${}^*\mathbb{R}_b$ is a subring of ${}^*\mathbb{R}$, and $\mu(0)$ is a maximal ideal in ${}^*\mathbb{R}_b$.*
ii) *The mapping $\mathrm{st}\colon {}^*\mathbb{R}_b \longrightarrow \mathbb{R}$ defined in Theorem 0.3.7 is a surjective homomorphism, and $\mathrm{Ker\ st} = \mu(0)$.*

PROOF. Lemma 0.3.6 gives us immediately that ${}^*\mathbb{R}_b$ is a subring of ${}^*\mathbb{R}$ and $\mu(0)$ is an ideal in ${}^*\mathbb{R}_b$. The equalities ${}^\circ(\xi + \eta) = {}^\circ\xi + {}^\circ\eta$ and ${}^\circ(\xi \cdot \eta) = {}^\circ\xi \cdot {}^\circ\eta$ also follow easily from this lemma. They show that $\mathrm{st}\colon {}^*\mathbb{R}_b \longrightarrow \mathbb{R}$ is a homomorphism. It is clear that $\mathrm{st}(t) = t$ $\forall t \in \mathbb{R}$, that is, st is surjective. The fact that $\mathrm{Ker\ st} = \mu(0)$ follows immediately from the definition of st, and the ideal $\mu(0)$ is maximal because $\mathbb{R}$ is a field.

PROPOSITION 0.3.9. *The bounded set $\mu(0)$ does not have a least upper bound, and hence no proper extension of $\mathbb{R}$ is isomorphic to $\mathbb{R}$.*

PROOF. It is obvious that $\mu(0)$ is bounded above, for example, by 1. Let $\xi = \sup \mu(0)$. If $\xi \in \mu(0)$, then $2\xi \in \mu(0)$ by virtue of Lemma 0.3.6, but $2\xi > \xi$, and we get a contradiction. Thus, $\xi \notin \mu(0)$, that is, $\exists w \in \mathbb{R}$ $(\xi > w > 0)$. But since $w \in \mathbb{R}$ and $w > 0$, it follows that $\forall \alpha \in \mu(0)$ $(\alpha < w)$, that is, $\sup \mu(0) = \xi < w$. Contradiction.

EXERCISE 0.3.10. 1) Show that the sets ${}^*\mathbb{R}_b$ and $\mathbb{R}$ are bounded above and below in ${}^*\mathbb{R}$, yet have neither a least upper bound nor a greatest lower bound.

2) Show that $\{\Omega \in {}^*\mathbb{R} \mid \Omega \sim +\infty\}$ (that is, $\Omega \sim \infty$ and $\Omega > 0$) is bounded below, yet does not have a greatest lower bound.

3°. Returning to the case when ${}^*\mathbb{R}$ is an elementary extension of the model $\mathbb{R}$, we remark first of all that the following result is proved in Example 0.3.4.

PROPOSITION 0.3.11. *Each internal subset of* ${}^*\mathbb{R}$ *that is bounded above (below) has a least upper (greatest lower) bound.*

The subsets of ${}^*\mathbb{R}^k$ and subclasses of ${}^*\mathcal{C}$ that are not internal are said to be *external.*

The next result now follows from Proposition 0.3.9 and Exercises 0.3.10.

PROPOSITION 0.3.12. *The following subsets of* ${}^*\mathbb{R}$ *are external:* $\mu(0)$, $\mathbb{R}$, ${}^*\mathbb{R}_b$, $\{\Omega \mid \Omega \sim \infty\}$.

On the other hand, Proposition 0.3.11 and the transfer principle yield some important properties of internal sets.

THEOREM 0.3.13.

i) (*Boundedness Principle*) *If an internal set* $B \subseteq \mathbb{R}$ *consists solely of bounded elements, then there exists a standard* $t \in \mathbb{R}$ *such that* $B \subseteq {}^*[-t, t]$ (*note that* ${}^*[-t,t] = \{\xi \in {}^*\mathbb{R} \mid -t \leq \xi \leq t\}$ *by the transfer principle; this set will be denoted simply by* $[-t;t]$).

ii) (*Permanence Principle*) *If an internal set* B *contains all the positive bounded numbers, then it contains also the interval* $[0;\Omega]$ *for some infinite* Ω.

iii) (*Cauchy Principle*) *If an internal set* B *contains all the infinitesimal numbers, then it contains also the interval* $[-a;a]$ *for some standard* $a \in \mathbb{R}$ (*the word "standard" can actually be dropped, of course, since all the elements of* $\mathbb{R}$ *are standard*).

iv) (*Robinson Principle*) *If an internal set* B *consists solely of infinitesimal numbers, then* B *is contained in the interval* $[-\varepsilon;\varepsilon]$ *for some infinitesimal* ε.

v) (*Robinson Lemma*) *If* $f\colon {}^*\mathbb{R} \longrightarrow {}^*\mathbb{R}$ *is an internal function and* $f(x) \approx 0$ $\forall x \in {}^*\mathbb{R}_b$, *then* $\exists \Omega \sim \infty$ $(f([-\Omega;\Omega]) \subseteq \mu(0))$.

PROOF. i) If $\lambda \sim +\infty$, then $|\xi| < \lambda$ $\forall \xi \in B$ by assumption, that is, B is bounded. Since $|B|$ is bounded above, there exists a positive $\mu = \sup|B|$. If $\mu \sim +\infty$, then $\mu - 1 \sim +\infty$, and $\exists \xi \in B$ $(\mu - 1 < |\xi| \leq \mu)$ by the definition of the supremum, which contradicts the boundedness of ξ. Consequently, μ is bounded, and $\exists t \in \mathbb{R}$ $(t > \mu)$. It is clear that $B \subseteq [-t;t]$. The proofs of ii)–iv) are left as exercises for the reader.

To prove v) we consider the set

$$B = \{r > 0 \mid \forall x\ (|x| \leq r \longrightarrow r \cdot |f(x)| < 1)\}.$$

This set is internal (see Example 0.3.3), since it is definable by a formula in $\mathcal{LR}$ in terms of the internal parameter f. By assumption, it contains all the bounded positive elements of ${}^*\mathbb{R}$, and $\exists \Omega \sim +\infty$ $(\Omega \in B)$ in view of the permanence principle. Consequently, $\forall x$ $(|x| \leq \Omega \longrightarrow |f(x)| < \Omega^{-1} \approx 0)$.

We now consider certain properties of the set ${}^*\mathbb{N}$. First of all, it follows from the transfer principle that

$$ {}^*\mathbb{R} \models \text{“}{}^*\mathbb{N} \text{ is the set of natural numbers”} $$

(in fact, the quotes enclose the formula $\mathrm{Nat}({}^*\mathbb{N})$ in §1, 2°, part e).

It is shown in Example 0.3.2 that ${}^*\mathbb{N}$ satisfies the induction principle for internal sets.

PROPOSITION 0.3.14. ${}^*\mathbb{N}\setminus\mathbb{N} \neq \emptyset$. *Furthermore, if* $\Omega \in {}^*\mathbb{N}$, *then* $\Omega \sim +\infty \iff \Omega \notin \mathbb{N}$.

PROOF. By the transfer principle,

$$ {}^*\mathbb{R} \models \forall \xi \in {}^*\mathbb{R}_+ \; \exists n \in {}^*\mathbb{N} \; (n \leq \xi < n+1). $$

If $\lambda \sim +\infty$, $\Omega \in {}^*\mathbb{N}$, and $\Omega \leq \lambda < \Omega + 1$, then clearly $\Omega \sim +\infty$, that is, $\Omega \in {}^*\mathbb{N} \setminus \mathbb{N}$. It remains to show that ${}^*\mathbb{N} \cap {}^*\mathbb{R}_b \subseteq \mathbb{N}$. Indeed, suppose that $n \in {}^*\mathbb{N} \cap {}^*\mathbb{R}_b$ and $t = {}^\circ n \in \mathbb{R}$. Then $t - 1 < n < t + 1$. Let $m = [t] \in \mathbb{N}$; then $m \leq t < m + 1$. Thus, $m - 1 < n < m + 2$, and by the transfer principle

$$ {}^*\mathbb{R} \models \forall x, y \in {}^*\mathbb{N} \; (x - 1 < y < x + 2 \longrightarrow (y = x \;\vee\; y = x + 1)). $$

Consequently, $n = m$ or $n = m + 1$. In both cases $n \in \mathbb{N}$.

COROLLARY. *The set* $\mathbb{N}$ *is external.*

EXERCISE. An internal set containing all infinite natural numbers contains all natural numbers beginning with some standard number.

It follows from the transfer principle that the following ‘internal’ Archimedean principle is valid in the ordered field ${}^*\mathbb{R}$:

$$ \forall x, y \in {}^*\mathbb{R}_+ \exists n \in {}^*\mathbb{N} \; (n \cdot y > x). $$

On the other hand, it follows from Lemma 0.3.6 that ${}^*\mathbb{R}$ is not an Archimedean field: if $\alpha \in \mu(0)$, then $\forall n \in \mathbb{N} \; (n \cdot \alpha < 1)$.

This is not in any way contradictory! The fact of the matter is that the symbol $*$ must be ‘hung’ on all the constants occurring in the formula when the transfer principle is used.

REMARK. If f is a function definable by a formula in $\mathcal{LR}$ (see §1, 2°, part d), then by virtue of the transfer principle its extension ${}^*f$ is defined in the model ${}^*\mathbb{R}$ by the same formula and has all the properties of f definable in $\mathcal{LR}$. The convention was made earlier to keep the traditional notation for such functions when there is a traditional notation. Therefore, in what follows we use the same notation also for the images of these functions in ${}^*\mathbb{R}$; for example, we shall simply write $\sin\alpha$ instead of ${}^*\sin\alpha$, and so on.

4°. The cardinality of an arbitrary set X is denoted by $\mathrm{Card}(X)$ in this book. For internal sets it is more natural to consider their internal cardinality. For example, as mentioned above, the model $\mathbb{R}$ has elementary extensions of arbitrarily large cardinality, that is, $\mathrm{Card}({}^*\mathbb{R})$ can be arbitrarily large. On the other hand, if B is an internal set and there exists an internal bijection between B and ${}^*\mathbb{R}$, then it is natural to assume that the internal cardinality of B is the cardinality of the continuum.

The theory of internal cardinalities is not treated here in general form. 'Internally finite' sets play a fundamental role for all the considerations of this book.

DEFINITION 0.3.15. An internal set B is said to be *hyperfinite* if for some $\nu \in {}^*\mathbb{N}$ there exists an internal bijection

$$\varphi: B \Longleftrightarrow \{n \in {}^*\mathbb{N} \mid n < \nu\} = \{0, 1, \ldots, \nu - 1\}$$

(the fact that $\{0, 1, \ldots, \nu - 1\}$ is an internal set follows from the arguments in Example 0.3.3). The transfer principle gives us that such a $\nu \in {}^*\mathbb{N}$ is unique. It is called the internal cardinality of the set B and is denoted by $|B|$. If $|B| \in \mathbb{N}$, then B is said to be *standardly finite*. One simply says 'cardinality' instead of 'internal cardinality' where this does not lead to confusion.

REMARK. If $|B| \in {}^*\mathbb{N} \setminus \mathbb{N}$, then as will be shown below, $\mathrm{Card}(B)$ is even uncountable.

PROPOSITION 0.3.16.
i) *If B is a hyperfinite set and $|B| \in \mathbb{N}$, then* $\mathrm{Card}(B) = |B|$.
ii) *If $B \subseteq {}^*\mathbb{R}^k$ and* $\mathrm{Card}(B) \in \mathbb{N}$, *then B is internal, and* $\mathrm{Card}(B) = |B|$.
iii) *If $B \subseteq \mathbb{R}^k$ and* $\mathrm{Card}(B) \in \mathbb{N}$, *then* ${}^*B = B$.

PROOF. i) is obvious, since the internal bijection φ in Definition 0.3.15 is a bijection in the ordinary sense, and $|B| \in \mathbb{N}$, that is, is a cardinal number.

ii) is proved by induction on $\mathrm{Card}(B)$. If $\mathrm{Card}(B) = 0$, then $B = \emptyset$, and $\emptyset$ is an internal set and $|\emptyset| = 0$ by the transfer principle. Suppose that the assertion has been proved for $n \in \mathbb{N}$, and assume that $\mathrm{Card}(B) = n + 1$. For brevity we assume that $k = 1$. Let $t \in B$. Then $B = B' \cup \{t\}$. By the induction hypothesis, B' is internal and $|B'| = n$. The set $\{t\} = \{x \in {}^*\mathbb{R} \mid x = t\}$ is internal (Example 0.3.3). Now the set B is internal, because the class of internal sets is closed under unions (the remark after Proposition 0.2.10). Further, by the transfer principle,

$${}^*\mathbb{R} \models \forall X \subseteq {}^*\mathbb{R}\ \forall t \in {}^*\mathbb{R}\ (|X| \in {}^*\mathbb{N}\ \wedge\ t \notin X \longrightarrow |X \cup \{t\}| = |X| + 1)$$

(it is clear that this formula can be written in $\mathfrak{LR}$; see §1, 2°, part e). This implies that $|B| = n + 1$.

iii) If $B = \{t_1, \ldots, t_n\}$, $n \in \mathbb{N}$, then $\mathbb{R} \models \forall x\ (x \in B \longleftrightarrow (x = t_1 \vee \ldots \vee x = t_n))$, and ${}^*\mathbb{R} \models \forall x\ (x \in {}^*B \longleftrightarrow (x = {}^*t_1\ \vee \ldots \vee\ x = {}^*t_n))$ by the transfer principle. The assertion now follows from the fact that ${}^*t_i = t_i$ for $i = 1, \ldots, n$.

REMARK. It can be proved similarly that: 1) if $\mathfrak{M} \subseteq {}^*\mathcal{C}_{k_1} \times \cdots \times {}^*\mathcal{C}_{k_n}$ and $\mathrm{Card}(\mathfrak{M}) \in \mathbb{N}$, then $\mathfrak{M}$ is an internal class; 2) if $\mathfrak{M} \subseteq \mathcal{C}_{k_1} \times \cdots \times \mathcal{C}_{k_n}$ and $\mathrm{Card}(\mathfrak{M}) \in \mathbb{N}$, then ${}^*\mathfrak{M} = \{{}^*\bar{a} \mid \bar{a} \in \mathfrak{M}\}$.

In view of the transfer principle, hyperfinite sets have all the properties of finite sets that can be formulated in $\mathfrak{LR}$. For example, we have the following result.

PROPOSITION 0.3.17.
i) *Each hyperfinite set $B \subseteq {}^*\mathbb{R}$ has largest and smallest elements.*
ii) *If B_1 and B_2 are hyperfinite sets and $B_1 \cap B_2 = \emptyset$, then $|B_1 \cup B_2| = |B_1| + |B_2|$.*

We now pass to a study of hyperfinite classes. Identifying functions with their graphs if necessary, we can confine ourselves to the case of classes of internal sets.

DEFINITION 0.3.18. A class $\mathfrak{M}$ of internal subsets of ${}^*\mathbb{R}^k$ is said to be hyperfinite if there exist a $\nu \in {}^*\mathbb{N}$ and an internal set $F \subseteq \{0, 1, \dots, \nu - 1\} \times {}^*\mathbb{R}^k$ such that:

(1) $\forall n < \nu\ F_n = \{\langle x_1, \dots, x_k \rangle \mid \langle n, x_1, \dots, x_k \rangle \in F\}$;
(2) $\forall m, n < \nu\ (m \neq n \longrightarrow F_m \neq F_n)$;
(3) $\forall B \in \mathfrak{M}\ \exists n < \nu\ (B = F_n)$.

It is easy to see that such a ν is unique. It is called the *internal cardinality* of $\mathfrak{M}$ and denoted by $|\mathfrak{M}|$. If $|\mathfrak{M}| \in \mathbb{N}$, then the class $\mathfrak{M}$ is said to be *standardly finite.*

The properties of standardly finite classes are mentioned in the remark after Proposition 0.3.16.

REMARK. In turn, every internal set F having the properties indicated in the last definition gives an internal hyperfinite class $\mathfrak{M}$. Thus, it is possible to identify hyperfinite classes with sets of the indicated form. Consequently, propositions can be expressed in $\mathcal{LR}$ containing quantifiers with respect to hyperfinite classes (cf. §1, 2°, part g). Therefore, instead of the term 'hyperfinite class' we shall use 'hyperfinite family', and sometimes even simply 'hyperfinite set'. Further, if an internal family is given by means of an internal set F satisfying Definition 0.3.18, then we call it an indexed internal family and write $\{F_n \mid n < \nu\}$.

An important role below is played by internal measures on finite Boolean algebras of sets. Let $\mathfrak{M} = \{F_n \mid n < \nu\}$ be an indexed hyperfinite family of sets that is a Boolean algebra, that is, there exists an internal set $B \in \mathfrak{M}$ such that $F_n \subseteq B$ for $n < \nu$ and $\mathfrak{M}$ is closed under unions, intersections, and the taking of complements. Then an internal mapping $\lambda\colon \{0, 1, \dots, \nu - 1\} \longrightarrow {}^*\mathbb{R}_+$ is called an internal measure on $\mathfrak{M}$ if

$$\forall n, m, k < \nu\ (n \neq m\ \wedge\ F_n \cap F_m = \emptyset\ \wedge\ F_k = F_n \cup F_m \longrightarrow \lambda(k) = \lambda(n) + \lambda(m)).$$

In what follows we write $\lambda(F_n)$ instead of $\lambda(n)$. Then the last condition is simply the additivity condition:

$$F_n \cap F_m = \emptyset \longrightarrow \lambda(F_n \cup F_m) = \lambda(F_n) + \lambda(F_m).$$

The most essential example here is the algebra of all internal subsets of an internal hyperfinite set B. This algebra is denoted below by ${}^*\mathcal{P}(B)$. It is defined by the formula ${}^*\mathcal{P}(B) = \{\mathcal{D} \mid \mathcal{D} \subseteq B\}$, that is, it is internal. By the transfer principle, all the elements of ${}^*\mathcal{P}(B)$ are hyperfinite, and $|{}^*\mathcal{P}(B)| = 2^{|B|}$, that is, ${}^*\mathcal{P}(B) = \{B_n \mid n < 2^{|B|}\}$. A canonical internal measure ν is given on ${}^*\mathcal{P}(B)$ by the formula $\nu(B_n) = |B_n| \cdot |B|^{-1}$, $n < 2^{|B|}$.

5°. Let us now consider nonstandard extensions of standard measure spaces, which will also be needed below.

Let $(X, \mathcal{Y}, \mu)$ be a measure space that can be imbedded in the model $\mathbb{R}$ (for this it suffices that $\mathrm{Card}(X), \mathrm{Card}(\mathcal{Y}) < \mathrm{Card}(\mathbb{R})$, of course), that is, $\mathcal{Y}$ is a σ-algebra of subsets of X, and μ is a countably additive measure on $\mathcal{Y}$. We assume that $X \subseteq \mathbb{R}^k$ and that $\mathcal{Y}$ is determined by a set $Y \subseteq \mathbb{R}^{k+1}$ as follows (see §1, 2°, part g and Exercise 10 in part h). Let $I = \mathrm{dom}^{(1)} Y$. Then

(1) $\forall i \in I\ (Y_i = \{\langle x_1, \dots, x_k \rangle \mid \langle i, x_1, \dots, x_k \rangle\} \in \mathcal{Y})$,
(2) $\forall i, j \in I\ (i \neq j \longrightarrow Y_i \neq Y_j)$,
(3) $\forall B \in \mathcal{Y}\ \exists i \in I\ (B = Y_i)$,

that is, $\mathcal{Y}$ is a family of subsets of X indexed by the elements of the set I. Furthermore, the condition that $\mathcal{Y}$ be a σ-algebra can be written in $\mathcal{LR}$.

For example, closedness under countable unions can be written as follows:

$$\forall \varphi\colon \mathbb{N} \longrightarrow I\ \exists i \in I\ \forall \overline{x} \in X\ (\overline{x} \in Y_i \longleftrightarrow \exists n \in \mathbb{N}\ (\overline{x} \in Y_{\varphi(n)})). \tag{0.3.3}$$

As in the preceding subsection, we can now regard μ as a function from I to $\mathbb{R}$, but it is natural to write $\mu(Y_i)$ instead of $\mu(i)$. We leave it as an exercise to write in $\mathcal{LR}$ the condition of countable additivity of the measure μ.

In going to the elementary extension $^*\mathbb{R}$ we can consider the triple $(^*X, {}^*\mathcal{Y}, {}^*\mu)$, where $^*\mathcal{Y}$ is defined in terms of $^*X$, that is, as a family of internal subsets of $^*X$ indexed by the elements of $^*I$. Further (Proposition 0.2.10), $I \subseteq {}^*I$ and $Y_i \subseteq {}^*Y_i$ $\forall i \in I$. But if $i \in {}^*I \setminus I$, then $^*Y_i$ is not the image of any standard set in $\mathcal{Y}$. We remark that $^*\mathcal{Y}$ is no longer a σ-algebra in general, because even though (0.3.3) is true in $^*\mathbb{R}$ in view of the transfer principle, this truth means only (as noted above) that $\bigcup_{n\in\mathbb{N}} {}^*Y_{\varphi(n)} \in {}^*\mathcal{Y}$ for any *internal* function $\varphi\colon {}^*\mathbb{N} \longrightarrow {}^*I$. But the sequences $\varphi\colon \mathbb{N} \longrightarrow {}^*I$ are not internal. On the other hand, $^*\mathcal{Y}$ is an algebra of sets, that is, it is closed under finite unions and intersections. This follows from the fact that a finite set of objects in $^*\mathcal{C}$ is always internal (see Proposition 0.3.16 and the remark after it). Thus, although

$$^*\mathbb{R} \models \text{“}{}^*\mathcal{Y} \text{ is a } \sigma\text{-algebra”},$$

the family $^*\mathcal{Y}$ of sets is not a σ-algebra.

This fact is usually expressed as follows: $^*\mathcal{Y}$ is an internal σ-algebra, but is not a σ-algebra from the external point of view.

In exactly the same way, although the measure $^*\mu$ is internally countably additive, it is not countably additive from the external point of view. However, the property of finite additivity is the same from both the external and internal points of view for the same reason as above, that is, $^*\mu\colon {}^*\mathcal{Y} \longrightarrow {}^*\mathbb{R}_+$ is a finitely additive set function. Note that in view of the transfer principle $^*\mu(i) = \mu(i) \in \mathbb{R}$ $\forall i \in I$. More precisely, this follows from Proposition 0.2.10 (3b).

Similarly, if $(X, \mathcal{Y})$ is a topological space, that is, $\mathcal{Y}$ is a base for a topology on X, then $\mathbb{R} \models \mathrm{Top}(X, \mathcal{Y})$ (see §1, 2°, part g), and also $^*\mathbb{R} \models \mathrm{Top}(^*X, {}^*\mathcal{Y})$ by virtue of the transfer principle. Only finite intersections appear in the formula $\mathrm{Top}(X, \mathcal{Y})$, so $^*\mathcal{Y}$ is a base for a topology on $^*X$ from the external point of view. It is easy to see, however, that even though $X \subset {}^*X$ the topological space $(X, \mathcal{Y})$ is not a subspace of $(^*X, {}^*\mathcal{Y})$. Indeed, if $x \in U \in \mathcal{Y}$, then $U = {}^*U \cap X$ (Proposition 0.2.10), where $^*U \in {}^*\mathcal{Y}$, but $^*\mathcal{Y}$ can (and does, as a rule) also contain other sets containing x.

For example, if $X = \mathbb{R}$ and $\mathcal{Y} = \{(a, b) \mid a \in \mathbb{R},\ b \in \mathbb{R},\ a < b\}$, then $^*X = {}^*\mathbb{R}$ and $^*\mathcal{Y} = \{(a, b) \mid a \in {}^*\mathbb{R},\ b \in {}^*\mathbb{R},\ a < b\}$. If $\alpha \approx 0$, then $(-\alpha, \alpha) \in {}^*\mathcal{Y}$, but for any standard open set U in $\mathbb{R}$ we have $U \neq (-\alpha, \alpha) \cap \mathbb{R}$.

In particular, if $(X, \mathcal{Y})$ is a separable (compact) space, then $^*\mathbb{R} \models$ “$(^*X, {}^*\mathcal{Y})$ is a separable (compact) space”, but $(^*X, {}^*\mathcal{Y})$ can be nonseparable and noncompact from the external point of view. Indeed, for example, compactness from the internal point of view means only that every *internal* open covering contains a *hyperfinite* subcovering.

Suppose now that X is an internal linear space over K, where $K = \mathbb{R}$ or $\mathbb{C}$. This means that two internal operations $+\colon X \times X \longrightarrow X$ and $\cdot\colon {}^*K \times X \longrightarrow X$ are given which satisfy the usual axioms of linear spaces. As a rule, an internal norm (inner product) is also given on X, that is, an internal mapping $\|\cdot\|\colon X \longrightarrow$

${}^*\mathbb{R}_+$ (respectively, $(\cdot,\cdot)\colon X^2 \longrightarrow {}^*K$) satisfying the usual axioms of a norm (inner product).

Since $K \subset {}^*K$, the internal linear space X is a linear space also from the external point of view; however, it is neither a normed nor a unitary space, not even if it is such a space from the internal point of view (that is, ${}^*\mathbb{R} \models$ "X is normed (unitary)"). Nevertheless, associated with each internal normed (unitary) space is a certain 'genuine' external Banach (Hilbert) space, which will be treated at length in the next section.

Internal hyperfinite-dimensional spaces will receive the most attention below. To define them it is necessary to make precise the sum of a hyperfinite set of elements of a linear space. To do this one should use the formula in $\mathfrak{L}\mathfrak{R}$ for the sum of the elements in a *finite* set as described in §1, 2°, part e, and then apply the transfer principle to it.

Thus, let X be an internal Abelian group, and Y a hyperfinite subset of X. We fix some bijection $\varphi\colon \{0,\dots,|Y|-1\} \Longleftrightarrow Y$, extend it to an internal sequence $f\colon {}^*\mathbb{N} \longrightarrow X$ by letting it be zero for $n > |Y|$, and define the sequence $g\colon {}^*\mathbb{N} \longrightarrow X$ by the formula $\sum(f,g)$ in §1, 2°, part e, that is, let $g \in {}^*\mathfrak{C}$ be such that ${}^*\mathbb{R} \models \sum(f,g)$. Then by virtue of the transfer principle, $g(|Y|-1)$ is independent of the choice of the internal bijection φ, and $g(|Y|-1) = \sum_{x\in Y} x$. By the transfer principle, the so-defined sum of a hyperfinite set has all the properties of an ordinary finite sum. For example, if $\{Y_m \mid m < \nu\}$ is an internal hyperfinite family of subsets of Y that is a partition of Y, then

$$\sum_{x\in Y} x = \sum_{i=0}^{\nu-1} \sum_{x\in Y_i} x,$$

and so on.

It is now easy to define internal hyperfinite-dimensional linear spaces. Let X be an internal linear space. An internal hyperfinite set $\{e_1,\dots,e_\Omega\}$, $\Omega \in {}^*\mathbb{N}$, is called a basis in X if for any $x \in X$ there exists a unique internal hyperfinite set $\{x_1,\dots,x_\Omega\}$ such that $x = \sum_{i=1}^{\Omega} x_i e_i$. A space having a hyperfinite basis is said to be hyperfinite-dimensional, and the internal cardinality of this basis is called the (internal) dimension of the space X: $\dim X = \Omega$. Below we keep the notation dim for the internal dimension of X, since its external dimension, that is, its dimension as a 'genuine' linear space, is not used in this book.

By the transfer principle, all the properties of finite-dimensional spaces and their finite bases carry over to hyperfinite-dimensional spaces and their hyperfinite bases. For example, $\dim X = \Omega$ if and only if there exists an internally linearly independent hyperfinite-dimensional subset $Y \subset X$ with internal cardinality Ω, and any hyperfinite set with internal cardinality $\Omega + 1$ is now internally linearly dependent. Here an internally hyperfinite set $\{y_1,\dots,y_\nu\}$ is said to be internally linearly independent if $\sum_{i=0}^{\nu} \lambda_i y_i \neq 0$ for any internal ν-sequence $\{\lambda_1,\dots,\lambda_\nu\}$ with at least one nonzero element, and it is said to be internally linearly dependent otherwise.

We remark that if a set $\{y_1,\dots,y_\nu\}$ is internally linearly independent, then it is linearly independent also from the external point of view. Indeed, its external linear dependence means the linear dependence over K of a standardly finite subset of it, and this is linear dependence also from the internal point of view, since a standardly finite set is internal (Proposition 0.3.16), and $K \subset {}^*K$.

On the other hand, an internally linearly dependent set can fail to be linearly dependent from the external point of view. For example, if $x \in X$, $x \neq 0$, and $\alpha \in {}^*K \setminus K$, then $\{x, \alpha x\}$ is an internally linearly dependent set, but it is externally linearly independent, since $\alpha \notin K$.

In referring to internal linear spaces below, we shall always have in mind internal linear dependence and independence, an internal basis, the internal dimension, and so on. Therefore, the adjective 'internal' itself will be omitted as a rule.

The most typical hyperfinite-dimensional space is ${}^*\mathbb{C}(X) = \{f\colon X \longrightarrow {}^*\mathbb{C}\}$, where X is some hyperfinite set. Of course, here we are concerned with the set of all internal mappings from X to $\mathbb{C}$. On this set we can consider the internal inner product

$$\langle f, g\rangle = \sum_{x \in X} f(x)\overline{g(x)}$$

and the norm induced by it:

$$\|f\| = \Big(\sum_{x \in X} |f(x)|^2\Big)^{1/2}.$$

It is also possible to consider other norms:

$$\|f\|_p = \Big(\sum_{x \in X} |f(x)|^p\Big)^{1/p}, \quad 1 \leq p \leq +\infty$$

(the index $p = 2$ is omitted in the notation for the norm, since it is the Euclidean norm we shall most often work with in this book).

From the internal point of view all these norms are equivalent by virtue of the transfer principle, that is, $\exists C_1, C_2 > 0$ $(C_1\|x\|_{p_1} \leq \|x\|_{p_2} \leq C_2\|x\|_{p_1})$. However, the constants C_1 and C_2 are in ${}^*\mathbb{R}$; in particular, they can be infinitesimal or infinite.

Finally, if X is a normed space, say over $\mathbb{C}$, then it is a topological space with $\mathcal{Y} = \{\{x \mid \|x - y\| < \varepsilon\} \mid y \in X,\ \varepsilon \in \mathbb{R}_+\}$ as a base for the topology. Then ${}^*X$ is an internally normed space over ${}^*\mathbb{C}$, and ${}^*\mathcal{Y} = \{\{x \mid {}^*\|x - y\| < \varepsilon\} \mid y \in {}^*X,\ \varepsilon \in {}^*\mathbb{R}_+\}$ is a base for its topology. In these cases the $*$ is usually dropped in the notation for the operations of addition and multiplication by scalars, as well as for the norm and inner product.

6°. In the conclusion of this section we translate to the language of nonstandard analysis the definitions of the basic concepts of mathematical analysis such as limits, continuity, and so on. The 'nonstandard' formulations obtained enable us to essentially simplify the proofs of many theorems in analysis. The corresponding material is presented in greater detail in the book [**15**]. Here we confine ourselves to the formulations needed in the exposition to follow.

THEOREM 0.3.19. *Let $\{x_n \mid n \in \mathbb{N}\}$ be a standard sequence. Then:*
i) $\lim_{n\to\infty} x_n = a \iff \forall \Omega \in {}^*\mathbb{N} \setminus \mathbb{N}\ ({}^*x_\Omega \approx a)$;
ii) *a is a limit point of x_n* $\iff \exists \Omega \in {}^*\mathbb{N} \setminus \mathbb{N}\ ({}^*x_\Omega \approx a)$;
iii) $\overline{\lim}_{n\to\infty} x_n = a \iff \forall \Omega \in {}^*\mathbb{N} \setminus \mathbb{N}\,({}^*x_\Omega \lesssim a) \ \wedge\ \exists \Omega \in {}^*\mathbb{N} \setminus \mathbb{N}\,({}^*x_\Omega \approx a)$;
iv) *x_n is Cauchy* $\iff \forall \Omega_1, \Omega_2 \in {}^*\mathbb{N} \setminus \mathbb{N}\ ({}^*x_{\Omega_1} \approx {}^*x_{\Omega_2})$.

PROOF. i) Let $\lim_{n\to\infty} x_n = a$. We fix an arbitrary standard $\varepsilon > 0$ and show that if $\Omega \sim +\infty$, then $|^*x_\Omega - a| < \varepsilon$. Indeed, it follows from the definition of a limit that there exists a standard $n_0 \in \mathbb{N}$ such that

$$\forall n \in \mathbb{N}\ (n > n_0 \longrightarrow |x_n - a| < \varepsilon).$$

By the transfer principle,

$$^*\mathbb{R} \models \forall n \in {}^*\mathbb{N}\ (n > n_0 \longrightarrow |^*x_n - a| < \varepsilon),$$

and if $\Omega \in {}^*\mathbb{N} \setminus \mathbb{N}$, then $\Omega > n_0 \in \mathbb{N}$ (Proposition 0.3.14), which proves what is required.

To prove the reverse implication we again fix a standard $\varepsilon > 0$ and consider the *internal* set $\{n \mid |^*x_n - a| < \varepsilon\}$. By the assertion on the right-hand side in the equivalence i), this set contains all infinite natural numbers, and in view of the exercise after Proposition 0.3.14 it also contains all the natural numbers from some point on. This proves that $\forall \varepsilon > 0\ \exists n_0 \in \mathbb{N}\ \forall n > n_0\ (|x_n - a| < \varepsilon)$, that is, $\lim_{n\to\infty} x_n = a$. The proofs of ii)–iv) are left to the reader as exercises.

Let $\overline{a} = \langle a_1, \ldots, a_n\rangle \in \mathbb{R}^n$, where n is a standard natural number. Then the external set $\mu(\overline{a}) = \mu(a_1) \times \cdots \times \mu(a_n) \subset {}^*\mathbb{R}^n$ is called the monad or halo of the point $\overline{a}$ (see Definition 0.3.5). Similarly, if $\overline{\xi} = \langle \xi_1, \ldots, \xi_n\rangle, \overline{\eta} = \langle \eta_1, \ldots, \eta_n\rangle \in {}^*\mathbb{R}^n$, then $\overline{\xi} \approx \overline{\eta} \Longleftrightarrow \xi_1 \approx \eta_1, \ldots, \xi_n \approx \eta_n$.

THEOREM 0.3.20. *Let $A \subseteq \mathbb{R}^n$. Then:*
i) *A is open* $\Longleftrightarrow \forall \overline{a} \in A\ (\mu(\overline{a}) \subset {}^*A)$;
ii) *A is closed* $\Longleftrightarrow \forall \overline{a} \in \mathbb{R}^n\ (\mu(\overline{a}) \cap {}^*A \neq \emptyset \longrightarrow a \in A)$;
iii) *A is bounded* $\Longleftrightarrow {}^*A \subset {}^*\mathbb{R}^n_b$.

PROOF. We confine ourselves for brevity to the case when $n = 1$.

i) If A is open and $a \in A$, then there exists a standard $\varepsilon > 0$ such that $(a - \varepsilon, a + \varepsilon) \subseteq A$. But $\mu(a) \subset {}^*(a - \varepsilon, a + \varepsilon) \subseteq {}^*A$, which is what was required. Conversely, if $\varepsilon \approx 0$, then $(a - \varepsilon, a + \varepsilon) \subset \mu(a)$, that is, $\{\varepsilon \mid {}^*(a - \varepsilon, a + \varepsilon) \subseteq {}^*A\}$ contains all the infinitesimal elements, so in view of the Cauchy principle (Theorem 0.3.13) it contains also some standard δ; therefore, ${}^*(a - \delta, a + \delta) \subseteq {}^*A$, and by the transfer principle (applied 'in the opposite direction'!) $(a - \delta, a + \delta) \subseteq A$, as required. The proofs of ii) and iii) are left to the reader as exercises.

THEOREM 0.3.21. *Suppose that $D \subseteq \mathbb{R}^n$ and $f\colon D \longrightarrow \mathbb{R}$. Then:*
i) $\lim_{x\to x_0} f(x) = A \Longleftrightarrow \forall \xi \in {}^*D\ (\xi \approx x_0 \longrightarrow {}^*f(\xi) \approx A)$;
ii) *f is continuous on D* $\Longleftrightarrow \forall a \in D\ \forall \xi \in {}^*D\ (\xi \approx a \Longrightarrow {}^*f(\xi) \approx f(a))$;
iii) *f is uniformly continuous on D* $\Longleftrightarrow \forall \xi \in {}^*D\ \forall \eta \in {}^*D\ (\xi \approx \eta \Longrightarrow {}^*f(\xi) \approx {}^*f(\eta))$.

PROOF. iii) Let f be uniformly continuous on D. We show that for any standard $\varepsilon > 0$ the inequality $|^*f(\xi) - {}^*f(\eta)| < \varepsilon$ holds if $\xi, \eta \in {}^*D$ and $\xi \approx \eta$. Since f is uniformly continuous, there is a standard $\delta > 0$ such that $\forall x < y \in D$ $(|x - y| < \delta \longrightarrow |f(x) - f(y)| < \varepsilon)$. By the transfer principle,

$$^*\mathbb{R} \models \forall x, y \in {}^*D\ (|x - y| < \delta \longrightarrow |^*f(x) - {}^*f(y)| < \varepsilon).$$

Since δ is standard and $\xi \approx \eta$, it follows that $|\xi - \eta| < \delta$, which proves what is required. Conversely, we again fix a standard $\varepsilon > 0$ and consider the internal set

$$\{\delta \mid \forall x, y \in {}^*D\ (|x - y| < \delta \longrightarrow |^*f(x) - {}^*f(y)| < \varepsilon)\}.$$

By virtue of the equivalence on the right-hand side of iii) this set contains all infinitesimals, that is, again by the Cauchy principle it contains also some standard δ. Using the transfer principle in the opposite direction again, we get what is required. The proofs of i) and ii) are left to the reader as exercises.

COROLLARY. *If $f\colon [a,b] \longrightarrow \mathbb{R}$ is a standard Riemann-integrable function, $\Delta \in \mathbb{R}$, $\Delta \approx 0$, and $N = \left[\frac{b-a}{\Delta}\right]$, then $\int_a^b f(x)\,dx = {}^\circ\!\left(\Delta \sum_{n=0}^{N-1} f(n\Delta)\right)$.*

PROOF. This follows from i) and the fact that

$$\lim_{\Delta \to 0} \Delta \sum_{n=0}^{N-1} f(n\Delta) = \int_a^b f(x)\,dx.$$

We illustrate the possibilities of using these theorems with the example of the proof of Cantor's theorem on the uniform continuity of a function that is continuous on a closed bounded set.

Let f be continuous on the closed bounded set $D \subset \mathbb{R}$, and let $\xi, \eta \in {}^*D$ be such that $\xi \approx \eta$. Since D is bounded, it follows from Theorem 0.3.20 iii) that ${}^*D \subset {}^*\mathbb{R}^n_b$; therefore, Theorem 0.3.7 gives us the existence of a standard $a \in \mathbb{R}^n$ such that $a \approx \eta$, and since $\xi \approx \eta$, it follows that $a \approx \xi$. By Theorem 0.3.20 ii) and the closedness of D, $a \in D$. It now follows from the continuity of f and Theorem 0.3.21 ii) that ${}^*f(\xi) \approx f(a)$ and ${}^*f(\eta) \approx f(a)$, that is, ${}^*f(\xi) \approx {}^*f(\eta)$, which proves the uniform continuity of f in view of Theorem 0.3.21 iii).

We return again to such examples in §1 of Chapter 1.

THEOREM 0.3.22. *Suppose that $f_n\colon D \longrightarrow \mathbb{R}$ is a sequence of functions, $D \subseteq \mathbb{R}^n$, and $f\colon D \longrightarrow \mathbb{R}$. Then:*

i) *f_n converges pointwise to $f \iff \forall a \in D\ \forall \Omega \in {}^*\mathbb{N} \setminus \mathbb{N}\ ({}^*f_\Omega(a) \approx f(a))$;*

ii) *f_n converges uniformly on $D \iff \forall \xi \in {}^*D\ \forall \Omega \in {}^*\mathbb{N} \setminus \mathbb{N}\ ({}^*f_\Omega(\xi) \approx {}^*f(\xi))$.*

PROOF. i) follows immediately from Theorem 0.3.19 i), and ii) is left to the reader as an exercise.

§4. Saturation and idealization

In this section we consider some properties of elementary extensions of the model $\mathbb{R}$ that play an important role in applications of nonstandard analysis. The existence of elementary extensions with these properties will be proved in the next section.

1°. If λ is a cardinal number (the cardinality of an infinite set), then λ^+ will denote the next cardinal number after it. The cardinality of a countable set is denoted by ω, ω^+ is denoted by ω_1, and so on (the continuum hypothesis says that $2^\omega = \omega_1$). In set theory ω is identified with $\mathbb{N}$. Therefore, the notation ω will sometimes be used instead of $\mathbb{N}$, and conversely.

Let $\{\mathfrak{M}_i \mid i \in I\}$ be a family of internal subclasses of ${}^*\mathcal{C}_{k_1} \times \ldots, \times {}^*\mathcal{C}_{k_n}$ (as above, we fix some elementary extension ${}^*\mathbb{R}$ of $\mathbb{R}$). As usual, this family is said to have the finite intersection property if any finite subfamily of it has nonempty intersection. We recall that each internal class $\mathfrak{M}$ of the model ${}^*\mathbb{R}$ is defined by a formula in the language $\mathcal{L}\mathcal{R}$ with elements of ${}^*\mathcal{C}$ substituted in place of certain free variables (see Definition 0.2.11). Consequently, the family of all internal classes has cardinality equal to $\mathrm{Card}({}^*\mathcal{C}) = \mathrm{Card}({}^*\mathbb{R}^{{}^*\mathbb{R}})$, and any subfamily of it does not have cardinality greater than this.

From this point the term 'nonstandard universe ${}^*\mathbb{R}$' will be used instead of the term 'elementary extension ${}^*\mathbb{R}$ of the model $\mathbb{R}$'. A nonstandard universe includes, of course, both the ordered extension ${}^*\mathbb{R}$ of the field $\mathbb{R}$ and the family ${}^*\mathcal{C}$ of all internal operations on ${}^*\mathbb{R}$.

DEFINITION 0.4.1. Let $\lambda > \omega$ be a cardinal number. A nonstandard universe ${}^*\mathbb{R}$ is said to be λ-saturated if $\bigcap_{i\in I} \mathfrak{M}_i \neq \emptyset$ for any family $\{\mathfrak{M}_i \mid i \in I\}$ of internal classes with $\mathrm{Card}(I) < \lambda$ having the finite intersection property. Further, a ω^+-saturated nonstandard universe is said to be countably saturated, and a $\mathrm{Card}({}^*\mathbb{R}^{{}^*\mathbb{R}})$-saturated nonstandard universe is simply said to be saturated.

REMARK. This definition is suitable also for arbitrary models of the language $\mathcal{LR}$, of course, since internal classes are defined in any model (see Definition 0.2.11).

THEOREM 0.4.2. *For any cardinal number $\lambda > \omega$ there exists a λ-saturated nonstandard universe.*

The countably saturated nonstandard universes will play the most important role for us in what follows. Their existence will be proved in the next section. The existence of saturated nonstandard universes requires the assumption of additional set-theoretic hypotheses, for example, hypotheses about the existence of an inaccessible cardinal. The next result follows immediately from the definition of countable saturation.

PROPOSITION 0.4.3. *If $\{\mathfrak{M}_n \mid n \in \mathbb{N}\}$ is a decreasing sequence of internal subclasses of a countably saturated nonstandard universe, that is, $\mathfrak{M}_0 \supset \mathfrak{M}_1 \supset \cdots \supset \mathfrak{M}_n \supset \cdots$, and if $\bigcap_{n\in\mathbb{N}} \mathfrak{M}_n = \emptyset$, then $\exists n_0 \in \mathbb{N}\ (\mathfrak{M}_{n_0} = \emptyset)$.*

PROPOSITION 0.4.4. *Suppose that $\{Y_n \mid n \in \mathbb{N}\}$ is a sequence of internal sets in a countably saturated universe. Then there exists an internal sequence extending the original one, that is, coinciding with it on standard n.*

PROOF. It is required to prove the existence of an internal set Y such that $\mathrm{dom}^{(1)} Y = {}^*\mathbb{N}$ and such that $\{\overline{x} \in {}^*\mathbb{R}^k \mid \langle n, \overline{x}\rangle \in Y\} = Y_n$ for any standard $n \in \mathbb{N}$.

We consider a sequence $\{\mathfrak{M}_s \mid s \in \mathbb{N}\}$ of internal classes such that

$$\xi \in \mathfrak{M}_s \iff \xi\colon {}^*\mathbb{N} \longrightarrow {}^*\mathcal{C}_k \ \wedge\ \forall i \le s\ (\xi_i = Y_i),$$

that is, the class $\mathfrak{M}_s$ is defined by the formula $A(\alpha^{(k+1)}, Y_0, \ldots, Y_s)$:

$$\mathrm{dom}^{(1)} \alpha^{(k+1)} = {}^*\mathbb{N} \ \wedge\ \mathrm{Set}(\alpha^{(k+1)}) \ \wedge\ \bigwedge_{i=0}^{s} \{\overline{x} \mid \langle i, \overline{x}\rangle \in \alpha^{(k+1)}\} = Y_i.$$

It is clear from the definition that this sequence is decreasing, and $\mathfrak{M}_s \neq \emptyset$ for any $s \in \mathbb{N}$. Indeed, $\mathfrak{M}_s$ contains, for example, the sequence ξ such that $\xi_i = Y_i$ for $i \le s$ and $\xi_i = 0$ for $i > s$. It follows from the countable saturation of the nonstandard universe that $\bigcap_{s\in\mathbb{N}} \mathfrak{M}_s \neq \emptyset$. Any Y in this intersection is the desired object.

REMARK. It is easy to see that the proof does not change if we require that each of the Y_n, $n \in \mathbb{N}$, satisfy some formula in $\mathcal{LR}$; then each of the Y_n, $n \in {}^*\mathbb{N}$, satisfies the same formula. For example, a sequence of internal mappings extends to an internal sequence of internal mappings, a sequence of elements of ${}^*\mathbb{R}^k$ (single-element internal sets) extends to an internal sequence of elements of ${}^*\mathbb{R}^k$, and so on.

2°. Using the property of λ^+-saturation of a nonstandard universe, we can extend the assertions obtained in the concluding subsection of the preceding section for the topological spaces $\mathbb{R}^k$, to the case of arbitrary topological spaces having a topological base with cardinality not exceeding λ. Before taking this up we make the following important remark. As shown in §1, 2°, part g, the standard model

$$\langle \mathbb{R}, \{0, 1, +, \cdot, <\}, \{\mathcal{C}_n \mid n \in \mathbb{N}\}\rangle$$

allows us to consider only standard objects (in particular, bases of a topology) with cardinality not exceeding that of the continuum. In this book the need does not arise to consider objects of greater cardinality as means of the language $\mathcal{LR}$. However, this restriction is not really fundamental and is adopted here only to make the exposition simpler.

Indeed, if we consider a standard model $\mathcal{M}$ of the language $\mathcal{LR}$ constructed on the basis of a set M of arbitrary cardinality exceeding that of the continuum, then we can assume that $\mathbb{R}$ is a subset of M, that is, an element of $\mathcal{C}_1(M)$, the operations of addition and multiplication on $\mathbb{R}$ are elements of $\mathcal{C}_3(M)$, and the order relation is an element of $\mathcal{C}_2(M)$. Consequently, the model $\mathbb{R}$ (see §3, 1°) occurs in $\mathcal{M}$, and its elementary extension (0.3.1) occurs in the elementary extension ${}^*\mathcal{M}$. Theorem 0.4.2 is valid for elementary extensions of arbitrary standard models. On the other hand, in a way analogous to what was described in §1, 2°, part g (see also §3, 5°) we can consider in $\mathcal{LR}$ standard families of sets having cardinality not exceeding $\mathrm{Card}(M)$, and hence also nonstandard extensions of them (images under elementary imbeddings). Thus, we can construct a nonstandard universe in which standard families of sets of arbitrarily large fixed cardinality are imbedded. In considering the language of set theory (see §6) we can also avoid the condition that the cardinality be fixed in advance. With the foregoing remarks in mind, we shall not stipulate below that the weight of the topological spaces considered does not exceed that of the continuum.

Up to the end of this subsection the nonstandard universe is assumed to be λ^+-saturated. We consider a standard topological space X whose weight does not exceed λ. Let $\mathcal{Y}$ be a base for the topology of X, $\mathrm{Card}(\mathcal{Y}) \leq \lambda$. If $x \in X$, then $\mathcal{Y}(x) = \{Y \mid x \in Y \in \mathcal{Y}\}$. The monad of a point $x \in X$ in the nonstandard extension ${}^*X$ of X is defined to be the (external, in general) set

$$\mu_X(x) = \bigcap\{{}^*Y \mid Y \in \mathcal{Y}(x)\}.$$

PROPOSITION 0.4.5. *For any $x \in X$ there exists an internal set $U \subseteq {}^*X$ such that $U \in {}^*\mathcal{Y}(x)$ and $U \subseteq \mu_X(x)$.*

PROOF. For each $W \in \mathcal{Y}(x)$ we consider the internal class $\mathfrak{M}_W = \{V \in {}^*\mathcal{Y}(x) \mid V \subseteq {}^*W\}$. The family $\{\mathfrak{M}_W \mid W \in \mathcal{Y}(x)\}$ of internal classes has the finite intersection property: if $W_1, \ldots, W_n \in \mathcal{Y}(x)$, then by the definition of a base of a topology, $\exists W_0 \in \mathcal{Y}(x)\ (W_0 \subseteq W_1 \cap \cdots \cap W_n)$, and then ${}^*W_0 \in \mathfrak{M}_{W_1} \cap \cdots \cap \mathfrak{M}_{W_n}$. Moreover, the cardinality of this family does not exceed λ. It follows from the λ^+-saturation of the nonstandard universe that $\cap\{\mathfrak{M}_W \mid W \in \mathcal{Y}(x)\} \neq \emptyset$. Any internal set in this intersection serves as the desired set.

If x is not an isolated point, then by the transfer principle any set satisfying the conditions of Proposition 0.4.5 contains elements different from x, that is, $\mu_X(x) \setminus \{x\} \neq \emptyset$.

COROLLARY. *A point $x \in X$ is not an isolated point if and only if $\mu_X(x) \setminus \{x\} \neq \emptyset$.*

If $\xi \in \mu_X(x)$, then we write $\xi \approx x$.

PROPOSITION 0.4.6.

i) *X is Hausdorff* $\Longleftrightarrow \forall x, y \in X\ (x \neq y \longrightarrow \mu_X(x) \cap \mu_X(y) = \emptyset)$;
ii) *$A \subseteq X$ is open* $\Longleftrightarrow \forall x \in A\ (\mu_X(x) \subseteq {}^*A)$;
iii) *$A \subseteq X$ is closed* $\Longleftrightarrow \forall x \in X\ (\mu_X(x) \cap {}^*A \neq \emptyset \longrightarrow x \in A)$;
iv) *$A \subseteq X$ is compact* $\Longleftrightarrow \forall \xi \in {}^*A\ \exists x \in A\ (\xi \approx x)$.

PROOF. i) If X is a Hausdorff space and $x \neq y$, then $\exists U \in \mathcal{Y}(x), V \in \mathcal{Y}(y)$ $(U \cap V = \emptyset)$. By the transfer principle, ${}^*U \cap {}^*V = \emptyset$, and by the definition of a monad, $\mu_X(x) \subseteq {}^*U$ and $\mu_X(y) \subseteq {}^*V$. Conversely, suppose that $x \neq y$, but $\forall U \in \mathcal{Y}(x), V \in \mathcal{Y}(y)\ (U \cap V \neq \emptyset)$. Then it is easy to see that the family $\mathcal{Y}(x) \cup \mathcal{Y}(y)$ has the finite intersection property, that is, the external family $\mathfrak{N} = \{{}^*W \mid W \in \mathcal{Y}(x) \cup \mathcal{Y}(y)\}$ has the finite intersection property, and $\mathrm{Card}(\mathfrak{N}) \leq \lambda$, therefore,

$$\emptyset \neq \bigcap \mathfrak{N} = \bigcap\{{}^*W \mid W \in \mathcal{Y}(x)\} \cap \bigcap\{{}^*W \mid W \in \mathcal{Y}(y)\} = \mu_X(x) \cap \mu_X(y),$$

which contradicts the assumption.

ii) If A is open and $x \in A$, then $\exists U \in \mathcal{Y}(x)\ (U \subseteq A)$, that is, ${}^*U \subseteq {}^*A$, and $\mu_X(y) \subseteq {}^*U$ by the definition of a monad. Conversely, if A is not open, then $\exists x \in A$ $\forall U \in \mathcal{Y}(x)\ (U \cap (X \setminus A) \neq \emptyset)$. Then the family $\{U \cap (X \setminus A) \mid U \in \mathcal{Y}(x)\}$ has the finite intersection property. Consequently, so does the family $\{{}^*U \cap ({}^*X \setminus {}^*A) \mid U \in \mathcal{Y}(x)\}$. Hence, the intersection of this family is not empty, though this intersection is equal to $\mu_X(x) \cap ({}^*X \setminus {}^*A)$, contradicting the fact that $\mu_X(x) \subseteq {}^*A$.

iii) This is left to the reader.

iv) Assume that $\forall x \in A\ (\xi \notin \mu_X(x))$. Then it follows from the definition of a monad that there is a $U_x \in \mathcal{Y}(x)$ such that $\xi \notin {}^*U_x$. The family $\{U_x \mid x \in A\}$ forms a covering of A. Since A is compact, $\exists x_1, \dots, x_n \in A\ (A \subseteq U_{x_1} \cup \dots \cup U_{x_n})$. By the transfer principle, ${}^*A \subseteq {}^*U_{x_1} \cup \dots \cup {}^*U_{x_n}$. If now $\xi \in {}^*A$, then $\exists i \leq n\ (\xi \in {}^*U_{x_i})$, which contradicts the definition of U_{x_i}. In the converse direction it suffices to show that a covering of A by elements of $\mathcal{Y}$ has a finite subcovering. Let $S \subseteq \mathcal{Y}$ be a covering of A that does not have a finite subcovering. Then it is easy to see that the family

$$\mathfrak{N} = \{{}^*A \setminus ({}^*W_1 \cup \dots \cup {}^*W_n) \mid W_i \in S,\ i = 1, \dots, n,\ n \in \mathbb{N}\}$$

has the finite intersection property. Consequently,

$$\emptyset \neq \bigcap \mathfrak{N} = {}^*A \setminus \bigcup\{{}^*W \mid W \in S\}. \tag{0.4.1}$$

Let $\xi \in \bigcap \mathfrak{N}$. By assumption, $\exists x \in A\ (\xi \in \mu_X(x))$. Then $\exists W \in S\ (x \in W)$ because S covers A. By the definition of a monad, $\mu_X(x) \subseteq {}^*W$, that is, $\xi \in {}^*W$, and this contradicts (0.4.1).

If $\xi \in {}^*X$ and $\exists x \in X$ $(\xi \approx x)$, then the element ξ will be said to be near-standard. The set of all near-standard elements of ${}^*X$ (in general this set is not internal, that is, it is external) will be denoted by $\mathrm{Ns}({}^*X)$, that is, $\mathrm{Ns}({}^*X) = \bigcup\{\mu_X(x) \mid x \in X\}$.

Proposition 0.4.6 (iv) immediately yields the

COROLLARY. *A topological space X is compact if and only if* $\mathrm{Ns}({}^*X) = {}^*X$.

If X is a Hausdorff space, then a mapping $\mathrm{st}\colon \mathrm{Ns}({}^*X) \longrightarrow X$ is defined such that

$$\mathrm{st}(\xi) = x \Longleftrightarrow \xi \approx x.$$

As in the case of $\mathbb{R}$ we sometimes write ${}^\circ\xi$ instead of $\mathrm{st}(\xi)$ and call this element the standard part or shadow of ξ.

PROPOSITION 0.4.7. *If $W \subseteq \mathrm{Ns}({}^*X)$ is an internal set, then $\mathrm{st}(W)$ is a closed subset of X.*

The proof is left to the reader as an exercise (see [**1**], Proposition 2.1.8).

PROPOSITION 0.4.8. *Let X and Y be topological spaces with weight not exceeding λ, and let $f\colon X \longrightarrow Y$. Then:*

i) $\lim_{x\to x_0} f(x) = y_0 \Longleftrightarrow \forall\xi \approx x_0\ ({}^*f(\xi) \approx y_0)$;

ii) *f is continuous* $\Longleftrightarrow \forall x \in X\ \forall\xi \in {}^*X\ (\xi \approx x \longrightarrow {}^*f(\xi) \approx f(x))$.

PROOF. We leave this to the reader.

If now X is a metric space, and ρ is a metric on X, then $\mu_X(x) = \{\xi \in {}^*X \mid {}^*\rho(\xi, x) \approx 0\}$ and $\forall\xi, \eta \in {}^*X$ $(\xi \approx \eta \rightleftharpoons {}^*\rho(\xi, \eta) \approx 0)$.

REMARK. In the case of a topological space the relation $\approx$ is defined on ${}^*X$ only for pairs with one of the elements standard. The relation of infinite closeness of two nonstandard elements is defined for uniform spaces, of which a special case is the case of metric spaces. As in the case of $\mathbb{R}^k$, for arbitrarily metric spaces we have Propositions 0.4.5 and 0.4.6 (i–iii) even without any assumptions about the saturation of the nonstandard universe (exercise), and we have Proposition 0.4.8 for a countably saturated universe. For Proposition 0.4.6 (iv) the assumption about saturation is required just as in the case of arbitrary topological spaces.

If (X, ρ) is a metric space, then an element $\xi \in {}^*X$ is said to be bounded if $\exists x \in X$ $({}^*\rho(x, \xi) \in {}^*\mathbb{R}_b)$. It is clear that in this case $\forall x \in X$ $({}^*\rho(x, \xi) \in {}^*\mathbb{R}_b)$. Obviously, every near-standard element is bounded. As shown above, the converse is also true for ${}^*\mathbb{R}^k$ with standard k. For an arbitrary metric space this is not true in general. For example, suppose that X is a Hilbert space, and $\{e_n \mid n \in \mathbb{N}\}$ is an orthonormal basis in X. If $\Omega \in {}^*\mathbb{N} \setminus \mathbb{N}$, then $({}^*e_\Omega, {}^*e_\Omega) = \|{}^*e_\Omega\|^2 = 1$ by the transfer principle, that is, ${}^*e_\Omega$ is a bounded element of ${}^*X$. We show that it is not near-standard. If $x = \sum_{i\in\mathbb{N}} x_i e_i \in {}^*X$, then $x_i \to 0$ as $i \to \infty$, and hence ${}^*x_\Omega \approx 0$. According to the transfer principle, ${}^*x_\Omega = (x, {}^*e_\Omega)$, that is, $(x, {}^*e_\Omega) \approx 0$. Consequently, $\|x - {}^*e_\Omega\|^2 = \|x\|^2 + \|{}^*e_\Omega\|^2 - 2(x, {}^*e_\Omega) \approx \|x\|^2 + 1 \neq 0$.

Moreover, note that in a metric space (X, ρ) the metric $\rho' = \frac{\rho}{\rho+1}$ determines the same topology. It is clear that any element is bounded with respect to ρ'.

The set of all bounded elements of a nonstandard extension ${}^*X$ of a metric space X is denoted below by ${}^*X_b$.

We recall that if X and Y are metric spaces, then a mapping $\varphi\colon X \longrightarrow Y$ is said to be completely continuous (compact) if the image $\varphi(A)$ of any bounded set A is precompact (a set A is bounded if $\sup\{\rho(x,y) \mid x,y \in A\} < +\infty$).

PROPOSITION 0.4.9. *If X and Y are metric spaces, then a continuous mapping $\varphi\colon X \longrightarrow Y$ is completely continuous if and only if $\varphi({}^*X_b) \subseteq \mathrm{Ns}({}^*Y)$.*

The proof is left to the reader as an exercise.

What assumption about the saturation of the universe is required here?

3°. In this and the next subsection it is assumed that the nonstandard universe is countably saturated.

Let E be an internal normed linear space over ${}^*K$, where $K = \mathbb{R}$ or $\mathbb{C}$ (see §1, 5°). Denote by E_b the set of all elements of E with bounded norm, that is,

$$E_b = \{e \in E \mid \|e\| \in {}^*\mathbb{R}_b\},$$

and by E_0 the set of all elements of E with infinitesimal norm, that is,

$$E_0 = \{e \in E \mid \|e\| \in \mu(0)\}.$$

It is clear that from the external point of view E_b and E_0 are linear spaces over K, and $E_0 \subset E_b$. Let us consider the quotient space $E^\# = E_b/E_0$. We can define a norm on $E^\#$, also denoted by $\|\cdot\|$, as follows.

For an $e \in E_b$ the class of e in $E^\#$ will be denoted by $e^\#$. Then $\|e^\#\|$ is defined in $E^\#$ by the formula

$$\|e^\#\| = {}^\circ\|e\|.$$

It is trivial to check that this definition is unambiguous and satisfies all the norm axioms. Thus, $E^\#$ is a normed space over K. In fact, we have

THEOREM 0.4.10.
i) *$E^\#$ is a Banach space.*
ii) *If ${}^*\mathbb{R} \models \dim E = n$, where n is standard, then $\dim E^\# = n$. In the opposite case $E^\#$ is not separable.*

PROOF. i) The completeness of $E^\#$ is a simple consequence of the countable saturation of the nonstandard universe (see [**1**], Proposition 2.2.1).

To prove ii) we show first that if $E^\#$ is separable, then it is finite dimensional. We introduce the following notation. If L is a normed space, then let $B(L)$ be the unit ball in L:

$$B(L) = \{l \in L \mid \|l\| \leq 1\}.$$

If $l_0 \in L$ and $\varepsilon > 0$, then

$$B^0_{l_0}(\varepsilon) = \{l \in L \mid \|l - l_0\| < \varepsilon\}, \quad B_{l_0} = \{l \in L \mid \|l - l_0\| \leq \varepsilon\}.$$

If $A \subseteq E$, then $A^\# = \{e^\# \mid e \in A\} \subseteq E^\#$. The last notation is used both for internal and external subsets of E.

It is easy to see that $B(E^\#) = B(E)^\#$. Indeed, the inclusion $B(E)^\# \subseteq B(E^\#)$ follows from the fact that $\forall \lambda \in {}^*\mathbb{R}\ (\lambda \leq 1 \longrightarrow {}^\circ\lambda \leq 1)$. Let $\xi \in B(E^\#)$, that is, $\|\xi\| \leq 1$. If $\|\xi\| < 1$, then $\exists e \in E\ (\xi = e^\# \wedge \|e\| < 1)$, that is, $e \in B(E)$, and so $\xi \in B(E^\#)$. If $\|\xi\| = 1$, then $\xi = e^\#$, where $\|e\| \approx 1$, and it is possible that $\|e\| > 1$. However, in this case $e' = \frac{e}{\|e\|} \approx e$, that is, $e'^\# = e^\#$, and $\|e'\| = 1$, so again $\xi \in B(E)^\#$. Similarly, if $e \subset E_b$, then $B^0_e(\varepsilon)^\# \subseteq B_{e^\#}(\varepsilon)$.

To prove that $E^\#$ is finite dimensional it suffices to show that $B(E^\#)$ is compact. Since $E^\#$ is separable, there is a countable dense subset $\{e_i^\# \mid i \in \mathbb{N}\}$ of $B(E^\#)$. By what was established above, it can be assumed that $\forall i \in \mathbb{N}$ $(e_i \in B(E))$. We fix an arbitrary $\varepsilon > 0$ and consider the increasing sequence of internal sets $M_n = \bigcup_{i=0}^n B_{e_i}^0(\varepsilon) \cap B(E)$. We show that $\bigcup_{n\in\mathbb{N}} M_n = B(E)$. Indeed, if $e \in B(E)$, then $\exists n \in \mathbb{N}$ $(\|e^\# - e_n^\#\| < \varepsilon)$, and thus also $\|e - e_n\| < \varepsilon$, that is, $e \in B_{e_n}^0(\varepsilon) \subseteq M_n$. It now follows from the countable saturation of the universe that $\exists n_0 \in \mathbb{N}$ $(M_{n_0} = B(E))$ (Proposition 0.4.3). Then

$$B(E)^\# = \left(\bigcup_{i=0}^{n_0} B_{e_i}^0(\varepsilon) \cap B(E)\right)^\# = \bigcup_{i=0}^{n_0} (B_{e_i}^0(\varepsilon) \cap B(E))^\# \subseteq \bigcup_{i=0}^{n_0} B_{e_i^\#}(\varepsilon) \cap B(E)^\#.$$

This inclusion shows that $\{e_0^\#, \dots, e_{n_0}^\#\}$ is an ε-net for $B(E^\#)$.

We now show that if $e_1, \dots, e_n \in E_b$, where $n \in \mathbb{N}$ and $e_1^\#, \dots, e_n^\#$ are linearly independent, then $e_1, \dots, e_n$ are linearly independent in E (over ${}^*K$). Indeed, suppose that $\lambda_1, \dots, \lambda_n \in {}^*K$ are not all zero and $\sum_{i=1}^n \lambda_i e_i = 0$. If $\lambda = \max_i |\lambda_i|$, then $\lambda \neq 0$ and $\sum_{i=1}^n \mu_i e_i = 0$, where $\mu_i = \frac{\lambda_i}{\lambda}$ and $|\mu_i| \leq 1$. Since $|\mu_i| \in {}^*\mathbb{R}_b$, ${}^\circ\mu_i$ exists. It is easily checked that in this case $(\sum_{i=1}^n \mu_i e_i)^\# = \sum_{i=1}^n {}^\circ\mu_i e_i^\# = 0$. But $|\mu_j| = 1$ for some j, that is, ${}^\circ\mu_j \neq 0$. This contradicts the linear independence of $e_1^\#, \dots, e_n^\#$.

This implies, in particular, that if ${}^*\mathbb{R} \models \dim E = n$ for some standard n, then $\dim E^\# \leq n$. To prove the converse inequality we follow [**39**] and use an Auerbach basis. A normalized basis $l_1, \dots, l_n$ in a normed space L is called an Auerbach basis if $\|\sum_{i=1}^n \alpha_i l_i\| \geq |\alpha_j|\ \forall j \leq n$. It is known (see, for example, [**60**]) that an Auerbach basis exists in any finite-dimensional normed space. Consequently, one exists also in an internal n-dimensional space E by virtue of the transfer principle. Suppose that ${}^*\mathbb{R} \models$ "$\{e_1, \dots, e_n\}$ is an Auerbach basis in E". We show that $e_1^\#, \dots, e_n^\#$ are linearly independent. If $\sum_{i=1}^n \lambda_i e_i^\# = 0$, then $\|\sum_{i=1}^n \lambda_i e_i\| \approx 0$, but $\|\sum_{i=1}^n \lambda_i e_i\| \geq |\lambda_i|\ \forall i \leq n$, contradicting the fact that all the λ_i are standard, and at least one of them is nonzero. This proves that if ${}^*\mathbb{R} \models \dim E = n$, then $\dim E^\# = n$. Suppose now that ${}^*\mathbb{R} \models \dim E > n$ for any standard n. Then for any standard n there exists an internal subspace $E_1 \subset E$ such that ${}^*\mathbb{R} \models \dim E_1 = n$. Obviously, $E_1^\#$ is imbedded in $E^\#$ with preservation of norm, that is, for any $n \in \mathbb{N}$ the space $E^\#$ contains an n-dimensional subspace, that is, $E^\#$ cannot be finite dimensional, hence also not separable.

Obviously, if E is an internal Hilbert space, then $E^\#$ is a Hilbert space in which the inner product is defined by the formula

$$(e_1^\#, e_2^\#) = {}^\circ(e_1, e_2).$$

If X is a standard normed space, then $({}^*X)^\#$ is called the *nonstandard hull* of X. It is clear that X is imbedded in $({}^*X)^\#$ with preservation of norm. If $x \in X$, then ${}^*x^\# \in ({}^*X)^\#$. The mapping $x \longrightarrow {}^*x^\#$ is the required imbedding.

EXERCISE. Prove that the imbedding described is an isomorphism if and only if X is finite-dimensional.

The above construction of $E^\#$ and especially the nonstandard hulls of standard spaces play an exceptionally important role in applications of nonstandard analysis to the theory of Banach spaces [**39**]. In this book we consider mainly $E^{\#}$ for the case when E is a hyperfinite-dimensional space.

4°. In this subsection we present one of the most important constructions in nonstandard analysis: the construction of Loeb measures, which are especially widely used in nonstandard probability theory. All the results here are well known, so the proofs are omitted. They can be found in the book [**1**]. Let $(X, \mathfrak{A}, \nu)$ be an internal measure space with a finitely additive nonnegative measure, that is, $\mathfrak{A}$ is an internal algebra of subsets of X, and $\nu\colon \mathfrak{A} \longrightarrow {}^*\mathbb{R}_+$ is an internal finitely additive measure on $\mathfrak{A}$. It is possible to define on $\mathfrak{A}$ a finitely additive (no longer internal) measure ${}^\circ\nu\colon \mathfrak{A} \longrightarrow \mathbb{R}_+ \cup \{+\infty\} = \overline{\mathbb{R}}_+$, where ${}^\circ\nu(A)$ is, as usual, the standard part of the number $\nu(A)$ when this number is finite, and $+\infty$ when $\nu(A)$ is infinite.

THEOREM 0.4.11. *The finitely additive measure* ${}^\circ\nu\colon \mathfrak{A} \longrightarrow \overline{\mathbb{R}}_+$ *has a unique countably additive extension* λ *to the* σ*-algebra* $\sigma(\mathfrak{A})$ *generated by the algebra* $\mathfrak{A}$*. Furthermore,* $\forall B \in \sigma(\mathfrak{A})$

$$\lambda(B) = \inf\{{}^\circ\nu(A) \mid B \subseteq A,\ A \in \mathfrak{A}\}.$$

If $\lambda(B) < +\infty$*, then*

$$\lambda(B) = \sup\{{}^\circ\nu(A) \mid A \subseteq B,\ A \in \mathfrak{A}\},$$

and there is an $A \in \mathfrak{A}$ *such that* $\lambda(A \Delta B) = 0$.

For an arbitrary $B \in \sigma(\mathfrak{A})$ *either there is an* $A \in \mathfrak{A}$ *such that* $A \subseteq B$ *and* ${}^\circ\nu(A) = +\infty$*, or there is a sequence* $\{A_n \mid n \in \mathbb{N}\}$ *of sets in* $\mathfrak{A}$ *such that* $B \subseteq \bigcup_{n\in\mathbb{N}} A_n$ *and* ${}^\circ\nu(A_n) < +\infty$ $\forall n \in \mathbb{N}$.

DEFINITION 0.4.12. The measure space $(X, S(\mathfrak{A}), \nu_L)$ with a σ-additive measure is called a Loeb space, where $S(\mathfrak{A})$ is the completion of $\sigma(\mathfrak{A})$ with respect to the measure λ, and ν_L is the extension of λ to $S(\mathfrak{A})$.

PROPOSITION 0.4.13. *If* $\nu_L(X) < +\infty$*, then* $B \in S(\mathfrak{A})$ *if and only if*

$$\sup\{{}^\circ\nu(A) \mid A \subseteq B,\ A \in \mathfrak{A}\} = \inf\{{}^\circ\nu(A) \mid B \subseteq A,\ A \in \mathfrak{A}\} = \nu_L(B).$$

A function $f\colon X \longrightarrow \overline{\mathbb{R}}$ is said to be *Loeb-measurable* if it is measurable with respect to the σ-algebra $S(\mathfrak{A})$. An internal function $F\colon X \longrightarrow {}^*\mathbb{R}$ is said to be $\mathfrak{A}$-measurable if $\{x \in X \mid F(x) \leq t\} \in \mathfrak{A}$ $\forall t \in {}^*\mathbb{R}$. A internal function F is said to be *simple* if $\mathrm{rng}(F)$ is a hyperfinite set. Obviously, a simple internal function F is $\mathfrak{A}$-measurable if and only if $F^{-1}(\{t\}) \in \mathfrak{A}$ $\forall t \in {}^*\mathbb{R}$. In this case the internal integral

$$\int_X F\,d\nu = \sum_{t\in\mathrm{rng}(F)} F(t)\nu(F^{-1}(\{t\}))$$

is defined for F. If $A \in \mathfrak{A}$, then, as usual, $\int_A F\,d\nu = \int_A f \cdot \chi_A\,d\nu$, where χ_A is the characteristic function of the set A.

DEFINITION 0.4.14. An internal simple $\mathfrak{A}$-measurable function $F\colon X \longrightarrow {}^*\mathbb{R}$ is said to be $\mathcal{S}$-integrable if $\int_{A_N} F\,d\nu \approx 0$ for any infinite $N \in {}^*\mathbb{N}$, where $A_N = \{x \mid |F(x)| \geq N\}$.

The two remaining theorems in this subsection relate to the case of Loeb spaces with a finite measure, that is, to the case when $\nu_L(X) < +\infty$.

THEOREM 0.4.15. *The following conditions are equivalent for any internal simple $\mathfrak{A}$-measurable function $F\colon X \longrightarrow {}^*\mathbb{R}$:*

a) *F is $\mathcal{S}$-integrable;*

b) $({}^\circ \int_X |F|\,d\nu < +\infty) \ \wedge\ \forall A \in \mathfrak{A}\ (\nu(A) \approx 0 \Longrightarrow \int_A |F|\,d\nu \approx 0)$;

c) $\int_X {}^\circ|F|\,d\nu_L = {}^\circ \int_X |F|\,d\nu$.

DEFINITION 0.4.16. An internal $\mathfrak{A}$-measurable function $F\colon X \longrightarrow {}^*\mathbb{R}$ is called a lifting of a function $f\colon X \longrightarrow \overline{\mathbb{R}}$ if $f(x) = {}^\circ F(x)$ for ν_L-almost all x.

THEOREM 0.4.17. a) *A function $f\colon X \longrightarrow \overline{\mathbb{R}}$ is measurable if and only if it has a lifting.*

b) *A function $f\colon X \longrightarrow \overline{\mathbb{R}}$ is integrable ($f \in L_1(\nu_L)$) if and only if it has an $\mathcal{S}$-integrable lifting $F\colon X \longrightarrow {}^*\mathbb{R}$. Furthermore, $\int_X f\,d\nu_L = {}^\circ \int_X F\,d\nu$.*

Only the case when X is a hyperfinite set, $\mathfrak{A} = {}^*\mathcal{P}(X)$, and $\nu(A) = \Delta \cdot |A|$ for any $A \in \mathfrak{A}$ will be considered in this book (obviously, Δ is the value of ν on one-element subsets of X). The corresponding Loeb space is denoted by $(X, S_\Delta, \nu_\Delta)$, and the measure ν_Δ is called a *uniform Loeb measure.* Unless otherwise specified, we shall always understand a Loeb space to be such a space in what follows. If necessary, we sometimes use the notation $(X, S_\Delta^X, \nu_\Delta^X)$.

If $\Delta = |X|^{-1}$, then such a Loeb space is said to be *canonical* and denoted by (X, S, ν_L) or (X, S^X, ν_L^X). Canonical Loeb spaces play an important role in nonstandard probability theory. In the case of uniform Loeb measures every internal function $F\colon X \longrightarrow {}^*\mathbb{R}$ is simple and $\mathfrak{A}$-measurable, and $\int_A F\,d\nu = \Delta \sum_{x\in A} F(x)$ for any $A \in \mathfrak{A}$. A Loeb measure ν_Δ is finite if $\Delta \cdot |X| \ll +\infty$. In the case of a finite Loeb measure if $F\colon X \longrightarrow {}^*\mathbb{R}$ is an $\mathcal{S}$-integrable lifting of the function $f\colon X \longrightarrow \overline{\mathbb{R}}$, then in view of Theorem 0.4.17 (b)

$$\int_X f\,d\nu_\Delta = {}^\circ\Big(\Delta \sum_{x\in X} F(x)\Big). \tag{0.4.2}$$

5°. To illustrate the possibilities of using Loeb measures in classical mathematics we consider one generalization of the well-known construction Furstenberg used [**69**] to prove the Szemerédi theorem on arithmetic progressions by ergodic theory methods. The history of the topic is presented at length in [**70**]. We recall that the upper density of a set $A \subseteq \mathbb{N}$ is defined to be the number

$$d(A) = \limsup_{n\to\infty} \frac{|A \cap \{1, 2, \ldots, n\}|}{n}.$$

The Szemerédi theorem asserts that any set $A \subseteq \mathbb{N}$ with positive upper density contains arbitrarily long finite arithmetic progressions. To prove this theorem Furstenberg used a special way of associating with each set of positive upper density a measure space with a finite countably additive measure and a measure-preserving transformation acting in this space. Then the Szemerédi theorem was easily derived from the following refinement (also due to Furstenberg) of the Poincaré recurrence theorem.

THEOREM 0.4.18. *If $(X, \mathcal{B}, \mu)$ is a finite measure space, $T\colon X \longrightarrow X$ is an automorphism of this space, $A \in \mathcal{B}$, and $\mu(A) > 0$, then for any $k \in \mathbb{N}$ there exists an $n \in \mathbb{N}$, $n \neq 0$, such that*

$$\mu(A \cap T^n(A) \cap T^{2n}(A) \cap \cdots \cap T^{(k-1)n}(A)) > 0.$$

Here we use Loeb spaces to carry out the construction of Furstenberg in the more general situation when an amenable group G is considered in place of $\mathbb{N}$. It is easy to see that in the definition of the upper density it is possible to pass from $\mathbb{N}$ to $\mathbb{Z}$ and consider

$$d_{\mathbb{Z}}(A) = \limsup_{n\to\infty} \frac{|A \cap \{-n, \dots, n\}|}{2n+1}$$

for $A \subseteq \mathbb{Z}$. Further, if $A \subseteq \mathbb{N}$ and $d(A) > 0$, then $d_{\mathbb{Z}}(A) > 0$.

We recall that a countable group G is said to be *amenable* if there is a finitely additive finite invariant measure on G that is not identically equal to zero. The following criterion for amenability of a group G is well known (see, for example, [**71**]).

FØLNER'S CRITERION. *A countable group G is amenable if and only if there exists a sequence $U = \{U_n \mid n \in \mathbb{N}\}$ of finite subsets of G such that $\forall g \in G$*

$$\lim_{n\to\infty} \frac{|U_n \Delta g U_n|}{|U_n|} = 0. \tag{0.4.3}$$

A sequence U satisfying (0.4.3) will be called a *Følner sequence.* With each Følner sequence U we can associate the U-upper density of a set $A \subseteq G$ by the formula

$$d_U(A) = \limsup_{n\to\infty} \frac{|A \cap U_n|}{|U_n|}.$$

It is easy to see that the sequence $U = \{\{-n, \dots, n\} \mid n \in \mathbb{N}\}$ is a Følner sequence in the additive group $\mathbb{Z}$, and the upper density introduced above is the U-upper density for this sequence. Using the apparatus developed in the preceding subsections, we now prove the following theorem, which, holding with [**70**], we call the *Furstenberg correspondence principle* for amenable groups.

THEOREM 0.4.19. *Suppose that G is a countable amenable group, and let $U = \{U_n \mid n \in \mathbb{N}\}$ be a Følner sequence of subsets of G. Then for any set $A \subseteq G$ with $d_U(A) > 0$ there exist a probability space $(X, \mathcal{B}, \mu)$, an invariant action T of G on X, and a Boolean algebra homomorphism $\varphi\colon \mathcal{P}(G) \longrightarrow \mathcal{B}$ such that:*

(1) $\forall B \subseteq G\ (\mu(\varphi(B)) \le d_U(B))$;
(2) $\forall B \subseteq G\ (\varphi(gB) = T_g(\varphi(B)))$;
(3) $\mu(\varphi(A)) = d_U(A)$.

REMARK. The homomorphism φ is required only to preserve finite Boolean operations, not necessarily countable ones.

PROOF. The condition that U be a Følner sequence means that

$$\forall \Omega \in {}^*\mathbb{N} \setminus \mathbb{N}\ \forall^{\mathrm{st}} g \in G \left(\frac{|gU_\Omega \Delta U_\Omega|}{|U_\Omega|} \approx 0 \right). \tag{0.4.4}$$

By Theorem 0.3.19 iii), for any $B \subseteq G$

$$\forall \Omega \in {}^*\mathbb{N} \setminus \mathbb{N} \left(\frac{|{}^*B \cap U_\Omega|}{|U_\Omega|} \lesssim d_U(B) \right), \tag{0.4.5}$$

and

$$\exists \Lambda \in {}^*\mathbb{N} \setminus \mathbb{N} \left(\frac{|{}^*A \cap U_\Lambda|}{|U_\Lambda|} \approx d_U(A) \right). \tag{0.4.6}$$

We fix some Λ satisfying (0.4.6) and consider the canonical Loeb space (see the preceding subsection) (U_Λ, S, ν_L), that is, S is the σ-algebra of internal subsets of U_Λ, and the measure ν_L is generated by the internal uniform measure on U_Λ. Thus, for an internal set $V \subseteq U_\Lambda$

$$\nu_L(V) = {}^\circ\left(\frac{|V|}{|U_\Lambda|}\right). \tag{0.4.7}$$

It follows from (0.4.4) that $\forall^{\mathrm{st}} g \in G\ (\nu_L(gU_\Lambda \cap U_\Lambda) = 1)$.

Let $X = \bigcap_{g \in G} U^{(g)}$, where $U^{(g)} = gU_\Lambda \cap U_\Lambda$. Since G is countable, it follows that $X \in S$ and $\nu_L(X) = 1$. If now $\mathcal{B} = \{C \cap X \mid C \in S\}$ and $\mu = \nu_L|_{\mathcal{B}}$, then $(X, \mathcal{B}, \mu)$ is a probability space. Let us show that it satisfies the conditions of the theorem. First of all we define an action T of G (G, and not ${}^*G$!) on X. It is clear from the definition of T that $\forall x \in X\ \forall^{\mathrm{st}} g \in G\ (g \cdot x \in X)$, that is, we can set $T_g(x) = g \cdot x$. We show that this action is invariant. It suffices to show this for a set of the form $C \cap X$, where C is an internal subset of U_Λ. We have that

$$\mu(C \cap X) = \nu_L(C \cap X) = \nu_L(C),$$
$$\mu(g(C \cap X)) = \mu(gC \cap gX) = \mu(gC \cap X) = \mu(gC \cap U_\Lambda \cap X) = \nu_L(gC \cap U_\Lambda).$$

Obviously, $|C| = |gC| = |gC \cap gU_\Lambda|$, and hence

$$\big|\,|C| - |gC \cap U_\Lambda|\,\big| = \big|\,|gC \cap gU_\Lambda| - |gC \cap U_\Lambda|\,\big| \le |U_\Lambda \Delta gU_\Lambda|.$$

Dividing both sides of this inequality by $|U_\Lambda|$ and using (0.4.4), we get that

$$\frac{|C|}{|U_\Lambda|} \approx \frac{|gC \cap U_\Lambda|}{|U_\Lambda|}.$$

Passing to standard parts, we see in view of (0.4.7) that $\mu(C \cap X) = \mu(g(C \cap X))$, and this proves that the action T is invariant. We now define a Boolean algebra homomorphism $\varphi \colon \mathcal{P}(G) \longrightarrow \mathcal{B}$ by setting

$$\varphi(B) = {}^*B \cap X$$

for an arbitrary standard $B \subseteq G$. First of all, ${}^*B \cap X = {}^*B \cap U_\Lambda \cap X \in \mathcal{B}$, because ${}^*B \cap U_\Lambda$ is an internal hyperfinite subset of U_Λ. Further, $\mu(\varphi(B)) = \mu({}^*B \cap U_\Lambda) = {}^\circ(|{}^*B \cap U_\Lambda| \cdot |U_\Lambda|^{-1}) \le d_U(B)$ by virtue of (0.4.5), and $\mu(\varphi(A)) = d_U(A)$ by virtue of (0.4.6). Moreover, if $g \in G$, then

$$\varphi(gB) = {}^*(gB) \cap X = g \cdot {}^*B \cap X = g \cdot {}^*B \cap g \cdot X = g \cdot ({}^*B \cap X) = T_g(B),$$

which proves (2).

COROLLARY. *Suppose that G is a countable amenable group. For any set $A \subseteq G$ of positive Følner upper density, any $g \in G$, and any $k \in \mathbb{N}$ there exist an $x \in A$ and an $n \in \mathbb{N}$ such that*

$$\{x, g^n x, g^{2n} x, \dots, g^{(k-1)n} x\} \subseteq A.$$

PROOF. Assume that $A \subseteq G$ and $d_U(A) > 0$, and $(X, \mathcal{B}, \mu)$, T, and φ satisfy the conditions of Theorem 0.4.19. Then $\mu(\varphi(A)) > 0$. Let $g \in G$. According to Theorem 0.4.18, we get from the properties 1)–3) of the homomorphism φ that for any $k \in \mathbb{N}$ there is an $n \in \mathbb{N}$ such that

$$\begin{aligned} 0 &< \mu(\varphi(A) \cap T_g^n(\varphi(A)) \cap \cdots \cap T_g^{(k-1)n}(\varphi(A))) \\ &= \mu(\varphi(A \cap g^n A \cap \cdots \cap g^{(k-1)n} A)) \le d_U(A \cap g^n A \cap \cdots \cap g^{(k-1)n} A), \end{aligned}$$

hence $(A \cap g^n A \cap \cdots \cap g^{(k-1)n} A) \neq \emptyset$.

If $G = \mathbb{Z}$, then this corollary is Szemerédi's theorem. Proofs of Theorem 0.4.19 and of the corollary are contained in [**70**] for certain concrete Abelian groups more general than the group $\mathbb{Z}$. These proofs use a fairly strong apparatus in functional analysis.

6°. The following notation is used below. If $\overline{k} = (k_1, \dots, k_n)$, then $\mathcal{C}_{\overline{k}} = \mathcal{C}_{k_1} \times \cdots \times \mathcal{C}_{k_n}$, and ${}^*\mathcal{C}_{\overline{k}} = {}^*\mathcal{C}_{k_1} \times \cdots \times {}^*\mathcal{C}_{k_n}$.

THEOREM 0.4.20 (Nelson idealization principle). *Suppose that the nonstandard universe is* $\mathrm{Card}(\mathbb{R}^{\mathbb{R}})^+$*-saturated,* $r, s \in \mathbb{N}$, $\overline{k} \in \mathbb{N}^r$, $\overline{l} \in \mathbb{N}^s$, *and* $\mathfrak{M} \subseteq {}^*\mathcal{C}_{\overline{k}} \times {}^*\mathcal{C}_{\overline{l}}$ *is an internal class with the property that*

$$\forall m \in \mathbb{N}\ \forall \{\overline{\varphi}_1, \dots, \overline{\varphi}_m\} \subseteq \mathcal{C}_{\overline{k}}\ \exists \overline{\psi} \in \mathcal{C}_{\overline{l}}\ \forall i \le m\ (\langle {}^*\overline{\varphi}_i, \overline{\psi} \rangle \in \mathfrak{M}). \tag{0.4.8}$$

Then

$$\exists \overline{\psi} \in {}^*\mathcal{C}_{\overline{l}}\ \forall \overline{\varphi} \in \mathcal{C}_{\overline{k}}\ (\langle {}^*\overline{\varphi}, \overline{\psi} \rangle \in \mathfrak{M}). \tag{0.4.9}$$

(Note that the $m, \overline{\varphi}_1, \dots, \overline{\varphi}_m, \overline{\varphi}$ in (0.4.8) and (0.4.9) are standard.)

PROOF. For each $\overline{\varphi} \in \mathcal{C}_{\overline{k}}$ we denote by $\mathfrak{M}_{\overline{\varphi}}$ the internal class

$$\mathfrak{M}_{\overline{\varphi}} = \{\overline{\psi} \in {}^*\mathcal{C}_{\overline{l}} \mid \langle {}^*\overline{\varphi}, \overline{\psi} \rangle \in \mathfrak{M}\}.$$

In view of (0.4.8) the family $\{\mathfrak{M}_{\overline{\varphi}} \mid \overline{\varphi} \in \mathcal{C}_{\overline{k}}\}$ of internal classes has the finite intersection property, and its cardinality does not exceed $\mathrm{Card}(\mathbb{R}^{\mathbb{R}})$ because $\mathcal{C}_{\overline{k}} \sim \mathbb{R}^{\mathbb{R}}$. Since the nonstandard universe is $\mathrm{Card}(\mathbb{R}^{\mathbb{R}})^+$-saturated, the given family now has nonempty intersection, and this is expressed by the relation (0.4.9).

REMARK 1. For the case when $\mathfrak{M}$ is a standard class, that is, a class definable by a formula of $\mathcal{LR}$ in terms of parameters of the form ${}^*\xi$ for $\xi \in \mathcal{C}$, this is still called the concurrence principle. We formulate this more precisely. Recall that if $\mathfrak{M}$ is a subclass of the standard model $\mathbb{R}$ definable by a formula in the language $\mathcal{LR}$ in terms of parameters $b_1, \dots, b_k \in \mathcal{C}$, then ${}^*\mathfrak{M}$ denotes the internal subclass of the nonstandard universe ${}^*\mathbb{R}$ definable by the same formula in terms of the parameters ${}^*b_1, \dots, {}^*b_k$. Subclasses of the form ${}^*\mathfrak{M}$ are called standard subclasses of the nonstandard universe. We can now give a precise formulation.

The concurrence principle. If $\mathfrak{M}$ is a subclass of the standard universe definable by a formula of $\mathcal{LR}$ in terms of parameters $b_1, \dots, b_k \in \mathcal{C}$ and is such that

$$\forall m \in \mathbb{N}\ \forall \{\overline{\varphi}_1, \dots, \overline{\varphi}_m\} \subseteq \mathcal{C}_{\overline{k}}\ \exists \overline{\psi} \in \mathcal{C}_{\overline{l}}\ \forall i \le m\ (\langle \overline{\varphi}_i, \overline{\psi} \rangle \in \mathfrak{M}), \tag{0.4.8′}$$

then

$$\exists \overline{\psi} \in {}^*\mathcal{C}_{\overline{l}}\ \forall \overline{\varphi} \in \mathcal{C}_{\overline{k}}\ (\langle {}^*\overline{\varphi}, \overline{\psi} \rangle \in {}^*\mathfrak{M}). \tag{0.4.9′}$$

It is easy to see that the concurrence principle is a special case of the idealization principle. The concurrence principle was introduced by Robinson in one of the first publications on nonstandard analysis [**61**], where it was proved that there is a nonstandard universe in which this principle is satisfied. At that time the saturation property had not yet been used in nonstandard analysis (it was apparently first employed systematically beginning with the papers [**42**] and [**43**] on Loeb measures), and assertions analogous to Propositions 0.4.6–0.4.8 were proved with the use of the concurrence principle.

EXERCISE. Prove Propositions 0.4.6–0.4.8 on the basis of the concurrence principle, that is, not assuming that the nonstandard universe has the saturation property, but rather assuming that the concurrence principle is satisfied in it.

REMARK 2. If we consider elementary extensions of a model $\mathcal{M}$ constructed on the basis of a set M of arbitrarily large cardinality (see the beginning of subsection 2° in this section), then the idealization principle for such elementary extensions follows from the condition that they are $\mathrm{Card}(M^M)^+$-saturated. The proof of this is analogous to that of Theorem 0.4.18.

REMARK 3. The idealization principle was actually formulated by Nelson for his internal set theory (IST; see the concluding section of this chapter). In the language of IST it is not possible to speak of arbitrary external sets and families of them, and hence the saturation property cannot even be formulated in the framework of this theory, which, of course, restricts its use in an essential way. However, Propositions 0.4.6–0.4.8 and propositions analogous to them, as well as most of the properties considered for the nonstandard universe that do not employ the concept of saturation, can be proved in internal set theory with the use of the idealization principle. There are also several nonstandard set theories that include external sets (see, for example, [**62**]–[**64**], along with [**65**], which treats BST—the theory of bounded internal sets, whose language coincides with that of IST, but which uses a clever device enabling one to refer to certain classes of external sets sufficient for most applications). In some of them it is possible to formulate also the saturation property (see, for example, the external set theory described in §6 of this chapter). However, it should be kept in mind that the existence of λ^+-saturated models can be proved without additional set-theoretic hypotheses even for arbitrary but previously fixed λ [**19**]. The idealization principle in nonstandard axiomatic set theories does not assume any restrictions on the cardinality of the set of standard elements ξ such that $\exists \eta\ (\langle {}^*\xi, \eta \rangle \in \mathfrak{M})$, where $\mathfrak{M}$ is an internal class, and its consistency does not require any additional set-theoretic hypotheses besides the system of Zermelo–Fraenkel axioms. That is, in the general case it is apparently impossible to assert that the Nelson idealization principle is a consequence of the saturation principle. However, if we confine ourselves to internal classes for which the cardinality of the above set of standard elements is bounded above by some sufficiently large cardinal number (the hypercontinuum is quite sufficient for most substantive propositions), then this restricted idealization principle follows from the corresponding saturation.

For most of the applications to be considered below, the countable saturation of the nonstandard universe is completely sufficient, and this is always assumed. In some cases it is required, moreover, that the nonstandard universe satisfy the idealization principle. The notation (IST) will indicate this. For example, the expression

"Proposition 1.2.3 (IST) ... " means that in Proposition 1.2.3 the nonstandard universe is assumed to satisfy the idealization principle, and the expression "§8 (IST) ... " means that this assumption holds in the course of the whole section.

In the next section we shall prove the existence of countably saturated nonstandard universes satisfying the idealization principle.

The following four propositions are consequences of the concurrence principle. It was noted in the remark after Proposition 0.3.16 that if $\operatorname{Card}\mathfrak{M} \in \mathbb{N}$, then ${}^*\mathfrak{M} = \{{}^*\bar{a} \mid \bar{a} \in \mathfrak{M}\}$.

PROPOSITION 0.4.21. *If the nonstandard universe satisfies the concurrence principle, and $\mathfrak{M}$ is a subclass of the standard model $\mathbb{R}$, then*

$${}^*\mathfrak{M} = \{{}^*\bar{a} \mid \bar{a} \in \mathfrak{M}\} \iff \operatorname{Card}\mathfrak{M} \in \mathbb{N}.$$

PROOF. In one direction the assertion has already been proved (see Proposition 0.3.16 and the remark after it). To prove the opposite direction we assume that $\operatorname{Card}\mathfrak{M} = \infty$ and consider the standard class

$$\mathfrak{N} = \{\langle \bar{a}, \bar{b}\rangle \mid \bar{a}, \bar{b} \in \mathfrak{M},\ \bar{a} \neq \bar{b}\}.$$

Since $\operatorname{Card}\mathfrak{M} = \infty$, it follows that ${}^*\mathfrak{N}$ satisfies (0.4.8′), and by virtue of (0.4.9′) there exists a $\bar{b} \in {}^*\mathfrak{M}$ such that $\forall \bar{a} \in \mathfrak{M}\ (\bar{b} \neq {}^*\bar{a})$. This contradiction proves what is required.

PROPOSITION 0.4.22. *Suppose that the nonstandard universe satisfies the concurrence principle, and let $k \in \mathbb{N}$. Then there exists a hyperfinite family $F = \{F_n \mid n < \nu \in {}^*\mathbb{N}\}$ of subsets of ${}^*\mathbb{R}^k$ such that ${}^*A \in F$ for any standard set $A \subseteq \mathbb{R}^k$ (see Definition 0.3.18 and the remark after it).*

PROOF. We consider the standard class

$$\mathfrak{M} = \{\langle A, F\rangle \mid A \subseteq \mathbb{R}^k\ \wedge\ \exists \nu \in \mathbb{N}\ (F = \{F_n \mid n < \nu\})\ \wedge\ A \in F\}.$$

Then $\mathfrak{M}$ obviously satisfies (0.4.8′), and

$${}^*\mathfrak{M} = \{\langle A, F\rangle \mid A \subseteq {}^*\mathbb{R}^k\ \wedge\ \exists \nu \in {}^*\mathbb{N}\ (F = \{F_n \mid n < \nu\})\ \wedge\ A \in F\}$$

satisfies (0.4.9′), that is, there exists an $F = \{F_n \mid n < \nu \in {}^*\mathbb{N}\}$ such that $\forall A \subseteq \mathbb{R}^k$ $({}^*A \in F)$.

COROLLARY. *For any internal class $\mathfrak{N}$ of subsets of ${}^*\mathbb{R}^k$ there exists a hyperfinite family $G = \{G_n \mid n < \lambda \in {}^*\mathbb{N}\}$ such that $G \subseteq \mathfrak{N}$ and*

$${}^*A \in \mathfrak{N} \iff {}^*A \in G$$

for any standard $A \subseteq \mathbb{R}^k$.

PROOF. Let $F = \{F_n \mid n < \nu \in {}^*\mathbb{N}\}$ be the hyperfinite family in Proposition 0.4.22. We consider the internal set $B = \{n < \nu \mid F_n \in \mathfrak{N}\}$. It is hyperfinite, as an internal subset of a hyperfinite set. Then there exists a $\lambda \in {}^*\mathbb{N}$ and an internal bijection $\varphi\colon \{0, 1, \dots, \lambda - 1\} \longrightarrow B$. Let $G_n = F_{\varphi(n)}$. Then $G = \{G_n \mid n < \lambda\}$ is the required hyperfinite family.

This leads easily to the following result, which is important in what follows.

THEOREM 0.4.23. *Suppose that the nonstandard universe satisfies the concurrence principle, and let* $(X, \mathcal{Y}, \mu)$ *be a standard measure space. Then there exists an element* $\xi \in {}^*X$ *such that*

$$\forall Y \in \mathcal{Y}\ (\mu(Y) = 0 \Longrightarrow \xi \in {}^*Y). \tag{0.4.10}$$

REMARK. In what follows, elements of ${}^*X$ satisfying (0.4.10) are said to be random (compare with the random elements in [**66**] and [**67**]; the random elements introduced in [**66**] are also described in [**68**]).

PROOF. We consider the internal class $\mathfrak{N} = \{Y \in \mathcal{Y} \mid {}^*\mu(Y) = 0\}$. Let $G = \{G_n \mid n < \lambda\}$ be a hyperfinite family satisfying the conditions of the preceding corollary. By the transfer principle, $Y_0 = \bigcup\{G_n \mid n < \lambda\} \in {}^*\mathcal{Y}$. Consequently, ${}^*\mu(Y_0) = 0$, that is, ${}^*\mu({}^*X \setminus Y_0) = {}^*\mu({}^*X) = \mu(X)$, and thus ${}^*X \setminus Y_0 \neq \emptyset$ (it is assumed that $\mu(X) > 0$). Obviously, any element in ${}^*X \setminus Y_0$ will work.

PROPOSITION 0.4.24. *If* X *is a standard linear space, then there exists an internal hyperfinite-dimensional subspace* $Y \subseteq {}^*X$ *(see §3, 5°) such that* $X \subset Y$ *(that is,* Y *contains all the standard elements of* ${}^*X$*).*

The proof is left to the reader as an exercise.

REMARK. It is this proposition that lies at the basis of most applications of nonstandard analysis in the theory of Banach spaces. In particular, the famous Bernstein–Robinson proof [**32**] of Halmos' conjecture about the existence of invariant subspaces of polynomially compact operators is based on it.

7°. In this subsection we consider an extension of the language $\mathfrak{LR}$, namely, the language $\mathfrak{LRS}$ intended for describing a nonstandard universe. It is obtained by adding to the language $\mathfrak{LR}$ the one-place predicate symbol $\mathrm{St}(\alpha)$, where α is a variable of arbitrary type. Moreover, we assume that in $\mathfrak{LRS}$ there are constants for arbitrary standard objects, and they have the same notation as the objects themselves. In the interpretation of the language $\mathfrak{LRS}$ in the universe ${}^*\mathbb{R}$ the symbols of $\mathfrak{LR}$ are interpreted in the usual way; for example, the variables of type k are interpreted as internal mappings from ${}^*\mathbb{R}^k$ to ${}^*\mathbb{R}$, and so on. If ξ is an internal object of type k, then $\mathrm{St}(\xi) = \mathbf{T}$ if and only if $\xi = {}^*f$ for some standard function $f\colon \mathbb{R}^k \longrightarrow \mathbb{R}$. If a is an object of type 0, then $\mathrm{St}(a) = \mathbf{T}$ if and only if a is a standard real number. We remark that the symbol "$*$" itself is absent in $\mathfrak{LRS}$, and the constants for denoting standard objects here denote the 'stars' of these objects, that is, their images under the imbedding in the nonstandard universe ${}^*\mathbb{R}$. For example, the constants $\mathbb{R}$ and $\mathbb{N}$ in the language $\mathfrak{LRS}$ denote the sets ${}^*\mathbb{R}$, ${}^*\mathbb{N}$, and so on. Formulas not containing the predicate symbol St, that is, in essence, the formulas of $\mathfrak{LR}$, are called internal formulas of $\mathfrak{LRS}$. The following abbreviations are also used below:

$$\forall^{\mathrm{st}}\alpha A \rightleftharpoons \forall\alpha\ (\mathrm{St}(\alpha) \longrightarrow A);$$
$$\exists^{\mathrm{st}}\alpha A \rightleftharpoons \exists\alpha\ (\mathrm{St}(\alpha) \wedge A);$$
$$\forall^{\mathrm{st\,fin}}X\ A \rightleftharpoons \forall X\ (\mathrm{St}(X) \wedge \mathrm{Fin}(X) \longrightarrow A);$$
$$\exists^{\mathrm{st\,fin}}X\ A \rightleftharpoons \exists X\ (\mathrm{St}(X) \wedge \mathrm{Fin}(X) \wedge A).$$

Here A is a formula, possibly containing the variables α and X, and $\mathrm{Fin}(X)$ means that X is a finite set. The corresponding formula in $\mathfrak{LR}$ can be found in §1, 2°, part e. In fact, for an arbitrary internal $X \subseteq {}^*\mathbb{R}^k$ the formula $\mathrm{Fin}(X) = \mathbf{T}$ if and only if X is a hyperfinite set, but $\mathrm{St}(X) \wedge \mathrm{Fin}(X)$ means that $X = \{\overline{a}_1, \dots, \overline{a}_n\}$ is a standard finite subset of $\mathbb{R}^k$.

The following formula in the language $\mathcal{LRS}$ expresses the fact that $X^{(k+1)}$ is a hyperfinite family of internal subsets of ${}^*\mathbb{R}^k$ (see Definition 0.3.18 and the remark after it):

$$\begin{aligned}\mathrm{Finn}\big(X^{(k+1)}\big) \rightleftharpoons \exists n \in \mathbb{N}\big(&\mathrm{dom}^{(1)}\big(X^{(k+1)}\big) = \{0, 1, \dots, n-1\}\\ &\wedge\ \forall m, l < n\big(m \neq l \longrightarrow \big\{\langle x_1, \dots, x_k\rangle \mid \langle m, x_1, \dots, x_k\rangle \in X^{(k+1)}\big\}\\ &\neq \big\{\langle x_1, \dots, x_k\rangle \mid \langle l, x_1, \dots, x_k\rangle \in X^{(k+1)}\big\}\big)\big).\end{aligned}$$

Since $\mathrm{Finn}\big(X^{(k+1)}\big)$ is regarded as a formula in $\mathcal{LRS}$, the constant $\mathbb{N}$ denotes the set ${}^*\mathbb{N}$, and

$$\begin{aligned}Y^{(k)} \in X^{(k+1)} \rightleftharpoons \exists m \in \mathrm{dom}^{(1)}&\big(X^{(k+1)}\big)\\ &\big(Y^{(k)} = \big\{\langle x_1, \dots, x_k\rangle \mid \langle m, x_1, \dots, x_k\rangle \in X^{(k+1)}\big\}\big).\end{aligned}$$

Now $\mathrm{St}\big(X^{(k+1)}\big) \wedge \mathrm{Finn}\big(X^{(k+1)}\big)$ means that $X^{(k+1)} = \{{}^*Y_1, \dots, {}^*Y_n\}$ for some standard natural number n and standard sets $Y_1, \dots, Y_n \subseteq \mathbb{R}^k$. The abbreviations $\forall^{\mathrm{st\,finn}} X$ and $\exists^{\mathrm{st\,finn}} X$ are used just as above.

For each internal formula $A(\alpha_1, \dots, \alpha_n)$ we can now write the following formula, which means that the transfer principle holds for A:

$$(0.4.11)\qquad Q_1^{st}\alpha_1 \dots Q_n^{st}\alpha_n A(\alpha_1, \dots, \alpha_n) \longleftrightarrow Q_1\alpha_1 \dots Q_n\alpha_n A(\alpha_1, \dots, \alpha_n).$$

Here each Q_i is either $\forall$ or $\exists$. We remark that it actually suffices to require the truth of the sentence

$$(0.4.12)\qquad \forall^{\mathrm{st}}\overline{\alpha}\ (\forall^{\mathrm{st}}\beta\ A(\overline{\alpha}, \beta) \longleftrightarrow \forall\beta\ A(\overline{\alpha}, \beta)),$$

because the formula (0.4.11) can be obtained formally from (0.4.12) by the rules of predicate logic.

The *idealization principle* can be formulated similarly. If $B(\overline{\alpha}, Y^{(k)}, \overline{\beta})$ is an internal formula, then the idealization principle for it is expressed by the following formula in $\mathcal{LRS}$:

$$\begin{aligned}(0.4.13)\quad \forall\overline{\beta}\ [\forall^{\mathrm{st\,finn}} X^{(k+1)}\ \exists\overline{\alpha}\ \forall Y^{(k)} \in X^{(k+1)}\ &B(\overline{\alpha}, Y^{(k)}, \overline{\beta})\\ &\longleftrightarrow \exists\overline{\alpha}\ \forall^{\mathrm{st}} Y^{(k)}\ B(\overline{\alpha}, Y^{(k)}, \overline{\beta})].\end{aligned}$$

The idealization principle is written as follows for an internal formula $B(\overline{\alpha}, \overline{x}, \overline{\beta})$ in which $\overline{x}$ is a k-tuple of variables of type 0:

$$(0.4.13')\qquad \forall\overline{\beta}\ [\forall^{\mathrm{st\,finn}} X^{(k)}\ \exists\overline{\alpha}\ \forall\overline{x} \in X^{(k)}\ B(\overline{\alpha}, \overline{x}, \overline{\beta}) \longleftrightarrow \exists\overline{\alpha}\ \forall^{\mathrm{st}}\overline{x}\ B(\overline{\alpha}, \overline{x}, \overline{\beta})].$$

If $A(\overline{x}, \overline{\alpha})$ is a formula in $\mathcal{LR}$, where $\overline{x}$ is a k-tuple of variables of type 0, then the following sentence is true in the model $\mathbb{R}$:

$$(0.4.14)\qquad \forall\overline{\alpha}\ \exists X^{(k)}\ \forall\overline{x}\ (\overline{x} \in X^{(k)} \longleftrightarrow A(\overline{x}, \overline{\alpha})).$$

This follows from the fact that the standard model contains all the subsets of $\mathbb{R}^k$, in particular, also the set $\{\overline{x} \mid \mathbb{R} \models A(\overline{x}, \overline{\alpha})\}$ for any $\overline{\alpha} \in \mathcal{C}$. By the transfer principle, the sentence (0.4.14) is true also in ${}^*\mathbb{R}$, that is, there exists the internal set $X \subseteq {}^*\mathbb{R}^k$ equal to $\{\overline{x} \in {}^*\mathbb{R}^k \mid {}^*\mathbb{R} \models A(\overline{x}, \overline{\alpha})\}$, where $\overline{\alpha} \in {}^*\mathcal{C}$.

The situation changes if A is a formula in $\mathcal{LRS}$ that is not internal (it is natural to call such formulas external in what follows). For example, the formula $A(x)$:

$$\forall^{\text{st}} t\ (t > 0 \longrightarrow |x| < t)$$

gives the monad of zero in ${}^*\mathbb{R}$, that is, an external set. But since the variables in $\mathcal{LRS}$ are interpreted only as internal objects, the sentence (0.4.14) with this formula A is false in ${}^*\mathbb{R}$.

On the other hand, any external formula $A(\overline{x}, \overline{\alpha})$ with an arbitrary $\overline{\alpha} \in {}^*\mathcal{C}$ determines the standard set $Y = \{\overline{x} \in \mathbb{R}^k \mid {}^*\mathbb{R} \models A(\overline{x}, \overline{\alpha})\}$ (it is possible that $Y = \emptyset$). Clearly, for each standard $\overline{x} \in \mathbb{R}^k$

$$\overline{x} \in {}^*Y \Longleftrightarrow {}^*\mathbb{R} \models A(\overline{x}, \text{-}a).$$

This proves that the sentence of $\mathcal{LRS}$

$$\forall \overline{\alpha}\ \exists^{\text{st}} X^{(k)} \forall^{\text{st}} \overline{x}\ (\overline{x} \in X^{(k)} \longleftrightarrow A(\overline{x}, \overline{\alpha})) \tag{0.4.15}$$

is true in ${}^*\mathbb{R}$. This sentence is called the *standardization principle.*

A standard set $X^{(k)}$ satisfying the formula (0.4.15) is called a standardization (external, in general) of the set $\{\overline{x} \mid A(\overline{x}, \overline{\alpha})\}$ and is denoted by ${}^S\{\overline{x} \mid A(\overline{x}, \overline{\alpha})\}$.

Let us consider some concrete examples of standardizations.

1. If B is a standard set, then it is easy to see that ${}^SB = {}^S\{x \mid x \in B\} = B$. Indeed, let $C = {}^SB$. Then B and C are standard sets, and by the definition of a standardization

$$\forall^{\text{st}} x\ (x \in C \longleftrightarrow x \in B),$$

but then $B = C$ by virtue of the transfer principle (because B and C are both standard!).

2. If $\Omega \in {}^*\mathbb{N} \setminus \mathbb{N}$, then

$${}^S\{n \in {}^*\mathbb{N} \mid n < \Omega\} = {}^*\mathbb{N}.$$

3. If $t \in {}^*\mathbb{R}_b$, then ${}^\circ t = \sup {}^S\{u \in {}^*\mathbb{R} \mid u < t\}$ (see the proof of Theorem 0.3.7).

We present an example of another true sentence of $\mathcal{LRS}$ that is needed in what follows.

Recall (Proposition 0.2.6, 11)) that for any formula $A(x,y)$ in the language $\mathcal{LR}$

$$\mathbb{R} \models \forall X\ \forall Y\ (\forall x \in X \exists y \in Y\ A(x,y) \longleftrightarrow \exists f \in Y^X\, \forall x \in X\ A(x, f(x))). \tag{0.4.16}$$

By the transfer principle, this sentence is true also in ${}^*\mathbb{R}$ for any formula A in the language $\mathcal{LR}$, or, what is the same, for any internal formula A in the language $\mathcal{LRS}$. But if A is an external formula in $\mathcal{LRS}$, then the equivalence (0.4.16) may no longer be true in ${}^*\mathbb{R}$. For example, suppose that $A(x,y) \rightleftharpoons \text{St}(y)\ \wedge\ y \approx x$. If we restrict the domain of the variable x to the interval ${}^*[0,1]$, then this formula determines a function $\varphi\colon {}^*[0,1] \longrightarrow {}^*[0,1]$ such that $\varphi(x) = {}^\circ x\ \forall x \in {}^*[0,1]$. However, this function is not *internal,* because $\text{rng}(f)$ is the set $[0,1]$ of standard elements of ${}^*[0,1]$, which is not internal. Nevertheless, we have

PROPOSITION 0.4.25. *For an arbitrary (including external) formula $A(x,y)$ in the language $\mathcal{LRS}$,*

$$\begin{aligned} {}^*\mathbb{R} \models \forall^{\text{st}} X\ \forall^{\text{st}} Y\ &(\forall^{\text{st}} x \in X\ \exists^{\text{st}} y \in Y\ A(x,y) \\ &\longleftrightarrow \exists^{\text{st}} f \in Y^X\ \forall^{\text{st}} x \in X\ A(x, f(x))). \end{aligned} \tag{0.4.17}$$

In the example considered in the preceding paragraph the standard function f whose existence is asserted in this sentence is simply the identity function on $[0,1]$.

PROOF. We prove only the implication $\longrightarrow$, since the converse implication is obvious. Let $T = {}^S\{\langle x,y\rangle \in X \times Y \mid A(x,y)\}$. Then the formula

$$^*\mathbb{R} \models \forall^{\mathrm{st}} x \in X\ \exists^{\mathrm{st}} y \in Y\ (\langle x,y\rangle \in T)$$

is true in $\mathfrak{LRS}$, and since X, Y, and T are standard and the formula $\langle x,y\rangle \in T$ is internal, it follows from the transfer principle (0.4.11) that

$$^*\mathbb{R} \models \forall x \in {}^*X\ \exists y \in {}^*Y\ (\langle x,y\rangle \in {}^*T)$$

(see the remark about the interpretation of standard constants in the language $\mathfrak{LRS}$ at the beginning of this subsection). Again by virtue of the transfer principle (now the usual one!)

$$\mathbb{R} \models \forall x \in X\ \exists y \in Y\ (\langle x,y\rangle \in T).$$

By Proposition 0.2.6 (11),

$$\mathbb{R} \models \exists h \in Y^X\ \forall x \in X\ (\langle x,h(x)\rangle \in T).$$

Setting $f = {}^*h$, we get what is required.

The language $\mathfrak{LRS}$ is convenient because there is an automatic procedure enabling us, given any sentence A in this language containing arbitrary standard constants, to construct a sentence B in $\mathfrak{LR}$ such that

$$^*\mathbb{R} \models A \Longleftrightarrow \mathbb{R} \models B.$$

In other words, for each nonstandard assertion about standard objects, a standard assertion about these objects that is equivalent to it is automatically constructed. This procedure was described in [**48**] for a more general situation, namely, for sentences of internal set theory (see §6 of this chapter), and it bears the name of Nelson's algorithm. This algorithm enables one to get all the assertions of Theorems 0.3.19–0.3.22 automatically. We describe the basic idea of Nelson's algorithm. A detailed description of it can be found in [**48**] or in [**20**]. First of all, the equivalence (0.4.11) shows that it suffices to reduce the sentence A to the form

$$Q_1^{\mathrm{st}}\alpha_1 \dots Q_n^{\mathrm{st}}\alpha_n\ C(\alpha_1,\dots,\alpha_n),$$

where C is an internal formula in $\mathfrak{LRS}$. For this one should first reduce A to prenex normal form (Proposition 0.2.7) by taking all the quantifiers of the form Q and Q^{st} out to the beginning of the formula, and then interchanging them to move all the quantifiers of the form Q^{st} to the left. Here quantifiers of the same name can be simply transposed in view of the logical equivalence

$$Q^{\mathrm{st}}\alpha Q\beta\ D(\alpha,\beta) \equiv Q\beta Q^{\mathrm{st}}\alpha\ D(\alpha,\beta),$$

where Q is either $\forall$ or $\exists$. The equivalences (0.4.13), (0.4.13′), and (0.4.17) should be used to transpose quantifiers of different name. Moreover, equivalences obtained from the indicated ones by passing to the negations may be needed. For example, the equivalence (0.4.13′) has the form

$$\forall^{\mathrm{st\,fin}} X\ \exists\overline{\alpha}\ \forall\overline{x} \in X\ D(\overline{\alpha},\overline{x}) \equiv \exists\overline{\alpha}\ \forall^{\mathrm{st\,fin}}\overline{x}\ D(\overline{\alpha},\overline{x}),$$

where D is an internal formula in $\mathcal{LRS}$. Replacing D by $\neg D$ here and using the fact that two formulas are equivalent if and only if their negations are equivalent, we get

$$\exists^{\mathrm{st\,fin}} X \;\forall \overline{\alpha}\; \exists \overline{x} \in X \; D(\overline{\alpha}, \overline{x}) \equiv \forall \overline{\alpha}\; \exists^{\mathrm{st}} \overline{x}\; D(\overline{\alpha}, \overline{x}). \tag{0.4.13''}$$

The equivalence

$$\exists^{\mathrm{st}} x \in X\; \forall^{\mathrm{st}} y \in Y\; A(x,y) \equiv \forall^{\mathrm{st}} f \in Y^X\; \exists^{\mathrm{st}} x \in X\; A(x, f(x)) \tag{0.4.17'}$$

is obtained from (0.4.17) in exactly the same way. We now consider two examples of the use of Nelson's algorithm.

EXAMPLE 0.4.26. Let A be an arbitrary standard subset of $\mathbb{R}$, and consider the following sentence in the language $\mathcal{LRS}$ (recall that a constant A in $\mathcal{LRS}$ is interpreted as the set ${}^*A \subseteq {}^*\mathbb{R}$):

$$\forall^{\mathrm{st}} x \in A\; \forall \xi\; (\forall^{\mathrm{st}} z > 0\; (|x - \xi| < z) \longrightarrow \xi \in A). \tag{0.4.18}$$

This sentence asserts that ${}^*A$ contains the monad of each standard point in it. By Theorem 0.3.20 i), this means that A is an open set. We get the last assertion by Nelson's algorithm. Using the basic logical equivalences (Propositions 0.2.4 and 0.2.6), we rewrite (0.4.18) in the form

$$\forall^{\mathrm{st}} x \in A\; \forall \xi\; \exists^{\mathrm{st}} z > 0\; (|x - \xi| < z \longrightarrow \xi \in A). \tag{0.4.19}$$

By virtue of (0.4.13″) the sentence (0.4.19) is equivalent to the sentence

$$\forall^{\mathrm{st}} x \in A\; \exists^{\mathrm{st\,fin}} B\; \forall \xi\; \exists z \in B\; (|x - \xi| < z \longrightarrow \xi \in A),$$

which, by the transfer principle (0.4.11), is equivalent to the standard sentence

$$\forall x \in A\; \exists^{\mathrm{fin}} B\; \forall \xi\; \exists z \in B\; (|x - \xi| < z \longrightarrow \xi \in A). \tag{0.4.20}$$

Again using logical transformations, we see that (0.4.20) is equivalent to the sentence

$$\forall x \in A\; \exists^{\mathrm{fin}} B\; \forall \xi\; (\forall z \in B\; (|x - \xi| < z) \longrightarrow \xi \in A). \tag{0.4.21}$$

Since B is a finite set, it follows that

$$\forall z \in B\; (|x - \xi| < z) \equiv |x - \xi| < \min B.$$

It is now obvious that (0.4.21) is equivalent to the sentence

$$\forall x \in A\; \exists z\; \forall \xi\; (|x - \xi| < z \longrightarrow \xi \in A),$$

which means that A is an open set.

EXAMPLE 0.4.27. We consider a sentence in $\mathcal{LRS}$ asserting that each point of the interval ${}^*[0,1]$ has a standard part:

$$\forall x \in [0,1]\; \exists^{\mathrm{st}} y \in [0,1]\; \forall^{\mathrm{st}} z > 0\; (|x - y| < z). \tag{0.4.22}$$

Below we write $z \in \mathbb{R}_+$ instead of $z > 0$. By (0.4.17′), the sentence (0.4.22) is equivalent to

$$\forall x \in [0,1]\; \forall^{\mathrm{st}} f \in \mathbb{R}_+^{[0,1]}\; \exists^{\mathrm{st}} y \in [0,1]\; (|x - y| < f(y)).$$

Transposing quantifiers of the same name, we arrive at the equivalent sentence

$$\forall^{\text{st}} f \in \mathbb{R}_+^{[0,1]}\ \forall x \in [0,1]\ \exists^{\text{st}} y \in [0,1]\ (|x-y| < f(y)).$$

Using (0.4.13″), we now get the sentence

$$\forall^{\text{st}} f \in \mathbb{R}_+^{[0,1]}\ \exists^{\text{st fin}} B\ \forall x \in [0,1]\ \exists y \in B\ (|x-y| < f(y)),$$

which, by the transfer principle (0.4.11), is equivalent to the standard sentence

$$\forall f \in \mathbb{R}_+^{[0,1]}\ \exists^{\text{fin}} B\ \forall x \in [0,1]\ \exists y \in B\ (|x-y| < f(y)). \tag{0.4.23}$$

We show that (0.4.23) is equivalent to the Heine–Borel lemma for $[0,1]$. Indeed, suppose that (0.4.23) is true, and let $\mathcal{A}$ be an arbitrary covering of $[0,1]$ by open intervals. With each $y \in [0,1]$ we associate some open interval $\Phi(y) \in \mathcal{A}$ containing y, and we denote by $f(y)$ the smaller of the distances from y to the endpoints of this interval. Then it follows from (0.4.23) that the finite family $\{\Phi(y) \mid y \in B\} \subseteq \mathcal{A}$ of open intervals covers $[0,1]$. Conversely, if $f \in \mathbb{R}_+^{[0,1]}$ and $\mathcal{A} = \{(y - f(y), y + f(y)) \mid y \in [0,1]\}$, then $\mathcal{A}$ is an open covering of $[0,1]$, and the truth of (0.4.23) follows immediately from the Heine–Borel lemma. It is clear that any standard closed interval can appear here instead of $[0,1]$.

EXERCISE. Prove all the assertions of Theorems 0.3.19–0.3.22 by using Nelson's algorithm.

§5. Construction of nonstandard models

1°. We briefly recall the main facts about ultrafilters. Proofs can be found, for example, in the book [**19**]. Let I be an infinite set. A family D of nonempty subsets of I is called a *filter* on I if: (1) it does not contain the empty set; (2) it is closed under finite intersections; (3) it contains all supersets of any set in it. Filters that are maximal with respect to inclusion are called *ultrafilters.* It follows immediately from Zorn's lemma that every filter on I is contained in an ultrafilter.

PROPOSITION 0.5.1. *The following conditions are equivalent:*
(1) *D is an ultrafilter*;
(2) $\forall A \in \mathcal{P}(I)\ (A \in D \ \vee\ \overline{A} \in D)$;
(3) $\forall A, B \in \mathcal{P}(I)\ (A \cup B \in D \longrightarrow (A \in D \ \vee\ B \in D))$.

The family $\mathcal{F} = \{A \subseteq I \mid 0 \leq |\overline{A}| < +\infty\}$ serves as an example of a filter on I. This filter is called the *Fréchet filter.* Ultrafilters majorizing the Fréchet filter are called *free* or *nonprincipal* ultrafilters. Ultrafilters that are not free are called *principal* ultrafilters. Proposition 0.5.1 immediately gives us

PROPOSITION 0.5.2. *An ultrafilter D on I is principal if and only if there exists an $a \in I$ such that $D = \{A \subseteq I \mid a \in A\}$.*

All the ultrafilters encountered below are assumed to be free unless otherwise stated. An ultrafilter D is said to be *countably complete* if $\bigcap_{n\in\mathbb{N}} A_n \in D$ for any sequence $\{A_n \mid n \in \mathbb{N}\} \subseteq D$. Otherwise D is said to be *countably incomplete.* Obviously, a principal ultrafilter is always countably complete. It is also easy to see that the converse is true in the case of a countable set I, that is, we have

PROPOSITION 0.5.3. *If I is a countable set, then an ultrafilter D on I is countably complete if and only if it is principal.*

It can be shown that there are countably incomplete ultrafilters on every infinite set I [**19**].

2°. In this subsection we fix an infinite set I and an ultrafilter D on it.

DEFINITION 0.5.4. i) The ultrapower of a set M with respect to the ultrafilter D is defined to be the quotient of the Cartesian power M^I by the equivalence relation $\sim_D$ defined by

$$\forall \mu, \nu \in M^I \ (\mu \sim_D \nu \rightleftharpoons \{i \mid \mu(i) = \nu(i)\} \in D).$$

The ultrapower of M with respect to D is denoted by $M^{(D)}$, that is, $M^{(D)} = M^I/{\sim_D}$. The class of an element $\mu \in M^I$ in $M^{(D)}$ is denoted by μ^D.

ii) There is a canonical imbedding $i_D \colon M \longrightarrow M^{(D)}$ defined by

$$\forall m \in M \ (i_D(m) = (\mu_m)^D),$$

where $\mu_m \in M^I$ is the mapping constantly equal to m. The elements m and $i_D(m)$ will be identified below, that is, it will be assumed that $M \subseteq M^{(D)}$.

iii) If $\rho \subseteq M^2$ is a binary relation on M, then ρ^D will denote the binary relation on $M^{(D)}$ such that

$$\mu_1^D \rho^D \mu_2^D \Longleftrightarrow \{i \mid \mu_1(i) \rho \mu_2(i)\} \in D$$

$\forall \mu_1, \mu_2 \in M^I$.

The properties of an ultrafilter imply immediately that this definition is unambiguous; see also the proof of Proposition 0.5.6. An analogous definition is used below for any n-place relation.

EXAMPLE 0.5.5. Let $\leq$ be a *linear* order relation on M. Then $\leq^D$ is a linear order relation on $M^{(D)}$. Indeed, let

$$A = \{i \mid \mu_1(i) \leq \mu_2(i)\}, \quad B = \{i \mid \mu_1(i) > \mu_2(i)\}.$$

Since the order on M is linear, $A = \overline{B}$, that is, precisely one of these sets is in D. If $A \in D$, then $\mu_1^D \leq^D \mu_2^D$, and if $B \in D$, then $\mu_2^D <^D \mu_1^D$. It is even simpler to verify the remaining properties of the order. We recall for comparison that the relation $<^I$ on M^I with

$$\mu_1 <^I \mu_2 \Longleftrightarrow \forall i \in I \ (\mu_1(i) < \mu_2(i))$$

is only a partial order. This example is a special case of a theorem of Łoś to be considered below, which is fundamental for us in this section.

PROPOSITION 0.5.6. *Suppose that $F = \langle f_i \mid i \in I \rangle$ is a family of functions from M^n to M, indexed by the elements of a set I. Then it determines a function $F^D \colon (M^{(D)})^n \longrightarrow M^{(D)}$ according to the rule*

$$F^D(\mu_1^D, \dots, \mu_n^D) = \langle f_i(\mu_1(i), \dots, \mu_n(i)) \mid i \in I \rangle^D \tag{0.5.1}$$

$\forall \mu_1, \dots, \mu_n \in M^I$. *Furthermore, if $H = \langle h_i \mid i \in I \rangle$ is another such family, then $F^D = H^D \Longleftrightarrow \{i \mid f_i = h_i\} \in D$.*

PROOF. We first prove that the definition (0.5.1) of the mapping F^D is unambiguous. Suppose that $\nu_1, \ldots, \nu_n \in M^I$ and that $\mu_k \sim_D \nu_k \ \forall k \leq n$. This means that the set $A_k = \{i \mid \mu_k(i) = \nu_k(i)\}$ is in D for any k. But then $A = \bigcap_{k=1}^n A_k \in D$. If now $\mu(i) = f_i(\mu_1(i), \ldots, \mu_n(i))$ and $\nu(i) = h_i(\nu_1(i), \ldots, \nu_n(i))$, then it is clear that $B = \{i \mid \mu(i) = \nu(i)\} \supseteq A \in D$. But then B is also in D, that is, $\mu \sim_D \nu$, which proves the required correctness of the definition.

Suppose now that $F^D = H^D$ but $\{i \mid f_i = h_i\} \notin D$. Then $B = \{i \mid f_i \neq h_i\} \in D$, and for any $i \in B$ there exist $m_1, \ldots, m_n \in M$ such that $f_i(m_1, \ldots, m_n) \neq h_i(m_1, \ldots, m_n)$. We determine $\mu_1, \ldots, \mu_n \in M^I$ such that

$$\forall i \in B \ (f_i(\mu_1(i), \ldots, \mu_n(i)) \neq h_i(\mu_1(i), \ldots, \mu_n(i))).$$

The functions μ_k are defined arbitrarily on the elements of $I \setminus B$. If now $A = \{i \mid f_i(\mu_1(i), \ldots, \mu_n(i)) = h_i(\mu_1(i), \ldots, \mu_n(i))\}$, then $A \subseteq \overline{B}$, that is, $A \notin D$. But this means that

$$F^D(\mu_1^D, \ldots, \mu_n^D) \neq H^D(\mu_1^D, \ldots, \mu_n^D).$$

The proof of the implication $\Longleftarrow$ is left to the reader.

If $F = \langle f_i \mid i \in I\rangle$ is such that $f_i = f \ \forall i \in I$, then we simply write f^D instead of F^D, that is, if $f \colon M^n \longrightarrow M$, then $f^D \colon (M^{(D)})^n \longrightarrow M^{(D)}$ is defined by

$$f^D(\mu_1^D, \ldots, \mu_n^D) = \langle f(\mu_1(i), \ldots, \mu_n(i)) \mid i \in I\rangle^D.$$

If $\mathcal{K}$ is a class of n-place operations on M, then $\mathcal{K}^{(D)}$ denotes the class of n-place operations on $M^{(D)}$ of the form F^D for $F \in \mathcal{K}^I$. The class $\mathcal{K}(D)$ will be called an *ultrapower of the class* $\mathcal{K}$.

Suppose now that

$$\mathcal{M} = \langle M; \{0, 1, +, \cdot, <\}; \{\mathcal{C}_n \mid n \in \mathbb{N}\}\rangle$$

is a model of the language $\mathcal{LR}$ (see Definition 0.2.1). We use it to construct a new model

$$\mathcal{M}^{(D)} = \langle M^{(D)}; \{0, 1, +^D, \cdot^D, <^D\}; \{\mathcal{C}_n^{(D)} \mid n \in \mathbb{N}\}\rangle.$$

Furthermore, for each n an imbedding $\varphi_n \colon \mathcal{C}_n \longrightarrow \mathcal{C}_n^{(D)}$ is defined such that $\varphi_n(f) = f^D \ \forall f \in \mathcal{C}_n$. The mapping i_D in Definition 0.5.4 ii) is taken as φ_0, or simply the identity imbedding in view of the convention adoped there. By using the fact that $I \in D$ it is easy to see that the sequence $\Phi = \{\varphi_n\}$ satisfies all the conditions of Definition 0.2.2, that is, it is an imbedding of $\mathcal{M}$ into $\mathcal{M}^{(D)}$, and $\mathcal{M}^{(D)}$ is thus an extension of $\mathcal{M}$.

PROPOSITION 0.5.7. *The imbedding Φ is an isomorphism of $\mathcal{M}$ onto $\mathcal{M}^{(D)}$ if and only if D is a principal ultrafilter.*

PROOF. The proof is left as an exercise for the reader.

Below we write simply $+$, $\cdot$, and $<$ instead of $+^D$, $\cdot^D$, and $<^D$. Moreover, for any $f \in \mathcal{C}_n$ we simply write f instead of f^D wherever this does not lead to confusion. The next theorem lies at the basis of the construction of all the models of nonstandard analysis.

THEOREM 0.5.8 (Łoś). *The model $\mathcal{M}^{(D)}$ is an elementary extension of the model $\mathcal{M}$.*

This theorem follows immediately from the next proposition.

PROPOSITION 0.5.9. *Let $\mathcal{A}(\alpha_1,\dots,\alpha_n)$ be a formula in $\mathcal{LR}$, and let $F_1,\dots,F_n$ be objects in $\mathcal{C}^I$ such that $F_k(i)$ has the same type as the variable α_k $\forall i \in I$ $\forall k \leq n$. Then*

$$\mathcal{M}^{(D)} \models \mathcal{A}(F_1^D,\dots,F_n^D) \iff \{i \mid \mathcal{M} \models \mathcal{A}(F_1(i),\dots,F_n(i))\} \in D.$$

PROOF. Let $t = t(\alpha_1,\dots,\alpha_n)$ be a term depending on the variables $\alpha_1,\dots,\alpha_n$. We show that

$$t(F_1^D,\dots,F_n^D) = \langle t(F_1(i),\dots,F_n(i)) \mid i \in I\rangle^D \tag{0.5.2}$$

(see Definition 0.2.5 and the remark after it).

If the term t is a variable α_m of type 0, then $t(F_m^D) = F_m^D$, and the equality (0.5.2) is obviously satisfied. This is true also in the case when the term is the constant 0 or 1. Suppose now that (0.5.2) has been proved for the terms $t_1,\dots,t_s$, and let $\langle t_m(F_1(i),\dots,F_n(i)) \mid i \in I\rangle^D = \mu_m^D$, where $\mu_m \in M^I$, $m = 1,\dots,s$. We assume that $t = \varphi(t_1,\dots,t_s)$, where φ is one of the variables $\alpha_1,\dots,\alpha_n$ having type s, or one of the function constants $+$ or $\cdot$ (the latter is possible for $s = 2$). It is assumed that an object $F^D \in \mathcal{C}^{(D)}$ is associated with the variable (or constant) φ under our interpretation. Then it follows from the definition of terms (Definition 0.2.5) that

$$t(F_1^D,\dots,F_n^D) = F^D(t_1(F_1^D,\dots,F_n^D),\dots,t_s(F_1^D,\dots,F_n^D)) = F^D(\mu_1^D,\dots,\mu_s^D)$$

because (0.5.2) holds for the terms t_m. It now follows from (0.5.1) that

$$t(F_1^D,\dots,F_n^D) = \langle F(i)(\mu_1(i),\dots,\mu_m(i)) \mid i \in I\rangle^D,$$

and this proves (0.5.2) for the term t.

The equality (0.5.2) and Definition 0.2.5 (3a) (of the truth of atomic formulas) immediately yield our proposition for atomic formulas. The rest of the proof is by induction with respect to the stucture of the formula $\mathcal{A}$ (that is, induction on the number of logical connectives and quantifiers appearing in $\mathcal{A}$). We leave it to the reader to carry out independently, or to read it in [**1**].

Proposition 0.5.7 and Theorem 0.5.8 show that an arbitrary ultrapower $\mathbb{R}^{(D)}$ with respect to a nonprincipal ultrafilter D on an infinite set I can be taken as a nonstandard universe $^*\mathbb{R}$, that is, as a proper elementary extension of the model $\mathbb{R}$. Further, if $\mathfrak{M} \subseteq \mathcal{C}_{n_1} \times \dots \times \mathcal{C}_{n_k}$ is a standard class, in particular, a standard set (see Definition 0.2.11), then the ultrapower $\mathfrak{M}^{(D)}$ is the nonstandard extension $^*\mathfrak{M}$ of it. To see what internal sets and internal classes in $\mathbb{R}^{(D)}$ are like, we recall that in the language $\mathcal{LR}$ sets are determined by characteristic functions (§1, 2°, part b). Therefore, every internal subset $A \subseteq (\mathbb{R}^{(D)})^n$ is determined by some function $F^D \in \mathcal{C}_n^{(D)}$ satisfying the condition

$$\{i \mid \forall \overline{x} \in \mathbb{R}^n \; ((F(i))(\overline{x}) = 0 \;\vee\; (F(i))(\overline{x}) = 1)\} \in D.$$

We give a more convenient description of internal sets in ultrapowers.

DEFINITION 0.5.10. i) Let $A = \{A_i \mid i \in I\}$ be a family of subsets of $\mathbb{R}^n$. The ultraproduct of this family is defined to be the set $A^{(D)} \subseteq (\mathbb{R}^{(D)})^n$ such that

$$\forall \mu \in (\mathbb{R}^n)^I \ (\mu^D \in A^{(D)} \iff \{i \mid \mu(i) \in A_i\} \in D).$$

ii) Let $\mathfrak{M} = \{\mathfrak{M}_i \mid i \in I\}$ be a family of subclasses of $\mathcal{C}_{n_1} \times \cdots \times \mathcal{C}_{n_k}$ for the standard model $\mathbb{R}$. The ultraproduct of this family is defined to be the subclass

$$\mathfrak{M}^{(D)} \subseteq \mathcal{C}_{n_1}^{(D)} \times \cdots \times \mathcal{C}_{n_k}^{(D)}$$

such that for any $F_m \in \mathcal{C}_{n_m}^I$ $(m = 1, \dots, k)$

$$\langle F_1^D, \dots, F_k^D \rangle \in \mathfrak{M}^{(D)} \iff \{i \mid (F_1(i), \dots, F_k(i)) \in \mathfrak{M}_i\} \in D.$$

PROPOSITION 0.5.11. i) *A set $A \subseteq (\mathbb{R}^{(D)})^n$ is internal if and only if it is an ultraproduct of some family of standard subsets of $\mathbb{R}^n$.*

ii) *Every internal subclass of $\mathbb{R}^{(D)}$ is an ultraproduct of standard subclasses of $\mathbb{R}$.*

REMARKS. 1. Definition 0.5.10 can be formulated also for ultrapowers of arbitrary models of the language $\mathcal{LR}$. Proposition 0.5.11 remains valid if "standard" is replaced by "internal" in it.

2. It follows from Propositions 0.5.9 and 0.5.11 that internal hyperfinite sets are ultraproducts of finite sets. Further, if $A^D \subseteq \mathbb{R}^{(D)}$ is a hyperfinite set, where $A = \{A_i \mid i \in I\}$ is a family of finite sets, then $\nu = \langle |A_i| \mid i \in I \rangle \in \mathbb{N}^I$, that is, $\nu^D \in \mathbb{N}^{(D)}$, a hypernatural number (an element of a nonstandard extension of the set $\mathbb{N}$) which is the internal cardinality of the hyperfinite set $A^{(D)}$. In exactly the same way, internal topological spaces are ultraproducts of topological spaces, internal linear (hyperfinite-dimensional) spaces are ultraproducts of linear (finite-dimensional) spaces, and so on.

3°. We now proceed to the investigation of the question of saturation of nonstandard universes that are ultrapowers. The results of this and the next subsections are special cases of general theorems in Chapter 6 of the book [**19**] on saturation of ultraproducts. We also adhere to the presentation in [**19**], after adapting it to our concrete case.

THEOREM 0.5.12. *If the nonstandard universe $^*\mathbb{R}$ is an ultrapower of a standard universe with respect to some countably incomplete ultrafilter D on an infinite set I, then it is countably saturated.*

PROOF. Since the ultrafilter D is countably incomplete, it follows (exercise) that there is a decreasing sequence of sets

$$I = I_0 \supset I_1 \supset \cdots \supset I_n \supset \cdots \tag{0.5.3}$$

such that $I_n \in D$ $\forall n \in \mathbb{N}$, and $\bigcap_{i=1}^{\infty} I_n = \emptyset$. Suppose now that $\{\mathfrak{M}_n \mid n \in \mathbb{N}_+\}$ is a countable sequence of internal subclasses of $\mathbb{R}^{(D)}$ with the finite intersection property. By Definition 0.2.11, each class $\mathfrak{M}_n$ is determined by some $\mathcal{LR}$-formula $\mathcal{A}_n(\overline{\alpha}, \overline{a}_n^D)$, where $\overline{\alpha}$ is a tuple of variables (possibly of different types), and $\overline{a}_n^D$ is a tuple of objects belonging to $^*\mathcal{C}$, which in our case is $\bigcup_{n\in\mathbb{N}} \mathcal{C}_n^{(D)}$, that is, each component of $\overline{a}_n$ belongs to $\mathcal{C}^I$ for any $n \in \mathbb{N}$. The finite intersection property of the system $\{\mathfrak{M}_n\}$ means that for any $n \in \mathbb{N}$

$$\mathbb{R}^{(D)} \models \exists \overline{\alpha}\ (\mathcal{A}_1(\overline{\alpha}, \overline{a}_1^D) \wedge \ldots \wedge \mathcal{A}_n(\overline{\alpha}, \overline{a}_n^D)).$$

By Proposition 0.5.9, this implies that

$$X_n = I_n \cap \{i \mid \mathbb{R} \models \exists \overline{\alpha}\ (\mathcal{A}_1(\overline{\alpha}, \overline{a}_1(i)) \wedge \ldots \wedge \mathcal{A}_n(\overline{\alpha}, \overline{a}_n(i)))\} \in D$$

for any n. Moreover, $X_n \supseteq X_{n+1}$ and $\bigcap_{n \in \mathbb{N}} X_n = \emptyset$. This means that for each $k \in I$ there exists a largest $n(i) \in \mathbb{N}$ such that $i \in X_{n(i)}$. Let $X_0 = I_0 = I$. We now construct an ordered tuple $\overline{b}$ consisting of elements of the same type as $\overline{\alpha}$ that belong to $\mathcal{C}^I$. If $n(i) = 0$, then we define $\overline{b}(i)$ arbitrarily, and if $n(i) > 0$, then we choose $\overline{b}(i)$ such that

$$\mathbb{R} \models \mathcal{A}_1(\overline{b}(i), \overline{a}_1(i)) \wedge \ldots \wedge \mathcal{A}_{n(i)}(\overline{b}(i), \overline{a}_{n(i)}(i)). \tag{0.5.4}$$

We now take an arbitrary $n \in \mathbb{N}$ and show that $\mathbb{R}^{(D)} \models \mathcal{A}_n(\overline{b}^D, \overline{a}_n^D)$. Indeed, $n \leq n(i)\ \forall i \in X_n$, that is, $\mathbb{R} \models \mathcal{A}_n(\overline{b}(i), \overline{a}_n(i))$ in view of (0.5.4). Thus, $\{i \mid \mathbb{R} \models \mathcal{A}_n(\overline{b}(i), \overline{a}_n(i))\} \supseteq X_n \in D$, which means what is required. Accordingly, $\overline{b}^D \in \bigcap_{n \in \mathbb{N}_+} \mathfrak{M}_n$.

4°. The proof of the existence of λ^+-saturated nonstandard universes for an arbitrary $\lambda > \omega$ is essentially more complicated. We need some special notation and definitions. As before, I is a set on which we are considering ultrafilters and ultrapowers. We fix some set Σ and take the set $\lambda(\Sigma)$ of all possible mappings from $\mathcal{P}^{\text{fin}}(\Sigma)$ to $\mathcal{P}(I)$. For any $f, g \in \lambda(\Sigma)$ we take $f \leq g$ to mean that $f(\sigma) \subseteq g(\sigma)$ $\forall \sigma \in \mathcal{P}^{\text{fin}}(\Sigma)$.

A function $f \in \lambda(\Sigma)$ is said to be monotone if

$$\forall \sigma_1, \sigma_2 \in \mathcal{P}^{\text{fin}}(\Sigma)\ (\sigma_1 \subseteq \sigma_2 \longrightarrow f(\sigma_1) \subseteq f(\sigma_2)),$$

and additive if

$$\forall \sigma_1, \sigma_2 \in \mathcal{P}^{\text{fin}}(\Sigma)\ (f(\sigma_1 \cup \sigma_2) = f(\sigma_1) \cup f(\sigma_2)).$$

It is clear that every additive function is monotone. If D is an ultrafilter on I, then we denote by $\Lambda(\Sigma, D)$ the set of functions in $\lambda(\Sigma)$ with values lying in D. The following concept is fundamental in this subsection. An ultrafilter D on I is said to be λ-good if for any set Σ with $|\Sigma| < \lambda$ and any monotone function $f \in \Lambda(\Sigma, D)$ there exists an additive function $g \in \Lambda(\Sigma, D)$ such that $g \leq f$. Every λ-good ultrafilter is obviously also ν-good for any cardinal $\nu < \lambda$. A proof of the next theorem can be found in [**19**] (Theorem 6.1.4).

THEOREM 0.5.13. *If I is a set of cardinality λ, then there exists a λ^+-good countably incomplete ultrafilter D on I.*

Theorem 0.4.2 on the existence of λ-saturated nonstandard universes for any cardinal $\lambda > \omega$ now follows from the next theorem.

THEOREM 0.5.14. *If D is a countably incomplete λ-good ultrafilter on I, then the nonstandard universe $\mathbb{R}^{(D)}$ is λ-saturated.*

PROOF. Suppose that $|\Sigma| < \lambda$ and $\{\mathfrak{M}_\gamma \mid \gamma \in \Sigma\}$ is a family of internal subclasses of $\mathbb{R}^{(D)}$ with the finite intersection property. It must be proved that $\bigcap_{\gamma\in\Sigma} \mathfrak{M}_\gamma \neq \emptyset$. As in the proof of Theorem 0.5.12, we assume that each class $\mathfrak{M}_\gamma$ is determined by a formula $\mathcal{A}_\gamma(\overline{\alpha}, \overline{a}_\gamma^D)$, where each component of $\overline{a}_\gamma$ lies in $\mathcal{C}^I$ for any γ. If $\sigma = \{\gamma_1, \dots, \gamma_n\} \in \mathcal{P}^{\text{fin}}(\Sigma)$, then $\bigwedge \sigma$ denotes the formula $\mathcal{A}_{\gamma_1} \wedge \dots \wedge \mathcal{A}_{\gamma_n}$, and σ_i denotes the formula $\mathcal{A}_{\gamma_1}(\overline{\alpha}, \overline{a}_{\gamma_1}(i)) \wedge \dots \wedge \mathcal{A}_{\gamma_n}(\overline{\alpha}, \overline{a}_{\gamma_n}(i))$. To say that $\{\mathfrak{M}_\gamma\}$ has the finite intersection property means that $\forall \sigma \in \mathcal{P}^{\text{fin}}(\Sigma)$ $(\mathbb{R}^{(D)} \models \exists \alpha \bigwedge \sigma)$, or, by Proposition 0.5.9,

$$\{i \mid \mathbb{R} \models \exists \alpha \bigwedge \sigma_i\} \in D. \tag{0.5.5}$$

We now use the countable incompleteness of D and choose a decreasing sequence $\{I_n\}$ of sets in (0.5.3) as in the proof of Theorem 0.5.12. Let the function $f \in \lambda(\Sigma)$ be defined by setting

$$f(\sigma) = I_{|\sigma|} \cap \{i \mid \mathbb{R} \models \exists \alpha \bigwedge \sigma_i\}$$

for any $\sigma \in \mathcal{P}^{\text{fin}}(\Sigma)$. It follows from (0.5.5) that $f(\sigma) \in D$, that is, $f \in \Lambda(\Sigma, D)$. Moreover, since the sequence $\{I_n\}$ is decreasing, and the implication $\exists x \bigwedge \tau \longrightarrow \exists x \bigwedge \sigma$ is logically valid in general for $\sigma \subseteq \tau$, the function f is monotone. Since D is a λ-good ultrafilter on I, there exists an additive function $g \in \Lambda(\Sigma, D)$ such that $g \le f$.

For each $i \in I$ we now define the set

$$\sigma[i] = \{\gamma \in \Sigma \mid i \in g(\{\gamma\})\}.$$

We first prove the following assertion:

$$\forall n \in \mathbb{N}\ \forall i\ [|\sigma[i]| \ge n \longrightarrow i \in I_n]. \tag{0.5.6}$$

Indeed, if $\gamma_1, \dots, \gamma_n$ are distinct elements of $\sigma[i]$, then by the additivity of g,

$$i \in g(\{\gamma_1\}) \cap \dots \cap g(\{\gamma_n\}) = g(\{\gamma_1, \dots, \gamma_n\}) \subseteq f(\{\gamma_1, \dots, \gamma_n\}) \subseteq I_n$$

(see (0.5.5)). From the fact that $\bigcap_{n\in\mathbb{N}} I_n = \emptyset$ and from (0.5.6) it now follows that $\sigma[i]$ is finite for any i. Further,

$$i \in \bigcap \{g(\{\gamma\}) \mid \gamma \in \sigma[i]\} = g(\sigma[i]) \subseteq f(\sigma[i]),$$

that is, $i \in f(\sigma[i])\ \forall i \in I$. For any $i \in I$ there now exists a $\overline{b}(i)$ such that

$$\mathbb{R} \models \bigwedge \sigma[i]_i\ (\overline{b}(i)) \tag{0.5.7}$$

(see the definition of f). We now show that for any $\gamma \in \Sigma$

$$\mathbb{R}^{(D)} \models \mathcal{A}_\gamma(\overline{b}^D, \overline{a}_\gamma^D). \tag{0.5.8}$$

This obviously concludes the proof of the theorem.

Since $g(\{\gamma\}) \in D$, to prove (0.5.8) it suffices to show that

$$\forall i \in g(\{\gamma\})\ \mathbb{R} \models \mathcal{A}_\gamma(\overline{b}(i), \overline{a}_\gamma(i)).$$

But if $i \in g(\{\gamma\})$, then $\gamma \in \sigma[i]$, that is , $\mathcal{A}_\gamma(\overline{b}(i), \overline{a}_\gamma(i))$ occurs in the conjuction $\bigwedge \sigma[i]_i\ (\overline{b}(i))$, which is true in $\mathbb{R}$ according to (0.5.7).

§6. The axiomatics of nonstandard set theory

1°. Despite the rich expressive possibilities of the language $\mathcal{LR}$, it does not suffice for the formalization of the whole of mathematics. The main reason is that in this language we can work with objects of cardinality λ that, though arbitrarily large, is fixed, and this does not permit us to define, for example, the set of all subsets of an arbitrary set X. Recall that we have dealt only with families of subsets of X indexed by the elements of some other set Y whose cardinality also cannot exceed λ. In particular, we could consider only topological spaces of a fixed weight, not exceeding the cardinality of the space itself (however, this is not an essential restriction in substantive problems).

The language Set of set theory is a universal language for the formalization of mathematics. Syntactically, this language is considerably simpler than $\mathcal{LR}$. It does not contain constants, and it contains variables of only one sort together with two symbols for binary relations: $=$ (equality) and $\in$ (membership). The remaining objects in it are the same as in the language $\mathcal{LR}$ (see Definition 0.1.1, and §§4, 5). Correspondingly, there are no terms other than variables in the language Set, and the atomic formulas of this language have the form $x = y$ and $x \in y$. The remaining formulas are defined just like the formulas in the language $\mathcal{LR}$ (see Definition 0.1.3, and §§2, 3). Of course, the syntactic simplicity of the language Set leads to very great complexity in the writing of substantive mathematical propositions in it, and, although everything expressible in $\mathcal{LR}$ can be expressed in Set, the formal writing of, for example, each of the sentences contained in the exercises of the first section occupies several pages. Since sets are the most primitive mathematical concepts, they can be defined only axiomatically. Here we consider two of the most widespread systems of set theory axioms: the Zermelo–Fraenkel (ZF) axiomatics, and the von Neumann–Bernays–Gödel (NBG) axiomatics. These axiomatics are equivalent—any substantive mathematical proposition provable in one of them is also provable in the other. The reasons for presenting both systems are connected with nonstandard analysis and will be explained below. Each of these systems is apparently an adequate formalization of mathematics. In any case all the known mathematical theorems can be obtained as formal consequences of the ZF axioms. On the other hand, for such sentences as, for instance, the continuum hypothesis, which had neither been proved nor disproved despite considerable efforts over a long period of time, it was rigorously proved that neither they themselves nor their negations are formal consequences of the ZF axioms. Of course, to prove this kind of assertion it is necessary to have a precise definition of a formal derivation from the axioms. Such a definition is formulated in mathematical logic, but here it will not be presented, since it is not used in what follows. It can be found in any textbook on mathematical logic (in the author's opinion, the best text for a cursory acquaintance is the book [**72**]). We confine ourselves to a brief description of this concept. A formal proof of a sentence $\mathcal{A}$ in some system of axioms is a finite sequence of sentences, each of which either is an axiom of this system or can be obtained from the preceding sentences by definite rules called rules of deduction, and the last of which is $\mathcal{A}$. We note that the concept of a formal proof in the system ZF, for example, is constructive in the sense that there is an algorithm that, given any finite sequence of sentences in the language Set, answers the question of whether or not this sequence is a proof of some sentence. The formal deducibility of a sentence in some system of axioms is commonly denoted by the symbol $\vdash$. For instance, the fact that a sentence $\mathcal{A}$ is a theorem in the theory ZF is written as follows: ZF $\vdash \mathcal{A}$.

We touch briefly on the question of the consistency of an axiomatics of set theory. An axiomatic system is said to be consistent if it is not possible to prove some sentence $\mathcal{A}$ and its negation $\neg\mathcal{A}$ simultaneously in it, or, what is the same, for every sentence $\mathcal{A}$ it is impossible to prove the sentence $\mathcal{A} \wedge \neg\mathcal{A}$. The property of consistency is very essential because in an inconsistent axiomatics it is possible to prove any sentence $\mathcal{A}$ at all. The consistency of this or that axiomatic system is usually proved by constructing a model in which all the axioms of this system are true. But since every model is a set, such a proof cannot be convincing for an axiomatics of set theory. For example, the assertion that the system ZF is consistent, like every mathematical proposition, can be formalized in the language Set, and therefore it is natural to try to construct a formal proof of it in ZF. It turns out that this is impossible. The famous Gödel incompleteness theorem asserts that the consistency of an axiomatic system including arithmetic cannot be formally proved in this system, provided only that the system itself is consistent (we noted above that anything desired can be proved in an inconsistent system). At the present time the consistency of ZF is taken on faith by the overwhelming majority of specialists on the foundations of mathematics. The proofs of independence discussed above bear a relative character: it is proved that this or that sentence $\mathcal{A}$ is not deducible in ZF if ZF is consistent, and this is proved finitistically, that is, an algorithm is produced that would transform a formal proof of $\mathcal{A}$ in ZF into a proof of some contradiction in ZF (a sentence of the form $\mathcal{B} \wedge \neg\mathcal{B}$). The systems of axioms considered below for nonstandard set theory are consistent relative to ZF (that is, are consistent if ZF is consistent). The system NBG is also equiconsistent with ZF.

2°. In the Zermelo–Fraenkel axiomatics it is assumed that all the objects involved are sets. Some sets can be elements of others. The expression $x \in y$ is read as "the set x is an element of (belongs to) the set y". Capital letters will sometimes be used in the notation for certain variables in formulas of ZF, strictly for reasons of clarity.

1. The *axiom of extensionality* asserts that two sets are equal if and only if they consist of the same elements:

$$\forall X, Y\ (\forall z(z \in X \longleftrightarrow z \in Y) \longleftrightarrow X = Y).$$

We express the concept of a subset in the language Set:

$$X \subseteq Y \rightleftharpoons \forall z\ (z \in X \longrightarrow z \in Y).$$

The axiom of extensionality can now be written in the form

$$X = Y \longleftrightarrow X \subseteq Y\ \wedge\ Y \subseteq X.$$

2. The *axiom of pairing* asserts that for any two sets a and b there exists a set u whose elements are a and b and only they:

$$\forall a, b\ \exists u\ \forall z\ (z \in u \longleftrightarrow (z = a\ \vee\ z = b)).$$

It follows from the axiom of extensionality that such a set u is unique. It is commonly denoted by $\{a, b\}$. Now we can define ordered pairs:

$$\langle a, b\rangle = \{\{a\}, \{a, b\}\}. \tag{0.6.1}$$

From these two axioms it is easy to derive the following characteristic property of an ordered pair:

$$\langle a, b\rangle = \langle c, d\rangle \longleftrightarrow a = b \ \wedge \ c = d.$$

It is this property that provides the basis for calling the set $\langle a, b\rangle$ the ordered pair of elements a, b. We can now easily define ordered triples, quadruples, and arbitrary n-tuples by induction:

$$\langle a, b, c\rangle = \langle\langle a, b\rangle, c\rangle,$$

and so on.

3. The *axiom of separation* asserts that from any set x it is possible to separate out the subset of those elements y with some property $\mathcal{A}$ that can be formulated in the language Set. Thus, there is actually a countable set of separation axioms: each formula $\mathcal{A}$ in the language Set has its own axiom, with structure of the form

$$\forall u_1, \ldots, u_n \ \forall x \ \exists y \ \forall z \ (z \in y \longleftrightarrow z \in x \ \mathcal{A}(z, u_1, \ldots, u_n)).$$

Such a collection of axioms is called a schema of axioms, that is, the axiom of separation is actually a schema of axioms. It follows from the axiom of extensionality that the set y is uniquely determined by the sets x and $u_1 \ldots, u_n$. It is denoted as follows:

$$y = \{z \in x \mid \mathcal{A}(z, u_1, \ldots, u_n)\}.$$

In naive set theory one assumes that any collection of objects with a certain property is a set. It is this assumption that leads to various paradoxes. In particular, it is well known that the assumption about the existence of the set $\{x \mid x = x\}$ of all sets leads to a contradiction. The axiom of extensionality asserts only that from an *already existing set* x it is possible to separate out the subset consisting of the elements having a given property.

We present examples of the construction of sets on the basis of the axiom of separation. The existence of at least one set is a consequence of the axioms of formal logic and thus is not included in the system of axioms of set theory. Let X be an arbitrary set. By the axiom of separation, there exists a set $\emptyset = \{z \in x \mid z \neq z\}$.

It is now easy to prove that $\forall z \ (z \notin \emptyset)$. And this and the axiom of extensionality imply that $\emptyset$ is unique, that is, it does not depend on x. This is the way the existence of the empty set is proved in ZF. If $x \neq \emptyset$ and $z \in x$, then the set $\{t \in z \mid \forall y \in x \ (t \in y)\}$ exists. It is easy to prove that this set is determined only by the set x and does not depend on z. It is called the intersection of x (we have in mind the intersection of all the sets belonging to x) and denoted by $\cap x$. In particular, $\cap\{a, b\}$ is denoted by $a \cap b$. The set-theoretic difference $a \setminus b = \{t \in a \mid t \notin b\}$ is now easily defined.

The existence of the union of x cannot be deduced from the preceding axioms. It is postulated in the Zermelo–Fraenkel axiomatics.

4. The *axiom of union*:

$$\forall x \ \exists y \ \forall z \ (z \in y \longleftrightarrow \exists w \in x \ (z \in w)).$$

It is easy to prove that the set y is uniquely determined by x. It is denoted by $\cup x$, and $\cup\{a, b\}$ is denoted by $a \cup b$.

5. The *axiom of the power set* asserts the existence of the set $\mathcal{P}(x)$ of all subsets of a set x:

$$\forall x \ \exists y \ \forall z \ (z \in y \longleftrightarrow z \subseteq x).$$

We are now in a position to prove the existence of the Cartesian product $X \times Y = \{\langle x, y\rangle \mid x \in X \ \wedge \ y \in Y\}$ and of the set Y^X of all mappings from X to Y. Indeed, it follows from (0.6.1) that if $x \in X$ and $y \in Y$, then $\langle x, y\rangle \subseteq \mathcal{P}(X \cup Y)$, hence, $\langle x, y\rangle \in \mathcal{P}(\mathcal{P}(X \cup Y))$, hence,

$$X \times Y = \{z \in \mathcal{P}(\mathcal{P}(X \cup Y)) \mid \exists x \in X \ \exists y \in Y \ (z = \langle x, y\rangle)\}.$$

It is easy to see that it is possible to write a formula in ZF analogous to $\mathrm{Map}(f, X, Y)$ (see §1, 2°, example b) asserting that f is a mapping from X to Y. We keep the notation Map for this formula. Obviously, $\mathrm{Map}(f, X, Y) \longrightarrow f \subseteq X \times Y$, that is, $f \in \mathcal{P}(X \times Y)$. Thus,

$$Y^X = \{f \in \mathcal{P}(X \times Y) \mid \mathrm{Map}(f, X, Y)\}.$$

6. The *axiom of infinity* asserts the existence of at least one infinite set:

$$\exists x \ (\emptyset \in x \ \wedge \ \forall y \in x \ (y \cup \{y\} \in x)).$$

We denote the formula after $\exists x$ by $\mathrm{Infin}(x)$. Then it is clear that any x satisfying the formula $\mathrm{Infin}(x)$ contains the elements

$$\emptyset, \{\emptyset\}, \{\emptyset, \{\emptyset\}\}, \{\emptyset, \{\emptyset\}, \{\emptyset, \{\emptyset\}\}\}, \ldots, \tag{0.6.2}$$

where the process of constructing these sets belonging to x does not break off at a finite step.

7. The *axiom of choice*:

$$\forall X \ (X \neq \emptyset \ \wedge \ \forall x \in X \ (x \neq \emptyset) \longrightarrow \exists f \ (\mathrm{Map}(f, X, \cup X) \ \wedge \ \forall x \in X \ (f(x) \in x))).$$

We note that the notation ZF is usually used for the system of Zermelo-Fraenkel axioms without the axiom of choice. The Zermelo–Fraenkel system with the axiom of choice is denoted by ZFC.

8. The *axiom of regularity*:

$$\forall S \ \exists x \in S \ (x \cap S = \emptyset).$$

This axiom implies the impossibility of the existence of inclusions like $x \in x$, as well as of arbitrary chains $x_1 \in x_2 \in \cdots \in x_n \in x_1$. It also implies that there is no infinite chain of sets that is decreasing with respect to membership, that is, of the form

$$x_1 \ni x_2 \ni \cdots \ni x_n \ni \cdots.$$

9. The *axiom of replacement.* This is also a countable sequence of axioms. Each axiom is constructed from a formula $\mathcal{A}(x, y, u_1, \ldots, u_n)$:

$$\forall u_1, \ldots, u_n \ (\forall x \ \exists! y \ \mathcal{A}(x, y, u_1, \ldots, u_n) \\ \longrightarrow \forall X \ \exists Y \ \forall y \ (y \in Y \longleftrightarrow \exists x \in X \ \mathcal{A}(x, y, u_1, \ldots, u_n))).$$

This axiom asserts that if the formula $\mathcal{A}$ determines a functional dependence of y on x, and the domain over which x varies is restricted to some set X, then the domain over which y varies is restricted to some set Y. It is easy to see that this axiom implies the axiom of separation. The latter was formulated separately for the following reasons. It is essentially simpler and more clear than the axiom of replacement, and it is completely sufficient for the construction of all of substantive mathematics. The axiom of replacement is needed only for constructing the whole theory of infinite cardinalities in full scope in the framework of the formal axiomatics. The Zermelo–Fraenkel system of axioms without the axiom of replacement is called the Zermelo axiomatics.

This concludes the list of axioms of ZFC.

From the axiom of infinity it is easy to deduce the existence of a set x that is smallest with respect to inclusion among the sets with the property $\mathrm{Infin}(x)$. This set is usually denoted by ω in axiomatic set theory. It is easy to see that a one-place operation $': \omega \longrightarrow \omega$ such that $n' = n \cup \{n\}$ $\forall n \in \omega$ is defined on ω.

By using the axiom of regularity it can be proved that the system $(\omega, ')$ satisfies all the Peano axioms for the natural numbers. The axiom of induction is an obvious corollary to the minimality of ω with respect to inclusion. It is clear from the above that in axiomatic set theory there are sufficient means to construct from the set of natural numbers the ring of integers, the fields of rational and real numbers, and all the remaining objects of classical mathematics. We remark that in this approach the sets in the sequence (0.6.2) are the natural numbers $0, 1, 2, 3, \dots$.

3°. In the von Neumann–Bernays–Gödel (NBG) axiomatics a somewhat different approach is taken. Here the variables are interpreted as classes of objects. Intuitively, a class is precisely the collection of all objects united by having some property. Some classes can be elements of other classes, which is denoted by the formula $X \in Y$ as above. If a class X occurs as an element in some other class, then it is called a set. Thus, the formula

$$M(X) \rightleftharpoons \exists Y \ (X \in Y)$$

asserts that X is a set. To shorten expressions in the NBG axiomatics we use the following method. Here variables are commonly denoted by capital letters, but variables satisfying the formula $M(X)$ are denoted by lower-case letters. Thus, the formula $x = Y$, for example, is simply an abbreviation of the formula $M(X) \wedge X = Y$, and $M(X) \equiv \exists y \ (X = y)$.

We list the basic axioms of NBG.

The *axiom of extensionality* has here exactly the same form as in ZF:

$$\forall X, Y \ (X = Y \longleftrightarrow \forall z \ (z \in X \longleftrightarrow z \in Y)).$$

The definition of the formula $X \subseteq Y$ (X is a subclass of Y) is also analogous to the definition in ZFC.

The *axioms of pairing, of union, of the power set, of infinity*, and *of choice* are formulated only for sets and have exactly the same form as in ZF. Ordered pairs, Cartesian products, and so on are defined in complete analogy.

In contrast to ZF, the *axiom of separation* is here not a countable collection of axioms, but a single axiom:

$$\forall X \ \forall x \ \exists y \ (y = x \cap X). \tag{0.6.3}$$

In particular, this implies that a subclass of a set is itself a set:

$$\forall X \ \forall x \ (X \subseteq x \longrightarrow \exists y \ (y = X)).$$

The *axiom of replacement* is also a single sentence. We leave its formulation to the reader.

The *axiom of regularity* has the form

$$\forall Y \ (Y \neq \emptyset \longrightarrow \exists x \in Y \ (x \cap Y = \emptyset)),$$

that is, formally it looks somewhat stronger than the corresponding axiom of ZF, since it concerns arbitrary classes, and not just sets; however, it can actually be deduced from the corresponding axiom for sets. The *axioms of existence of classes* are essentially new here in comparison with ZF. We list them:

Cl 1 . $\exists X\ \forall z\ (z \in X \longleftrightarrow \exists u, v\ (z = \langle u, v\rangle\ \wedge\ u \in v))$ (the $\in$-relation);
Cl 2 . $\forall X, Y\ \exists Z\ \forall u\ (u \in Z \longleftrightarrow u \in X\ \wedge\ u \in Y)$ (intersection);
Cl 3 . $\forall X\ \exists Z\ \forall u\ (u \in Z \longleftrightarrow u \notin X)$ (complementation);
Cl 4 . $\forall X\ \exists Z\ \forall u\ (u \in Z \longleftrightarrow \exists v\ (\langle u, v\rangle \in X))$ ($\operatorname{dom} X$);
Cl 5 . $\forall X\ \exists Z\ \forall u\ \forall v\ (\langle u, v\rangle \in Z \longrightarrow u \in X)$ (cylinder);
Cl 6 . $\forall X\ \exists Z\ \forall u, v, w\ (\langle u, v, w\rangle \in Z \longleftrightarrow \langle v, w, u\rangle \in X)$;
Cl 7 . $\forall X\ \exists Z\ \forall u, v, w\ (\langle u, v, w\rangle \in Z \longleftrightarrow \langle u, w, v\rangle \in X)$.

By using the axioms Cl 4–Cl 7 it can easily be shown that for any permutation $\sigma \in S_n$

$$\forall X\ \exists Z\ \forall u_1, \ldots, u_n\ (\langle u_1, \ldots, u_n\rangle \in Z \longleftrightarrow \langle u_{\sigma(1)}, \ldots, u_{\sigma(n)}\rangle \in X).$$

A formula is said to be *predicative* if all the quantifiers in it relate only to set variables (more precisely, can be reduced to such a form after use of the abbreviations described above). The next result can be deduced from the axioms of existence of classes (see, for example, [**72**]).

THEOREM 0.6.1. *If $\varphi(x_1, \ldots, x_n, X_1, \ldots, X_m)$ is a predicative formula, then*

$$\forall X_1, \ldots, X_m\ \exists Y\ \forall x_1, \ldots, x_n\ (\langle x_1, \ldots, x_n\rangle \in Y \longleftrightarrow \varphi(x_1, \ldots, x_n, X_1, \ldots, X_m)).$$

For arbitrary formulas that are not necessarily predicative formulas, an analogous assertion can no longer be proved in NBG, provided that NBG is consistent, of course. The question of the consistency of the theory obtained if this assertion is added to NBG as an axiom is apparently open. Another interesting strengthening of NBG is obtained if the axiom of choice is strengthened by requiring that there be a universal choice function, that is, a class function that selects an element from each nonempty set. The reader is left to write this sentence independently as a formula in the language Set (using the abbreviations adopted in NBG). As already noted, NBG and ZFC are equiconsistent. Any sentence in which only set variables appear is a theorem in NBG if and only if it is provable in ZFC. Intuitively, one can think of classes in NBG as collections $\{x \mid \varphi(x)\}$, where φ is a formula in ZFC. As mentioned above, such collections are by no means always sets.

4°. In this subsection we consider Nelson's [**48**] axiomatization of nonstandard set theory—internal set theory (IST). This axiomatization is historically the first and most natural. The premises at the basis of Nelson's axiomatization were described at length in the treatment of the language $\mathfrak{LRS}$ (see §4, 6°), which is in essence a fragment of IST that suffices for most applications. As in the case of $\mathfrak{LRS}$, the language of IST is obtained from the language Set by adding the one-place predicate symbol St (St(x) is read "x is a standard set"). Formulas of IST not containing this predicate (that is, in fact, formulas of ZF) are called *internal* formulas. Intuitively, one should think of a sentence of IST as a proposition about internal sets of the nonstandard universe. Furthermore, the proposition St(B) ("B is standard") means that $B = {}^*A$, where A is a set of the standard universe. Thus, standard sets can contain also nonstandard elements in IST. Recall that in the treatment of the language $\mathcal{LRS}$ we identified a standard set A with its nonstandard extension $^*A$.

Of course, here we have in mind nonstandard extensions of a considerably broader standard universe than that in which the language $\mathcal{LRS}$ was interpreted. Putting aside logical subtleties, we can assume that we are dealing with the universe of all sets. In a nonstandard extension of it there are other internal sets besides the images of standard sets. Thus, the language IST describes the collection of all internal sets of a nonstandard extension, especially singling out the images of standard sets with the help of the predicate St.

The system of axioms IST contains all the axioms of ZFC and the three schemata of axioms for the predicate St: the transfer principle (T), the idealization principle (I), and the standardization principle (S), which are generalizations of the corresponding principles formulated in $\mathcal{LRS}$ and discussed in detail in §4, 6°. They have the following form.

The *transfer principle*:

$$\text{(T)} \qquad \forall^{\mathrm{st}} t_1, \dots, t_k \ (\forall^{\mathrm{st}} x \ \mathcal{A}(x, t_1, \dots, t_k) \longrightarrow \forall x \ \mathcal{A}(x, t_1, \dots, t_k)),$$

where $\mathcal{A}(x, t_1, \dots, t_k)$ is an internal formula of the language IST.

The *idealization principle*

$$\text{(I)} \qquad \forall^{\mathrm{st\,fin}} z \ \exists x \ \forall y \in z \ \mathcal{B}(x, y) \longleftrightarrow \exists x \ \forall^{\mathrm{st}} y \ \mathcal{B}(x, y),$$

where $\mathcal{B}$ is an internal formula possibly containing, in addition to x and y, also some free variables.

The *standardization principle*:

$$\text{(S)} \qquad \forall^{\mathrm{st}} x \ \exists^{\mathrm{st}} y \ \forall^{\mathrm{st}} z \ (z \in y \longleftrightarrow z \in x \ \wedge \ \mathcal{C}(z)),$$

where $\mathcal{C}(z)$ is an arbitrary (not necessarily internal) formula of IST, possibly containing besides z also some free variables.

Let $\mathcal{A}$ be an arbitrary internal sentence. We denote by $\mathcal{A}^{\mathrm{st}}$ the sentence obtained by replacing each quantifier Q occurring in $\mathcal{A}$ by Q^{st}. Then it follows from the transfer principle (T) and Propositions 0.2.6 and 0.2.7 that the sentence $\mathcal{A} \longleftrightarrow \mathcal{A}^{\mathrm{st}}$ is provable in IST for any internal sentence $\mathcal{A}$. In fact, $\mathcal{A} \longleftrightarrow \mathcal{A}^{\mathrm{st}}$ is a formal analogue of the transfer principle formulated in §3, 1°.

Since all the axioms of ZFC occur in IST, all the theorems of ZFC are also theorems in IST. It was proved in [**48**] that the converse also holds, all *internal* theorems of IST are theorems of ZFC. This implies that ZFC and IST are equiconsistent.

Moreover, the algorithm of Nelson considered in §4, 6° enables us to construct for any sentence in IST an internal sentence equivalent to it.

From the idealization principle (I) it is easy to deduce [**48**] the following two propositions, which essentially generalize and strengthen Propositions 0.4.21 and 0.4.22.

PROPOSITION 0.6.2. *All the elements of a set X are standard if and only if X is a standard finite set.*

PROPOSITION 0.6.3. *There exists an internal hyperfinite set containing all standard elements.*

This proposition shows, in particular, that the axiom of separation is valid in IST only for internal formulas (then it is an axiom of ZFC), since it implies that the class $\{x \mid \mathrm{St}(x)\}$ of all standard sets (which is not an internal set, of course) is a subclass of some internal set. An essentially weaker assertion—the standardization principle (S)—is an analogue of the axiom of separation for external formulas.

5°. The expressive possibilities of IST are limited to a considerable extent by the absence in it of variables with external sets as values. Because of this, the saturation principle cannot be formulated in IST, and the constructions of nonstandard hulls and Loeb measures are not formalized in it. At present there are several axiomatic systems ([**62**]–[**64**]) for nonstandard set theory that include the means for formalizing assertions about external sets.

The theory considered below—nonstandard class theory (NCT)—extends NBG just as IST extends ZFC. The language of NCT is just like the language of ZFC, but the variables of NCT, by analogy with NBG, take classes as values, and a class is a set when it occurs as an element of some other class. The classes (including sets) on which the one-place predicate St is true are called standard classes, of course. As in NBG (see the concluding paragraph of 3°), the classes of NCT can be thought of as collections of sets of IST definable by formulas of IST. Furthermore, the standard classes are collections of sets definable by internal formulas with standard parameters. Internal classes *are defined* as sections of standard class relations by sets. In this approach all sets turn out to be internal. The role of external sets is played by classes that are not internal. Further, in contrast to NBG, there are subclasses of sets that are not sets in NCT (the semisets of Vopěnka!). The basic facts about NCT are not used in the exposition to follow, so they are given here without proof. Detailed proofs can be found in [**73**].

We proceed to a formulation of the axioms of NCT. As in NBG, there are finitely many of them—there are no schemata of axioms. We make the same conventions as in NBG about capital letters for denoting variables in the general case, and lower-case letters for variables taking sets as values (instead of using the formula $\exists Y\ (X \in Y)$).

The *axioms of extensionality, pairing, union, the power set, infinity,* and *choice* are formulated precisely as in NBG.

The *axiom of existence of the empty set* is added: $\exists x\ \forall y\ (y \notin x)$.

The *axiom of separation* is formulated here only for standard classes:

$$\forall x\ \forall^{\mathrm{st}} X\ \exists z\ \forall u\ (u \in z \longleftrightarrow u \in x\ \wedge\ u \in X),$$

that is, the intersection of a *standard* class and a set is a set.

The same applies to the *axiom of replacement*:

$$\forall^{\mathrm{st}} F\ (\mathrm{Fnc}(F) \longrightarrow \forall x\ \exists y\ \forall z\ (z \in y \longleftrightarrow \exists t \in x\ (\langle t, z\rangle \in F))).$$

Here $\mathrm{Fnc}(F)$ is the formula expressing the fact that the class F is a functional relation (see §1, 2°, part b).

The *axiom of regularity* is changed in precisely the same way: $\forall^{\mathrm{st}} S\ \exists x\ (x \cap S = \emptyset)$.

In addition to the axioms of existence of classes in the theory NBG there are axioms asserting that the classes constructed from standard classes are themselves standard. Thus, the *axioms of existence of classes* in NCT have the following form:

(1) SCl 1. $\exists^{\mathrm{st}} X\ \forall z\ (z \in X \longleftrightarrow \exists u, v\ (\langle u, v\rangle = z\ \wedge\ u \in v))$ (the relation of membership is a standard class);
(2) the axioms Cl 2–Cl 7 are the same as in NBG;

(3) the axioms SCl 2–SCl 7, which differ from Cl 2–Cl 7 only in that the quantifiers relating to the variables X, Y, and Z are relativized to standard classes, that is, for example, the axiom SCl 2 has the form $\forall^{\mathrm{st}} X\ \forall^{\mathrm{st}} Y\ \exists^{\mathrm{st}} Z\ \forall u\ (u \in Z \longleftrightarrow u \in X \ \wedge\ u \in Y)$, and the axioms Cl 3–Cl 7 are modified similarly;

(4) $\exists X\ \forall x\ (x \in X \longleftrightarrow \mathrm{St}(x))$ (this axiom asserts that the class Stan of all standard sets exists).

It follows immediately from these axioms that the set $\emptyset$ is standard. Indeed, if X is some standard class (the existence of at least one standard class follows from the axiom SCl 1), then $\emptyset = X \cap \overline{X}$ is a standard class in view of the axioms SCl 2 and SCl 3. The existence of the empty set can actually be proved by applying the axiom of separation to some standard class, for example, the relation of membership. It follows from the axioms SCl 2 and SCl 3 only that the empty *class* exists. Now the class of all sets $V = \overline{\emptyset}$ is also standard.

Just as in NBG, a formula of NCT is said to be predicative if in it only variables taking sets as values are bound by quantifiers. By analogy with IST, a formula of NCT is said to be internal if it does not contain the predicate symbol St.

PROPOSITION 0.6.4. *Suppose that $\varphi(x_1, \dots, x_n, Y_1, \dots, Y_m)$ is a predicative formula of NCT. Then the following sentence is true in NCT*:

$$\forall Y_1, \dots, Y_m\ \exists Z\ \forall x_1, \dots, x_n\ (\langle x_1, \dots, x_n\rangle \in Z \longleftrightarrow \varphi(x_1, \dots, x_n, Y_1, \dots, Y_m)).$$

Furthermore, if φ is an internal formula, then the following sentence is also true:

$$\forall^{\mathrm{st}} Y_1, \dots, Y_m\ \exists^{\mathrm{st}} Z\ \forall x_1, \dots, x_n\ (\langle x_1, \dots, x_n\rangle \in Z \longleftrightarrow \varphi(x_1, \dots, x_n, Y_1, \dots, Y_m)).$$

The proof of this sentence is completely analogous to that of Theorem 0.6.1 (see, for example, [**72**]). In particular, it implies that any class definable in ZFC is standard.

COROLLARY. *Let $\varphi(x)$ be a formula of ZFC (that is, an internal formula containing only variables with sets as values) with a single free variable x. Then*

$$\exists^{\mathrm{st}} Z\ \forall x\ (x \in Z \longleftrightarrow \varphi(x)).$$

This implies, for example, that the set ω of natural numbers is standard, since it is definable in ZFC (see the last paragraph of 2°).

We define the following formula of NCT:

$$\mathrm{Int}(X) \rightleftharpoons \exists^{\mathrm{st}} Y\ \exists y\ \forall x\ (x \in X \longleftrightarrow \langle x, y\rangle \in Y).$$

A class X satisfying this formula will be called an *internal* class.

PROPOSITION 0.6.5. i) *Every set is internal.*
ii) *Any standard class is internal.*

PROOF. Let $E = \{\langle x, y\rangle \mid x \in y\}$. By axiom SCl 1, E is a standard class. An arbitrary set y now satisfies the equality $y = \{x \mid \langle x, y\rangle \in E\}$, which proves what is required. If X is a standard class, then it follows easily from the axioms of existence of classes that $Y = X \times \{\emptyset\}$ is also a standard class. The assertion ii) now follows the fact that $x \in X \longleftrightarrow \langle x, \emptyset\rangle \in Y$.

The obvious abbreviations $\forall^{\mathrm{int}}$ and $\exists^{\mathrm{int}}$ are used below.

It follows easily by Propositions 0.6.4 and 0.6.5 that if $\varphi(x_1, \dots, x_n, Y_1, \dots, Y_m)$ is an internal predicative formula, and $A_1, \dots, A_m$ are internal classes (in particular, arbitrary internal sets), then the class

$$Z = \{\langle x_1, \dots, x_n\rangle \mid \varphi(x_1, \dots, x_n, A_1, \dots, A_m)\}$$

is internal. The transfer, idealization, and standardization principles are now formulated in NCT as follows.

The *transfer principle*:

$$\text{(T)} \qquad \forall^{\text{st}} X\ (X \neq \emptyset \longrightarrow \exists^{\text{st}} x\ (x \in X)).$$

The *idealization principle*:

$$\text{(I)} \qquad \forall^{\text{int}} X\ (\forall^{\text{st fin}} z\ \exists x\ \forall y \in z\ (\langle x, y\rangle \in X) \longleftrightarrow \exists x\ \forall^{\text{st}} y\ (\langle x, y\rangle \in X)).$$

The *standardization principle*:

$$\text{(S)} \qquad \forall X\ \forall^{\text{st}} x\ \exists^{\text{st}} y\ \forall^{\text{st}} z\ (z \in y \longleftrightarrow z \in x\ \wedge\ z \in X).$$

We note that in NCT a set z is said to be finite if there exists an internal bijection between z and a set $\{0, \dots, n\}$ with n a natural number, possibly nonstandard (recall that in NCT, as in IST, no distinction is made between standard sets and their nonstandard extensions, so standard sets can contain nonstandard elements), that is, hyperfinite sets are said to be finite. And if the number n is standard, then the set z is said to be standardly finite. It follows from the transfer principle that a standard hyperfinite set z is standardly finite.

The NCT theory is equiconsistent with IST. What is more, we have

THEOREM 0.6.6. *Any sentence of NCT in which there are only variables with sets as values is a theorem of NCT if and only if it is a theorem of IST.*

In the terminology of mathematical logic this means that NCT is a conservative extension of IST.

We dwell in somewhat greater detail on the properties of the set ω of natural numbers in NCT. It follows from the preceding theorem that the induction principle is true in NCT in the following form:

$$\forall x \subseteq \omega\ (0 \in x\ \wedge\ \forall n\ (n \in x \longrightarrow n+1 \in x) \longrightarrow x = \omega),$$

where $n + 1 = n \cup \{n\}$ (see the end of 2°).

If in this sentence we replace x by X, then it ceases to be true. Let us prove this. For an arbitrary class A let ${}^\circ A$ denote the class of all standard elements of A, that is, the class $A \cap \text{Stan}$. Then it is obvious that the class ${}^\circ\omega$ satisfies the premises of the induction axiom: $0 \in {}^\circ\omega\ \wedge\ \forall n(n \in {}^\circ\omega \longrightarrow n+1 \in {}^\circ\omega)$. On the other hand, it follows from the idealization principle that there exist nonstandard natural numbers, that is, ${}^\circ\omega \neq \omega$. This shows that the indicated strengthening of the induction principle is false and, moreover, that ${}^\circ\omega$, though a subclass of a set, is not itself a set.

Subclasses of internal sets that are not sets will (following Vopěnka) be called *semisets*.

Notation: $\text{Smst}(X) \rightleftharpoons \exists x\ (X \subseteq x)$ (that is, $\text{Smst}(X)$ is true if and only if X is either a set or a semiset).

From the usual mathematical point of view, the semiset ${}^\circ\omega$ is the 'genuine set' of natural numbers. It does satisfy the induction principle for classes.

PROPOSITION 0.6.7. *In NCT the following sentence is true:*

$$\forall X \subseteq {}^\circ\omega \; (0 \in X \;\wedge\; \forall n(n \in X \longrightarrow n+1 \in X) \longrightarrow X = {}^\circ\omega) \tag{0.6.4}$$

PROOF. If this sentence is false, then there exists a class $X \subseteq {}^\circ\omega$ such that $0 \in X \;\wedge\; \forall n(n \in X \longrightarrow n+1 \in X)$, but $Y = {}^\circ\omega \setminus X \neq \emptyset$. By the standardization principle,

$$\exists^{\mathrm{st}} x \; \forall^{\mathrm{st}} n \; (n \in x \longleftrightarrow n \in Y), \tag{0.6.5}$$

and $x \subseteq \omega$ by the transfer principle. Consequently, if $x \neq \emptyset$, then it has a smallest element m, which by the transfer principle is also standard, because x is a standard set. Then the standard number $m-1$ is not in x, and $m-1 \notin Y$ by virtue of (0.6.5), that is, $m-1 \in X$; but then $m \in X$ by the property of X. The contradiction shows that $x = \emptyset$. However, then also $Y \cap {}^\circ\omega = \emptyset$, that is, $X = {}^\circ\omega$.

By using the induction principle (0.6.4) it is easy to prove in NCT the analogue of Proposition 0.3.16 iii) (see also the remark after that proposition) asserting that every class whose cardinality is a standard natural number is a set. Before formulating it we note that in NCT we can state that classes are externally of the same cardinality; namely, classes A and B are externally of the same cardinality if $\exists F \; (F\colon A \Longleftrightarrow B)$. Further, sets a and b are externally of the same cardinality if $\exists f \; (f\colon a \Longleftrightarrow b)$. If λ is some standard cardinal, then $\mathrm{Card}(X) = \lambda$ means that X is externally of the same cardinality as ${}^\circ\lambda$, and $|x| = \lambda$ means that x is externally of the same cardinality as λ.

PROPOSITION 0.6.8. *In NCT the following sentence is true:*

$$\forall X \; (\exists^{\mathrm{st}} n \in \omega \; (\mathrm{Card}(X) = n) \longrightarrow \exists x \; (x = X \;\wedge\; |x| = n)).$$

We can now formulate the following beautiful criterion for a natural number to be standard.

PROPOSITION 0.6.9. *A natural number n is standard if and only if the set $\{0, \dots, n\}$ does not contain subsemisets.*

REMARK. In alternative set theory, where the predicate of standardness is absent, this proposition is taken as the definition of standard natural numbers, which are called finite natural numbers there.

PROOF. By using the induction principle (0.6.4) it is easy to prove that if n is standard, then the indicated set does not contain subsemisets. By using the same, it easy to prove that if n is infinite, then all the standard natural numbers are less than n, that is, ${}^\circ\omega \subseteq \{0, \dots, n\}$, which proves what is required.

COROLLARY. *A set is standardly finite if and only if it does not contain subsemisets.*

We formulate an open question concerning the axiomatics introduced.

It follows from the axioms of NCT and the definitions that the axiom of separation is valid also for internal classes, that is, the intersection of an internal class and a set is a set. Is the converse valid, that is, is it true that if the intersection of some class with an arbitrary set is a set, then this class is internal?

There is a basis for conjecturing that this question has a negative answer. The fact is that classes are considered in alternative set theory that have this property (that is, their intersections with any set is a set). There they are called sharp classes. As shown by Sochor [**74**], the family of sharp classes is not closed under projection, that is, in AST it is possible to construct a sharp class X such that the class $\operatorname{dom}(X)$ is no longer sharp. Obviously, the family of internal classes is closed under projection. On the other hand, in Sochor's proof essential use is made of very specific axioms of AST—for example, the axiom of two infinite cardinals, which is not present in NCT.

In the language NCT the saturation principle can be formulated in various forms. For this we say that a class A has the finite intersection property (denoted by $\mathrm{FIP}(X)$) if each nonempty standard finite subset of it has nonempty intersection. It is clear that the predicate $\mathrm{FIP}(X)$ is defined by a formula in NCT. Then a stronger form of the saturation principle is

$$\text{(SSat)} \qquad \forall X\ (\mathrm{FIP}(X) \longrightarrow \bigcap X \neq \emptyset).$$

The strong saturation principle (SSat) implies the idealization principle in a somewhat weakened form. A class A is said to be weakly bounded (denoted by $\mathrm{Boun}(A)$) if for any standard y the class $\{x \mid \langle x, y\rangle \in A\}$ is a semiset or a set. The weakly bounded idealization principle (I^{wb}) now has the form

$$\exists^{\mathrm{int}} X\ (\mathrm{Boun}(X) \longrightarrow (\forall^{\mathrm{st\,fin}} z\ \exists x\ \forall y \in z\ (\langle x, y\rangle \in X) \longleftrightarrow \exists x\ \forall^{\mathrm{st}} y\ (\langle x, y\rangle \in X))).$$

It is easy to see that $\mathrm{SSat} \longrightarrow \mathrm{I}^{\mathrm{wb}}$. Indeed, if a class X satisfies the conditions of the principle I^{wb}, then the class $A = \{z \mid \exists^{\mathrm{st}} y\ \{x \mid \langle x, y\rangle \in X\} \subseteq z\}$ obviously has the finite intersection property, and hence has nonempty intersection. Any element x in this intersection satisfies the conclusion of the principle I^{wb}.

We can easily see that the principle (SSat) is too strong to be true, because it contradicts NCT. To prove this consider, for example, the class $X = \{[\alpha, +\infty) \mid \alpha \in {}^*\mathbb{R}\}$. Obviously, $\mathrm{FIP}(X)$ is true, but $\cap X = \emptyset$.

The following saturation principle, which will be denoted simply by Sat, seems more relevant to the content of §4, though also too strong. We say that a class X has standard size if it is externally of the same cardinality as the semiset ${}^\circ x$ of all standard elements of some standard set x (notation: $\mathrm{Ssize}(X)$). Then the principle Sat asserts that each semiset or set of standard size with the finite intersection property has nonempty intersection:

$$\forall X\ (\mathrm{FIP}(X) \wedge \mathrm{Smst}(X) \wedge \mathrm{Ssize}(X) \longrightarrow \bigcap X \neq \emptyset).$$

As above, this principle implies a certain weak form of the idealization principle: the principle $\mathrm{I}^{(\mathrm{b})}$ obtained from I by requiring that the internal class X appearing in I be a subset of some standard set:

$$\begin{aligned}\forall u, v\ (\exists^{\mathrm{st}} w\ (u, v \subseteq w) \longrightarrow (\forall^{\mathrm{st\,fin}} z \subseteq v\ \exists x\ \forall y \in z\ (\langle x, y\rangle \in u) \\ \longleftrightarrow \exists x\ \forall^{\mathrm{st}} y \in v\ (\langle x, y\rangle \in u))).\end{aligned}$$

The question of the interrelation between Sat and I in full scope, and, in particular, the question of the consistency of the theory $\mathrm{NCT} + \mathrm{Sat}$, remains open. For any cardinal λ that is definable in ZFC, and hence standard, we denote by Sat_λ the λ-saturation principle, which is obtained from Sat by replacing the condition $\mathrm{Ssize}(X)$

by the condition $\mathrm{Card}(X) < \lambda$. It can be shown that the theory $\mathrm{NCT} + \mathrm{Sat}_\lambda$ is consistent if ZFC is consistent.

In order to have the possibility of formalizing the construction of nonstandard hulls it is necessary to strengthen somewhat the theory $\mathrm{NCT} + \mathrm{Sat}_\lambda$, replacing the axiom of choice by a stronger axiom asserting that every semiset A can be well ordered, that is, there exists a semiset $B \subseteq A \times A$ that is a linear order relation on A for which each nonempty subclass of A has a smallest element in this order. The theory obtained again turns out to be consistent, and in it one can now construct a semiset of representatives of the equivalence classes of the semiset of elements of bounded norm of an internal linear space with respect to infinite closeness. Here it is rather more complicated to formalize the construction of Loeb measures, but is also possible by using a method analogous to that employed in §1, 5°.

The approach sketched in this subsection enables us to axiomatize the theory of hyperfinite sets. For this one should consider the theory FNCT obtained by replacing the axiom of infinity by its negation. This theory can be interpreted in NCT, that is, it is consistent. On the basis of it one can develop a mathematics analogous to that constructed in alternative set theory.

CHAPTER 1

Nonstandard Analysis of Operators Acting in Spaces of Measurable Functions

§1. Relatively standard elements in internal set theory

1°. A large number of the applications of nonstandard analysis are based on equivalences like the nonstandard definitions of limit, continuity, compactness, and so on (see Theorems 0.3.19–0.3.21). In particular, it is such equivalences that enable us to give proofs that are simpler and more natural than the standard proofs for many classical theorems of analysis. There are numerous examples of such proofs in the books [**15**] and [**28**]. However, an essential limitation here is that all these equivalences hold only for standard objects.

Even in the cases when nonstandard analysis is applied to the study of standard objects there are often difficulties connected with this limitation. We look at some typical examples.

1) Let $f\colon \mathbb{R}^2 \longrightarrow \mathbb{R}$ be a standard function. The question arises of how to get an equivalent nonstandard definition of the condition that $\lim_{x\to 0}\lim_{y\to 0} f(x,y) = 0$ (of course, any standard numbers can appear in place of zero).

Using Theorem 0.3.21, we obtain immediately that

$$\lim_{x\to 0}\lim_{y\to 0} f(x,y) = 0 \Longleftrightarrow \forall \alpha \approx 0\ (\lim_{y\to 0} {}^*f(\alpha,y) \approx 0). \tag{1.1.1}$$

It is not possible to use the nonstandard definition of limit any further, because the function $\varphi(y) = {}^*f(\alpha,y)$ is internal but nonstandard. This difficulty arises at once when we try to get an assertion analogous to the corollary to Theorem 0.3.21 for improper integrals.

2) Let us attempt to apply nonstandard analysis to the proof of the 'difficult' part of l'Hôpital's rule. Suppose that f and g are standard functions differentiable in a neighborhood of a standard point a, $\lim_{x\to a} f(x) = \lim_{x\to a} g(x) = \infty$, $g'(x) \neq 0$ in a neighborhood of a, and

$$\lim_{x\to a} \frac{f'(x)}{g'(x)} = d.$$

It is required to show that $\lim_{x\to a} \frac{f(x)}{g(x)} = d$. We take arbitrary points y, $z \approx a$. For definiteness, let $a < z < y$. By Cauchy's theorem, there is an $\eta \in [z,y]$ such that

$$\frac{{}^*f(y) - {}^*f(z)}{{}^*g(y) - {}^*g(z)} = \frac{{}^*f'(\eta)}{{}^*g'(\eta)} \approx d, \tag{1.1.2}$$

because $\eta \approx a$. We now consider the equality

$$\frac{{}^*f(y) - {}^*f(z)}{{}^*g(y) - {}^*g(z)} = \frac{{}^*f(z)}{{}^*g(z)} \cdot \left(1 - \frac{{}^*f(y)}{{}^*f(z)}\right) \cdot \left(1 - \frac{{}^*g(y)}{{}^*g(z)}\right)^{-1}. \tag{1.1.3}$$

If y were standard and $z \approx a$, then in view of the conditions imposed on f and g we would have the relations $\frac{{}^*f(y)}{{}^*f(z)} \approx \frac{{}^*g(y)}{{}^*g(z)} \approx 0$, that is, $\frac{{}^*f(z)}{{}^*g(z)} \approx d$, which would prove what is required. But in our case y itself is $\approx a$, that is, it is not standard. Thus, to obtain the necessary result we must take z essentially closer to a than y. However, the question then arises of how to give the preceding statement a precise meaning that takes into account the fact that z and y are already infinitely close to a.

3) We consider the following proposition: "A uniform limit of a sequence $\{f_n\}$ of bounded uniformly continuous functions on a uniform space is a uniformly continuous function." Let us try to apply nonstandard analysis to prove it.

Thus, the standard sequence $\{f_n\}$ converges uniformly to a standard function f on a set X. By using familar criteria it is easy to deduce that ${}^*f_N(x) \approx {}^*f(x)$ for infinite N and for any $x \in {}^*X$, because $\sup\{|{}^*f_N(x) - {}^*f(x)| \mid x \in {}^*X\} \approx 0$. By well-known results, it is enough to prove that $\forall x', x'' \in {}^*X\ (x' \approx x'' \Longrightarrow {}^*f(x') \approx {}^*f(x''))$. We have that ${}^*f(x') \approx {}^*f_N(x')$ and ${}^*f(x'') \approx {}^*f_N(x'')$. However, we cannot assert that ${}^*f_N(x') \approx {}^*f_N(x'')$ because, though ${}^*f_N$ is uniformly continuous by the transfer principle, the nonstandard criterion does not hold for it since N is an infinite number and ${}^*f_N$ is not a standard function.

In view of this obstacle the 'nonstandard' proof of the proposition turns out to be even more complicated than the familiar standard proof.

To overcome this kind of difficulty in situations that are analogous but less elementary we introduce and investigate here the concept of relative standardness. All the arguments in this section are carried out in IST, but all the results obtained are valid also in any nonstandard universe ${}^*\mathbb{R}$ satisfying the Nelson principle. We recall that in working in IST it is considered that standard sets themselves, and not their nonstandard extensions (since such a concept is simply absent in IST), contain nonstandard elements in the case when they are infinite; for example, $\mathbb{R}$ itself contains infinite and infinitesimal elements, and ${}^*\mathbb{R}$ does not. But if we are working with ${}^*\mathbb{R}$, then elements of the form ${}^*X$ with $X \in \mathbb{R}$ are standard elements of ${}^*\mathbb{R}$ (${}^*X$ is still called the nonstandard extension of X).

2°. Along with the abbreviations $\mathrm{Fnc}(f)$, $\mathrm{dom}\, f$, $\mathrm{rng}\, f$, and $\mathrm{Fin}(x)$ already introduced we use also the notation

$$\mathrm{Ffin}(f) \Longleftrightarrow \mathrm{Fnc}(f) \wedge \forall x \in \mathrm{rng}(f)\ \ \mathrm{Fin}(x),$$

that is, the values taken by f are finite sets. We recall that $\mathrm{Fin}(x)$ means only that the cardinality of x is an element of ω, that is, a natural number, possibly infinite if x is a nonstandard set.

We call a set x bounded ($\mathrm{Bnd}(x)$) if $\exists^{\mathrm{st}} X\ (x \in X)$.

The relation $x \,\mathrm{st}\, y$ (read "x is standard relative to y" or "x is y-standard") is defined in IST as follows:

$$x \,\mathrm{st}\, y \Longleftrightarrow \exists^{\mathrm{st}} \varphi\ (\mathrm{Ffin}(\varphi)\ \wedge\ y \in \mathrm{dom}\, \varphi\ \wedge\ x \in \varphi(y)). \tag{1.1.4}$$

PROPOSITION 1.1.1.

1. $x \,\mathrm{st}\, y \Longrightarrow \mathrm{Bnd}(x) \wedge \mathrm{Bnd}(y)$.
2. $x \,\mathrm{st}\, y \wedge y \,\mathrm{st}\, z \Longrightarrow x \,\mathrm{st}\, z$.
3. $x \,\mathrm{st}\, y \wedge \mathrm{Fin}(x) \Longrightarrow \forall z \in x\ (z \,\mathrm{st}\, y)$.
4. $\mathrm{Bnd}(y) \wedge \mathrm{st}(x) \Longrightarrow x \,\mathrm{st}\, y$.

To prove the proposition we need

LEMMA 1.1.2. *Suppose that $A(x, y)$ is a formula in* ZFC, *and that* ZFC $\vdash \forall x\, \exists! y\ A(x, y)$. *Then*

$$\mathrm{IST} \vdash \forall^{\mathrm{st}} x\ \forall y\ (A(x, y) \Longrightarrow \mathrm{st}(y)).$$

(Recall that $\vdash$ means formal deducibility in a theory.)
The proof of the lemma follows immediately from the transfer principle (T).

PROOF OF PROPOSITION 1.1.1. Assertion 1 is obvious. To prove assertion 2 we assume that $x \in \varphi_1(y)$ and $y \in \varphi_2(z)$, where the φ_i are standard functions such that $\varphi_i(t)$ is a finite set $\forall t \in \operatorname{dom} \varphi_i$. We construct a function h such that $\operatorname{dom} h = \operatorname{dom} \varphi_2$ and $h(t) = \{\varphi_1(u) \mid u \in \varphi_2(t) \cap \operatorname{dom} \varphi_1\}$ $\forall t \in \operatorname{dom} \varphi_2$. Obviously, $h(t)$ is finite $\forall t \in \operatorname{dom} \varphi_2$ (since $\varphi_2(t)$ is finite), and $x \in h(z)$. By the lemma, h is a standard function, and this proves assertion 2.

To prove assertion 3 we assume that $x \in \psi(y)$, where ψ is a standard function, and $\psi(t)$ is finite $\forall t \in \operatorname{dom} \psi$. We construct a new standard function g such that $\operatorname{dom} g = \operatorname{dom} \psi$ and $g(t) = \{v \mid v \in \psi(t)\ \wedge\ v \text{ is finite}\}$ $\forall t \in \operatorname{dom} \psi$ (g is standard, again by Lemma 1.1.2). As a union of a finite set of finite sets, $g(t)$ is finite $\forall t \in \operatorname{dom} \psi$, and $z \in g(y)$ by assumption.

To prove assertion 4 we assume that $y \in X$, where X is standard (the existence of such an X follows from the fact that $\mathrm{Bnd}(y)$ is true). We define a standard function φ such that $\varphi(t) = \{x\}$ $\forall t \in X$ (this function is standard because x is standard). Then $x \in \varphi(y)$. □

We introduce abbreviations $\forall^{\mathrm{st}\, y} x$ and $\exists^{\mathrm{st}\, y} x$ analogous to the abbreviations $\forall^{\mathrm{st}}$ and $\exists^{\mathrm{st}}$ in §4 of Chapter 0.

THEOREM 1.1.3 (relativized transfer principle T_τ). *If $A(x, t_1, \ldots, t_k)$ is an internal formula ($k \geq 0$), then for any bounded τ*

$$\forall^{\mathrm{st}\tau} t_1 \ldots \forall^{\mathrm{st}\tau} t_k\ (\forall^{\mathrm{st}\tau} x\ A(x, t_1, \ldots, t_k) \longrightarrow \forall x\ A(x, t_1, \ldots, t_k)).$$

PROOF. For brevity assume that $k = 1$. We must prove in IST the sentence

$$\forall \tau\ (\mathrm{Bnd}(\tau) \longrightarrow \forall^{\mathrm{st}\tau} t\ (\forall^{\mathrm{st}\tau} x\ A(x, t) \longrightarrow \forall x\ A(x, t))). \tag{1.1.5}$$

Since $\neg\,\mathrm{Bnd}(t) \longrightarrow \neg t \operatorname{st} \tau$ in view of Proposition 1.1.1, (1.1.5) is equivalent to

$$\forall \tau\ \forall^{\mathrm{st}\tau} t\ (\forall^{\mathrm{st}\tau} x\ A(x, t) \longrightarrow \forall x\ A(x, t)). \tag{1.1.6}$$

We rewrite (1.1.6) solely in terms of IST. Further, in the following formulas we assume that φ and ψ run through functions with finite sets as values, and Φ and Ψ are sets of such functions (this convention applies also to the proof of Theorem 1.1.4):

$$\begin{aligned} &\forall \tau\ \forall^{\mathrm{st}} \varphi\ \forall t\ (\tau \in \operatorname{dom} \varphi\ \wedge\ t \in \varphi(\tau) \\ &\qquad \longrightarrow [\forall^{\mathrm{st}} \psi\ \forall x\ (\tau \in \operatorname{dom} \psi\ \wedge\ x \in \psi(\tau) \longrightarrow A(x, t)) \longrightarrow \forall y\ A(y, t)]). \end{aligned} \tag{1.1.7}$$

We shall apply Nelson's algorithm [**48**] to the formula (1.1.7). The latter is equivalent (in the predicate calculus) to the formula

$$\begin{aligned} &\forall^{\mathrm{st}} \varphi\ \forall \tau, t\ \exists^{\mathrm{st}} \psi\ (\tau \in \operatorname{dom} \varphi\ \wedge\ t \in \varphi(\tau) \\ &\qquad \wedge\ \forall x\ (\tau \in \operatorname{dom} \psi\ \wedge\ x \in \psi(\tau) \longrightarrow A(x, t)) \longrightarrow \forall y\ A(y, t)). \end{aligned} \tag{1.1.8}$$

In view of the idealization principle, (1.1.8) is equivalent to the formula

$$\forall^{\mathrm{st}}\varphi\ \exists^{\mathrm{st\,fin}}\Phi\ \forall\tau\ \forall t\ (\tau\in\operatorname{dom}\varphi\ \wedge\ t\in\varphi(\tau)$$
$$\wedge\ \forall\psi\in\Phi\ \forall x\ (\tau\in\operatorname{dom}\psi\ \wedge\ x\in\psi(\tau)\Longrightarrow A(x,t))\Longrightarrow\forall y\ A(y,t)).$$

By the transfer principle, the sign $\operatorname{st} y$ of the first two quantifiers can be omitted, that is, we need to prove in ZFC the formula

$$\forall\varphi\ \exists^{\mathrm{fin}}\Phi\ \forall\tau,t\ (\tau\in\operatorname{dom}\varphi\ \wedge\ t\in\varphi(\tau)$$
$$\wedge\ \forall\psi\in\Phi\ \forall x\ (\tau\in\operatorname{dom}\psi\ \wedge\ x\in\psi(\tau)\Longrightarrow A(x,t))\Longrightarrow\forall y\ A(y,t)).\tag{1.1.9}$$

We fix an arbitrary function φ and construct a singleton set $\Phi=\{\psi\}$, where the function ψ is defined as follows. Let $M=\bigcup\operatorname{rng}\varphi$ and $M_1=\{t\in M\mid\exists y\ \neg A(y,t)\}$. It follows from the ZFC axioms that there exists a function h such that $\operatorname{dom}h=M_1$ and $\forall t\in\operatorname{dom}h\ (\neg A(h(t),t))$. Now let $\operatorname{dom}\psi=\operatorname{dom}\varphi$, and let $\psi(\alpha)=\{h(\nu)\mid\nu\in\varphi(\alpha)\cap M_1\}\ \forall\alpha\in\operatorname{dom}\varphi$ (if $\varphi(\alpha)\cap M_1=\emptyset$, then $\psi(\alpha)=\emptyset$). Note that $\psi(\alpha)$ is finite because $\varphi(\alpha)$ is finite. We now fix a $\tau\in\operatorname{dom}\varphi$ and a $t\in\varphi(\tau)$. To prove (1.1.9) it then suffices to show that the formula

$$\forall x\in\psi(\tau)\ A(x,t)\Longrightarrow\forall y\ A(y,t)\tag{1.1.10}$$

is true. If $t\in M\setminus M_1$, then the conclusion of the implication (1.1.10) is true, and if $t\in M_1$, then its premise is false. Indeed, if $x=h(t)$, then $x\in\psi(\tau)$ because $t\in\varphi(\tau)\cap M_1$, but $A(h(t),t)$ is false. □

THEOREM 1.1.4 (relativized idealization principle I_τ). *If $B(x,y)$ is an internal formula, possibly containing free variables in addition to x and y, then for any bounded τ*

$$\forall^{\mathrm{st}\,\tau\,\mathrm{fin}}z\ \exists x\ \forall y\in z\ B(x,y)\longleftrightarrow\exists x\ \forall^{\mathrm{st}\tau}y\ B(x,y).$$

We remark that to prove I_τ it suffices to prove the implication $\longrightarrow$, since the implication $\longleftarrow$ follows from Proposition 1.1.1 (3).

PROOF. Assume that the formula B contains another free variable t. Then we must prove the formula

$$\forall\tau\ (\mathrm{Bnd}(\tau)\longrightarrow\forall t\ (\forall^{\mathrm{st}\,\tau\,\mathrm{fin}}x\ \exists y\ \forall z\in x\ B(z,y,t)\longrightarrow\exists u\ \forall^{\mathrm{st}\tau}v\ B(v,u,t))).\tag{1.1.11}$$

Again using Proposition 1.1.1, we conclude that (1.1.11) is equivalent to

$$\forall\tau\ \forall t\ (\forall^{\mathrm{st}\,\tau\,\mathrm{fin}}x\ \exists y\ \forall z\in x\ B(z,y,t)\longrightarrow\exists u\ \forall^{\mathrm{st}\tau}v\ B(v,u,t)).$$

We rewrite this sentence in terms of IST (that is, we write the predicate $x\operatorname{st}\tau$ in terms of IST). This gives the formula in IST

$$\forall\tau\ \forall t\ (\forall^{\mathrm{st}}\varphi\ \forall^{\mathrm{fin}}x\ (\tau\in\operatorname{dom}\varphi\ \wedge\ x\in\varphi(\tau)\longrightarrow\exists y\ \forall z\in x\ B(z,y,t))$$
$$\longrightarrow\exists u\ \forall^{\mathrm{st}}\Psi\ \forall v\ (\tau\in\operatorname{dom}\psi\ \wedge\ v\in\psi(\tau)\longrightarrow B(v,u,t)))$$
$$\sim\forall\tau\ \forall t\ (\forall^{\mathrm{st}}\varphi\ \forall^{\mathrm{fin}}x\ (\tau\in\operatorname{dom}\varphi\ \wedge\ x\in\varphi(\tau)\longrightarrow\exists y\ \forall z\in x\ B(z,y,t))$$
$$\longrightarrow\forall^{\mathrm{st\,fin}}\Psi\ \exists u\ \forall\psi\in\Psi\ \forall v\ (\tau\in\operatorname{dom}\psi\ \wedge\ v\in\psi(\tau)\longrightarrow B(v,u,t)))$$
$$\sim\forall^{\mathrm{st\,fin}}\Psi\ \exists^{\mathrm{st\,fin}}\Phi\ \forall\tau\ \forall t\ (\forall\varphi\in\Phi\ \forall^{\mathrm{fin}}x\ (\tau\in\operatorname{dom}\varphi\ \wedge\ x\in\varphi(\tau)$$
$$\longrightarrow\exists y\ \forall z\in x\ B(z,y,t))\longrightarrow\exists u\ \forall\psi\in\Psi\ \forall v\ (\tau\in\operatorname{dom}\psi\ \wedge\ v\in\psi(\tau)$$
$$\longrightarrow B(v,u,t)))$$

(the sign $\sim$ means equivalence of formulas).

In the last formula, as in the proof of Theorem 1.1.3, it is possible to omit the sign st on the first two quantifiers in view of the transfer principle, and to prove the sentence obtained in ZFC. We fix an arbitrary finite set Ψ of functions and define Φ by setting $\Phi = \{\varphi\}$, where $\operatorname{dom}\varphi = \bigcup_{\psi\in\Psi}\operatorname{dom}\psi$ and $\varphi(\alpha) = \{\cup\psi(\alpha) \mid \psi \in \Psi, \alpha \in \operatorname{dom}\psi\}$. Note that $\varphi(\alpha)$ is a singleton, and thus finite. We fix arbitrary τ and t. If $\tau \notin \operatorname{dom}\varphi$, then $\forall\psi \in \Psi\ (\tau \notin \operatorname{dom}\psi)$, that is, our formula is true. If $\tau \in \operatorname{dom}\varphi$, then we must establish the truth of the implication

$$(1.1.12)\quad \exists y\ \forall z \in \cup\{\psi(\tau) \mid \psi \in \Psi,\ \tau \in \operatorname{dom}\psi\}\ B(z, y, t) \\ \longrightarrow \exists u\ \forall\psi \in \Psi\ \forall v\ (\tau \in \operatorname{dom}\psi\ \wedge\ v \in \psi(\tau) \longrightarrow B(v, u, t)).$$

We transform the premise of this implication:

$$\exists y\ \forall z \in \cup\{\psi(\tau) \mid \psi \in \Psi,\ \tau \in \operatorname{dom}\psi\}\ B(z, y, t) \\ \sim \exists y\ \forall z\ (\exists\psi \in \Psi\ (\tau \in \operatorname{dom}\psi\ \wedge\ z \in \psi(\tau) \longrightarrow B(z, y, t)) \\ \sim \exists y\ \forall z\ \forall\psi \in \Psi\ (\tau \in \operatorname{dom}\psi\ \wedge\ z \in \psi(\tau) \longrightarrow B(z, y, t)).$$

It is now obvious that in the implication (1.1.12) the premise is equivalent to the conclusion. □

We present two simple corollaries to the theorems proven.

PROPOSITION 1.1.5. *Under the conditions of Lemma* 1.1.2 *if x is τ-standard and $A(x, y)$, then y is τ-standard.*

The proposition follows immediately from the relativized transfer principle T_τ.

PROPOSITION 1.1.6. *If x is τ-standard, then:*
1) $\operatorname{Fin}(x) \Longleftrightarrow \forall u \in x\ (u \operatorname{st} \tau)$;
2) $\operatorname{Fin}(x) \Longrightarrow |x| \operatorname{st} \tau$.

PROOF. Assertion 2 follows immediately from the preceding proposition. The implication $\Longrightarrow$ in assertion 1 is Proposition 1.1.1 (3). Let us prove the implication $\Longleftarrow$. To do this we rewrite the right-hand side in the form $\forall u \in x\ \exists^{\operatorname{st}\tau} v\ (u = v)$. By the idealization principle I_τ, this sentence is equivalent to the sentence $\exists^{\operatorname{st}\tau\operatorname{fin}} V\ \forall u \in x\ \exists v \in V\ (u = v)$, which means that $x \subseteq V$, and x is finite because V is.

3°. Let τ be an arbitrary bounded element, and X a τ-standard topological space. For any τ-standard point $a \in X$ we can define its τ-*monad*:

$$\mu^\tau(a) = \cap\{u \subseteq X \mid u \text{ is open},\ a \in u,\ u \operatorname{st} \tau\}.$$

If $x \in \mu^\tau(a)$, then we say that x *is τ-infinitely close to* a ($x \overset{\tau}{\approx} a$). If X is a uniform space with uniformity Σ (which is also τ-standard, of course), then for any two points x and y in it we shall say that x is *τ-infinitely close to* y ($x \overset{\tau}{\approx} y$) if $\langle x, y\rangle \in \cap\{\sigma \in \Sigma \mid \sigma \operatorname{st} \tau\}$.

It follows immediately from the idealization principle that if a is not an isolated point, then $\mu^\tau(a) \setminus \{a\} \neq \emptyset$.

PROPOSITION 1.1.7. *If a topological space X and a point $a \in X$ are τ-standard, and $\tau \operatorname{st} \lambda$, then $\mu^\lambda(a) \subseteq \mu^\tau(a)$ and $\forall x, y \in X$ $(x \overset{\lambda}{\approx} y \Longrightarrow x \overset{\tau}{\approx} y)$. In particular, if X and a are standard, then $\mu^\lambda(a) \subseteq \mu(a)$ and $\forall x, y \in X$ $(x \overset{\lambda}{\approx} y \Longrightarrow x \approx y)$ for any bounded λ.*

The proof follows from Proposition 1.1.1 (2). The same proposition gives us that $X \operatorname{st} \lambda$, that is, $\mu^\lambda(a)$ is defined. In the case of standard X and a one should take $\tau = \emptyset$ (or any other standard set), since $\operatorname{st}(X) \Longleftrightarrow X \operatorname{st} \emptyset$. □

THEOREM 1.1.7. *Suppose that the topological spaces X and Y, the mapping $f\colon X \longrightarrow Y$, the subset $A \subseteq X$, and the points $a \in X$ and $b \in Y$ are τ-standard. Then:*

1) *A is open* $\Longleftrightarrow \forall^{\operatorname{st}\tau} x \in A$ $(\mu^\tau(x) \subseteq A)$;
2) *A is closed* $\Longleftrightarrow \forall^{\operatorname{st}\tau} x \in X\ \forall \eta \in A$ $(\eta \in \mu^\tau(x) \longrightarrow x \in A)$.

The following statements hold for any bounded λ such that $\tau \operatorname{st} \lambda$:

3) $\lim_{x \to a} f(x) = b \Longleftrightarrow \forall \xi \in X$ $(\xi \overset{\lambda}{\approx} a \longrightarrow f(\xi) \overset{\tau}{\approx} b)$;
4) *when X and Y are uniform spaces, f is uniformly continuous* $\Longleftrightarrow \forall \xi, \eta \in X$ $(\xi \overset{\lambda}{\approx} \eta \Longrightarrow f(\xi) \overset{\tau}{\approx} f(\eta))$.

This theorem is a generalization of the corresponding criteria to the case of τ-standard objects, but for the compactness criterion the corresponding generalization is not true (see §1, 5°).

PROOF. Assertions 1 and 2 are proved just as in the case of standard objects. For completeness we present a proof of assertion 1.

Let $\mathbb{T}$ be the topology on X, and let $\mathbb{T}(a)$ be the collection of all open neighborhoods of a. It follows from the τ-standardness of the topological space X that $\mathbb{T}$ is τ-standard and $\mathbb{T}(a)$ is τ-standard $\forall a \operatorname{st} \tau$ (Proposition 1.1.6). Suppose that A is open, $x \operatorname{st} \tau$ and $x \in A$. Then by the definition of the τ-monad, $\mu^\tau(x) = \cap\{u \operatorname{st} \tau \mid u \in \mathbb{T}(x)\} \subseteq A$, because $A \in \mathbb{T}(x)$. To prove the converse assertion assume that $\forall^{\operatorname{st}\tau} x \in A$ $(\mu^\tau(x) \subseteq A)$, but A is not open. Since $X, A \operatorname{st} \tau$,

$$\exists^{\operatorname{st}\tau} x \in A\ \forall^{\operatorname{st}\tau} U \in \mathbb{T}(x)\ \exists^{\operatorname{st}\tau} y\ (y \in U\ \wedge\ y \notin A)$$

by the principle T_τ. We consider the binary relation $\mathcal{R} \subseteq \mathbb{T}(x) \times X$ with

$$\langle u, z \rangle \in \mathcal{R} \Longleftrightarrow z \in U\ \wedge\ z \notin A.$$

Since $\forall^{\operatorname{st}\tau\, \operatorname{fin}} I \subseteq \mathbb{T}(x)$ $(\cap I \in \mathbb{T}(x) \wedge \cap I \operatorname{st} \tau)$, the relation $\mathcal{R}$ satisfies the condition of the idealization principle. But this means that $\exists y\ \forall^{\operatorname{st}\tau} U \in \mathbb{T}(x)$ $(y \in U\ \wedge\ y \notin A)$, that is, $y \in \mu^\tau(x) \setminus A$. Contradiction.

We prove assertion 3. Again because of the τ-standardness of f, a, and b and the transfer principle T_τ,

$$\lim_{x \to a} f(x) = b \Longleftrightarrow \forall^{\operatorname{st}\tau} W \in \mathbb{T}_Y(b)\ \exists^{\operatorname{st}\tau} u \in \mathbb{T}_X(a)\ (f(u) \subseteq W).$$

Let $\lim_{x \to a} f(x) = b$. It is required to prove that $f(\mu^\lambda(a)) \subseteq \mu^\tau(b)$. Since $\mu^\lambda(a) \subseteq \mu^\tau(a)$ (Proposition 1.1.7), it suffices to show that $f(\mu^\tau(a)) \subseteq \mu^\tau(b)$, that is, $\forall^{\operatorname{st}\tau} W \in \mathbb{T}_Y(b)$ $(f(\mu^\tau(a)) \subseteq W)$. Suppose that $u \operatorname{st} \tau$, $u \in \mathbb{T}_X(a)$, and $f(u) \subseteq W$. Then $\mu^\tau(a) \subseteq u$, that is, $f(\mu^\tau(a)) \subseteq W$, which is what was required. To prove the converse implication assume that $f(\mu^\lambda(a)) \subseteq \mu^\tau(b)$. We fix an arbitrary $W \operatorname{st} \tau$, $W \in \mathbb{T}_Y(b)$, and note that, by Proposition 1.1.1 (2), $W \operatorname{st} \lambda$ and $f(\mu^\lambda(a)) \subseteq W$.

We prove first that $\exists^{\mathrm{st}\,\lambda} U \in \mathbb{T}_X(a)\ (f(U) \subseteq W)$. If this is not so, then the relation $\mathcal{R}_1 \subseteq \mathbb{T}_X(a) \times Y$ with $\langle U, y\rangle \in \mathcal{R}_1 \iff y \in U \ \wedge\ f(y) \notin W$ satisfies the condition of the idealization principle I_λ. Then using I_λ, we get that $\exists y\ \forall^{\mathrm{st}\,\lambda} U \in \mathbb{T}_X(a)$ $(y \in U \ \wedge\ f(y) \notin W)$. This contradicts the inclusion $f(\mu^\lambda(a)) \subseteq W$. Thus, $\exists U \in \mathbb{T}_X(a)\ (f(U) \subseteq W)$. Since all the parameters in the last sentence are τ-standard, the transfer principle T_τ gives us that $\exists^{\mathrm{st}\,\tau} U \in \mathbb{T}_X(a)\ (f(U) \subseteq W)$. Assertion 4 is proved similarly. □

REMARKS. 1. Of course, we can take τ as λ in assertions 3 and 4. However, it is the case $\lambda \neq \tau$ that is used in applications (see the next subsection).

2. Theorem 1.1.7 is now applicable for arbitrary bounded objects because $x \,\mathrm{st}\, x$ is obviously true for every such x. For example, $X, A \,\mathrm{st}\, \tau = \langle X, A\rangle$ in assertions 1 and 2, and $x, y, f, a, b \,\mathrm{st}\, \tau = \langle x, y, f, a, b\rangle$ in assertions 3 and 4.

We dwell on the case of the field $\mathbb{R}$ in greater detail. As before, let τ be an arbitrary bounded internal set. A number $x \in \mathbb{R}$ is said to be *τ-infinitesimal* if $x \overset{\tau}{\approx} 0$, *$\tau$-infinite* ($x \overset{\tau}{\sim} \infty$) if $x^{-1} \overset{\tau}{\approx} 0$, and *$\tau$-finite* ($x \overset{\tau}{\ll} \infty$) if x is not τ-infinite. It follows directly from these definitions that

$$
\begin{aligned}
x \overset{\tau}{\approx} 0 &\iff \forall^{\mathrm{st}\,\tau} y \in \mathbb{R}_+\ (|x| \leq y),\\
x \overset{\tau}{\sim} \infty &\iff \forall^{\mathrm{st}\,\tau} y \in \mathbb{R}_+\ (|x| > y), \qquad (1.1.13)\\
x \overset{\tau}{\ll} \infty &\iff \exists^{\mathrm{st}\,\tau} y \in \mathbb{R}_+\ (|x| \leq y).
\end{aligned}
$$

PROPOSITION 1.1.8. *A number $x \in \mathbb{R}_+$ is τ-infinitesimal if and only if $|x| < \varphi(\tau)$ for any $\mathbb{R}_+$-valued standard function φ such that $\tau \in \operatorname{dom}\varphi$.*

PROOF. The proposition is obvious in one direction (if x is τ-infinitesimal). Suppose that $y \,\mathrm{st}\, \tau$ and $y \in \mathbb{R}_+$. We show that $|x| < y$. It follows from the conditions for y that there exists a standard function ψ such that $\tau \in \operatorname{dom}\psi$, $\operatorname{rng}\psi \subseteq \mathcal{P}^{\mathrm{fin}}(\mathbb{R}_+)$ ($\mathcal{P}^{\mathrm{fin}}(A)$ is the set of finite subsets of A), and $y \in \psi(\tau)$. We define the standard (in view of Lemma 1.1.2) function $\varphi\colon \operatorname{dom}\psi \longrightarrow \mathbb{R}_+$ by setting $\varphi(\alpha) = \min \psi(\alpha)$. Then it follows from the conditions of the proposition that $|x| < \varphi(\tau)$, and $y \geq \varphi(\tau)$ by the construction of φ. □

PROPOSITION 1.1.9. *$n \,\mathrm{st}\, x$ for any $x \in \mathbb{R}_+$ and any natural $n \leq x$.*

PROOF. Let $m = [x]$ (the integer part); then $m \,\mathrm{st}\, x$, and $n \leq m$. Consider the set $\overline{m} = \{0, 1, \dots, m\}$. It is clear that $\overline{m} \,\mathrm{st}\, m$ (Proposition 1.1.5). Moreover, $\overline{m}$ is finite, that is, $\mathrm{Fin}(\overline{m})$. By Proposition 1.1.6 (1), $n \,\mathrm{st}\, \overline{m}$ if $n \in \overline{m}$. □

PROPOSITION 1.1.10. *If $\lambda \,\mathrm{st}\, \tau$ and $x \overset{\tau}{\approx} 0$ ($x \overset{\tau}{\sim} \infty$), then $x \overset{\lambda}{\approx} 0$ ($x \overset{\lambda}{\sim} \infty$). If x is λ-finite, then it is τ-finite.*

PROOF. This follows immediately from the relations in (1.1.13). □

PROPOSITION 1.1.11. *The boundedness, permanence, Cauchy, and Robinson principles remain valid if instead of infinitesimal, infinite, and finite numbers we consider in them τ-infinitesimal, τ-infinite, and τ-finite numbers, respectively.*

The proof does not differ from the usual one.

PROPOSITION 1.1.12. *For any infinite (infinitesimal) number $x \in \mathbb{R}$ there exists a nonstandard number η such that $x \overset{\eta}{\sim} \infty$ ($x \overset{\eta}{\approx} 0$). This η can be chosen to be finite, or infinitesimal, or infinite.*

PROOF. We consider the internal relation $\sigma \subseteq \mathbb{R}_+^{\mathbb{R}} \times \mathbb{R} \times \mathbb{R}$ with

$$\langle f, \xi, \eta \rangle \in \sigma \Longleftrightarrow f(\eta) < |x| \ \wedge \ \eta \neq \xi.$$

By using the fact that x is infinite it is easy to see that σ satisfies the conditions of the idealization principle, that is,

$$\forall^{\mathrm{st\,fin}} M \subseteq \mathbb{R}_+^{\mathbb{R}} \times \mathbb{R} \ \exists \eta \ \forall \langle f, \xi \rangle \in M \ (\langle f, \xi, \eta \rangle \in \sigma).$$

Using the idealization principle, we get that

$$\exists \eta \ \forall^{\mathrm{st}} f \in \mathbb{R}_+^{\mathbb{R}} \ \forall^{\mathrm{st}} \xi \in \mathbb{R} \ (f(\eta) < |x| \ \wedge \ \eta \neq \xi).$$

Such an η obviously works. If the η obtained is finite, then $\eta' = \eta - {}^{\circ}\eta \approx 0$ also satisfies the requirements in view of Proposition 1.1.10. □

4°. Let us consider some examples illustrating ways of using these concepts for overcoming difficulties analogous to those discussed in the introduction. We begin with iterated limits.

a) THEOREM 1.1.13. *Suppose that $f\colon \mathbb{R}^2 \longrightarrow \mathbb{R}$ and $a \in \mathbb{R}$ are standard, and $\lim_{y\to 0} f(x,y)$ exists for each x in some neighborhood of 0. Then*

$$\lim_{x\to 0} \lim_{y\to 0} f(x,y) = a \Longleftrightarrow \forall \alpha \approx 0 \ \forall \beta \overset{\alpha}{\approx} 0 \ (f(\alpha,\beta) - a \approx 0).$$

PROOF. Assume $\lim_{x\to 0} \lim_{y\to 0} f(x,y) = a$, and let $\varphi(x) = \lim_{y\to 0} f(x,y)$. Then $\varphi(\alpha) \approx a \ \forall \alpha \approx 0$. We remark that φ is a standard function in view of Lemma 1.1.2, so $\varphi(\alpha) \operatorname{st} \alpha$. By Theorem 1.1.7 and Proposition 1.1.1 (2), the assertion

$$\varphi(\alpha) = \lim_{y\to 0} f(x,y)$$

is equivalent to the assertion

$$\forall \beta \overset{\alpha}{\approx} 0 \ (f(\alpha,\beta) \overset{\alpha}{\approx} \varphi(\alpha)).$$

According to Proposition 1.1.7, $f(\alpha,\beta) \overset{\alpha}{\approx} \varphi(\alpha) \implies f(\alpha,\beta) \approx \varphi(\alpha)$, and $f(\alpha,\beta) \approx a$ because $\varphi(\alpha) \approx a$.

We prove the converse. It obviously suffices to prove the sentence

$$\forall \varepsilon > 0 \ \exists \delta > 0 \ \forall x \ (|x| < \delta \Longrightarrow \exists \gamma > 0 \ \forall y \ (|y| < \gamma \Longrightarrow |f(x,y) - a| < \varepsilon))$$

(recall that in the theorem it is assumed that $\lim_{y\to 0} f(x,y)$ exists for any x in some neighborhood of 0).

We fix an arbitrary standard ε and consider the internal set

$$M = \{\delta > 0 \mid \forall x \ (|x| < \delta \Longrightarrow \exists \gamma > 0 \ \forall y \ (|y| < \gamma \Longrightarrow |f(x,y) - a| < \varepsilon))\}.$$

It is easy to see that M contains all infinitesimals. Indeed, if $\delta \sim 0$ and $|x| < \delta$, then $x \sim 0$. If $\gamma \overset{x}{\approx} 0$, then $\forall y \ (|y| < \gamma \Longrightarrow y \overset{x}{\approx} 0)$, and hence $|f(x,y) - a| \approx 0$, so that $|f(x,y) - a| < \varepsilon$. It is now obvious that M also contains some standard element. □

This theorem is true also in the case when $x \to b$ and $y \to c$ for arbitrary standard b and c. It can be carried over without difficulty to the case of an infinite limit and a limit at infinity, as well as to the case of arbitrary topological spaces.

COROLLARY. *If $f\colon \mathbb{R} \longrightarrow \mathbb{R}$ is Riemann-integrable in each finite interval and $\int_{-\infty}^{\infty} f(x)\,dx$ exists at least in the principal value sense, then*

$$\int_{-\infty}^{\infty} f(x)\,dx = {}^{\circ}\left(\Delta \sum_{-[\frac{a}{\Delta}]}^{[\frac{a}{\Delta}]} {}^{*}f(k\Delta)\right)$$

for any $a \sim \infty$ and any $\Delta \overset{a}{\approx} 0$.

b) To conclude the proof of l'Hôpital's rule in part 2 of the introduction to this section it is required to take $z \overset{y}{\approx} a$. In this case $\frac{{}^*f(y)}{{}^*f(z)} \approx 0$ and $\frac{{}^*g(y)}{{}^*g(z)} \approx 0$ in view of Theorem 1.1.7 (3) (more precisely, the obvious modification of it relating to infinite limits), and hence $\frac{{}^*f(z)}{{}^*g(z)} \approx d$ by virtue of (1.1.3). Thus, $\frac{{}^*f(z)}{{}^*g(z)} \approx d\ \forall z \overset{y}{\approx} a$. According to Theorem 1.1.7 (3), this means that $\lim_{x\to a} \frac{f(x)}{g(x)} = d$.

Similarly, to conclude the proof of the assertion in part 3 of the introduction it is required to take x' and x'' so that $x' \overset{N}{\approx} x''$. Then by Theorem 1.1.7 (4), ${}^*f_N(x') \approx {}^*f_N(x'')$, and we get immediately that ${}^*f(x') \approx {}^*f(x'')$, that is, we have proved that $\forall x', x''\ (x' \overset{N}{\approx} x'' \Longrightarrow {}^*f(x') \approx {}^*f(x''))$. This is equivalent to the uniform continuity of f in view of Theorem 1.1.7 (4).

c) As an example of the use of Proposition 1.1.12 we consider the following assertion ([**1**], Proposition 1.3.2).

Let $(a_{m,n})_{m\in\omega, n\in\omega}$ be a standard sequence such that $\lim_{n\to\infty}\lim_{m\to\infty} a_{m,n} = a$ and $\lim_{m\to\infty}\lim_{n\to\infty} a_{m,n} = b$. Then

$$\forall m \sim \infty\ \exists n_1, n_2 \sim \infty\ (\forall n < n_1\ (n \sim \infty \longrightarrow a_{m,n} \approx a)\ \wedge\ \forall n > n_2\ (a_{m,n} \approx b)).$$

To prove this take $n_1 \sim \infty$ such that $m \overset{n_1}{\sim} \infty$ (this can be done in view of Proposition 1.1.12), and $n_2 \overset{m}{\sim} \infty$.

5°. We show that for the standardization principle (S) we cannot obtain a result analogous to Theorems 1.1.3 and 1.1.4.

As in the case of the principles T and I, we write the relativized standardization principle (S_τ):

$$\forall^{\mathrm{st}\,\tau} x\ \exists^{\mathrm{st}\,\tau} y\ \forall^{\mathrm{st}\,\tau} z\ (z \in y \longleftrightarrow z \in x\ \wedge\ C(z)), \tag{1.1.14}$$

where $C(z)$ is a formula of IST that possibly contains free variables besides z. We show that (1.1.14) is false, even if we impose the following restriction on $C(z)$: every occurrence of the predicate st in $C(z)$ has the form $\ldots \mathrm{st}\,\tau$, and the one-place predicate $\mathrm{st}(x)$ does not occur in $C(z)$.

The fact of the matter is that the existence of the standard part ${}^{\circ}t$ for every finite real number $t \in \mathbb{R}$ follows in IST from the standardization principle [**48**]. By repeating the corresponding argument it is easy to see that (1.1.14), with a formula $C(z)$ satisfying the restriction described above, implies the sentence

$$\forall \tau\ \forall t \in \mathbb{R}\ (\exists^{\mathrm{st}\,\tau} u \in \mathbb{R}\ (|t| < u) \longrightarrow \exists^{\mathrm{st}\,\tau} v \in \mathbb{R}\ (|t - v| \overset{\tau}{\approx} 0)). \tag{1.1.15}$$

The fact that (1.1.15) is false, and with it (1.1.14), now follows from the next proposition.

PROPOSITION 1.1.14. *There exist an infinite natural number N and an $x \in [0,1]$ such that if y is N-infinitely close to x, then y is not N-standard.*

PROOF. Assume that the assertion is false. Then the sentence

$$\forall N \in \omega\ \forall x \in [0,1]\ \exists^{\mathrm{st}} \varphi \in (\mathcal{P}^{\mathrm{fin}}(\mathbb{R}_+)^\omega)\ \exists z \in \mathbb{R}_+\ \forall^{\mathrm{st}} \psi \in \mathbb{R}_+^\omega \quad (z \in \varphi(N)\ \wedge\ |x-z| < \psi(N)) \tag{1.1.16}$$

is true in IST. Here we employ Proposition 1.1.8. Nelson's algorithm is applicable to (1.1.16). We shall assume that the variables N, x, φ, z, and ψ run through the sets indicated in (1.1.16), and we shall not indicate this in the formulas. Using the idealization principle, we arrive at the sentence

(1.1.17)

$$\forall N\ \forall x\ \exists^{\mathrm{st}} \varphi\ \forall^{\mathrm{st}} \Xi\ \exists z\ \forall^{\mathrm{st}} \psi \in \Xi\ (z \in \varphi(N)\ \wedge\ |x-z| < \psi(N)), \text{ where } \Xi \in \mathcal{P}^{\mathrm{fin}}(\mathbb{R}_+^\omega).$$

Applying to (1.1.17) the consequence (0.4.17) of the standardization principle, we get that (1.1.17) is equivalent to

$$\forall N\ \forall x\ \forall^{\mathrm{st}} \widetilde{\Xi}\ \exists^{\mathrm{st}} \varphi\ \exists z\ \forall \psi \in \widetilde{\Xi}(\varphi)\ (z \in \varphi(N)\ \wedge\ |x-z| < \psi(N)), \tag{1.1.18}$$

where $\widetilde{\Xi} \colon \mathcal{P}^{\mathrm{fin}}(\mathbb{R}_+)^\omega \longrightarrow \mathcal{P}^{\mathrm{fin}}(R_+^\omega)$. Transposing the quantifiers $\forall$ on the left, using again the idealization principle, and then using the transfer principle, we get that (1.1.18) is equivalent to the internal sentence

$$\forall \widetilde{\Xi}\ \exists \Phi\ \forall N\ \forall x\ \exists \varphi \in \Phi\ \exists z\ \forall \psi \in \widetilde{\Xi}(\varphi)\ (z \in \varphi(N)\ \wedge\ |x-z| < \psi(N)), \tag{1.1.19}$$

where $\Phi \in \mathcal{P}^{\mathrm{fin}}(\mathcal{P}^{\mathrm{fin}}(\mathbb{R}_+)^\omega)$.

It remains to prove the falseness of (1.1.19) in ZFC.

We take the function $\widetilde{\Xi}$ such that $\widetilde{\Xi}(\varphi) = \{\psi\}\ \forall \varphi \in \mathcal{P}^{\mathrm{fin}}(\mathbb{R}_+)^\omega$, where

$$\psi(n) = a_n = \frac{1}{2n|\varphi(n)|}$$

($|\varphi(n)|$ is the number of elements in $\varphi(n)$). Let $M_{\varphi,n} = \bigcup_{z \in \varphi(n)} (z - a_n, z + a_n)$. The Lebesgue measure $\mu(M_{\varphi,n})$ of $M_{\varphi,n}$ is $\leq 1/n$. If $\Phi \in \mathcal{P}^{\mathrm{fin}}(\mathcal{P}^{\mathrm{fin}}(\mathbb{R}_+)^\omega)$, then we set $M_{\Phi,n} = \bigcup_{\varphi \in \Phi} M_{\varphi,n}$. It is clear that $\mu(\overline{M}_{\Phi,n}) \leq \frac{|\Phi|}{n}$. If (1.1.19) is true, then we clearly get by applying it to the function constructed that

$$\exists \Phi \in \mathcal{P}^{\mathrm{fin}}(\mathcal{P}^{\mathrm{fin}}(\mathbb{R}_+)^\omega)\ \forall n \in \omega\ ([0,1] \subseteq \overline{M}_{\Phi,n}). \tag{1.1.20}$$

To see that (1.1.20) is false it suffices to take $n > |\Phi|$, because here $\mu(\overline{M}_{\Phi,n}) < 1$.□

It also follows from the proposition proved that the criterion for compactness does not generalize to the case of τ-standard objects (see Theorem 1.1.7).

6°. Let us consider another variant of the definition of the predicate of relative standardness. For bounded x and y we set

$$x \,\mathrm{sst}\, y \iff \exists^{\mathrm{st}} \varphi\ (\mathrm{Fnc}(\varphi)\ \wedge\ y \in \mathrm{dom}\,\varphi\ \wedge\ x = \varphi(y))$$

(read "x is strongly standard relative to y" or "x is strongly y-standard"). It is clear that $x \,\mathrm{sst}\, y \Longrightarrow x \,\mathrm{st}\, y$. Below we shall prove that the converse is false.

PROPOSITION 1.1.15. *Assertions* 1, 2, *and* 4 *of Proposition* 1.1.1, *along with the relativized transfer principle* T_τ, *remain true if all the occurrences of the predicate* $\ldots\mathrm{st}\ldots$ *are replaced in them by* $\ldots\mathrm{sst}\ldots$. *In the relativized idealization principle* I_τ *the implication* $\Longrightarrow$ *is preserved.*

The proof is exactly like the proof of the corresponding assertions and is thus omitted. We remark that the implication $\Longleftarrow$ in the idealization principle is a consequence of Proposition 1.1.1, assertion 3. It will be proved below that both these assertions fail to carry over to the case of the predicate sst.

To define τ-infinitesimal (τ-infinite, τ-finite) numbers relative to the predicate sst, just replace st by sst in the relations (1.1.13). Proposition 1.1.8 shows that the concept of a τ-infinitesimal (τ-infinite, τ-finite) number relative to the predicate $\ldots\mathrm{st}\,\tau$ coincides with the corresponding concept relative to the predicate $\ldots\mathrm{sst}\,\tau$. A simple modification of the proof of Proposition 1.1.8 shows that the concepts of τ-infinite closeness in arbitrary topological spaces and uniform spaces relative to the predicates st and sst coincide. This implies that Theorems 1.1.7 and 1.1.13 and Propositions 1.1.10–1.1.12 remain true when the predicate $\ldots\mathrm{st}\,\tau$ is replaced in them by $\ldots\mathrm{sst}\,\tau$. The proposition below shows that Proposition 1.1.9 is not preserved here. This implies also the fact that neither Proposition 1.1.1, assertion 3 nor the implication $\Longleftarrow$ in the principle I_τ are preserved upon making this substitution.

PROPOSITION 1.1.16. $\exists N \in \omega\ \exists n < N\ (\neg n\,\mathrm{sst}\,N)$.

PROOF. It suffices to prove that the sentence

$$\forall N \in \omega\ \forall n < N\ \exists^{\mathrm{st}}\varphi \in \omega^\omega\ (n = \varphi(N)) \tag{1.1.21}$$

is false in IST.

Applying the idealization principle and the transfer principle to (1.1.21), we get the internal sentence

$$\exists \Phi \in \mathcal{P}^{\mathrm{fin}}(\omega^\omega)\ \forall n, N \in \omega\ (n < N \Longrightarrow \exists \varphi \in \Phi\ (n = \varphi(N))). \tag{1.1.22}$$

We show that the negation of (1.1.22) is true in ZFC.

Indeed, if $\Phi = \{\varphi_1, \ldots, \varphi_k\}$ is an arbitrary finite set of functions $\varphi_i\colon \omega \longrightarrow \omega$, then for $N > k$ there clearly exists an $n \in \{0, 1, \ldots, N-1\} \setminus \{\varphi_1(N), \ldots, \varphi_k(N)\}$. The pair n, N of numbers satisfies the formula $n < N\ \wedge\ \forall \varphi \in \Phi\ (n \neq \varphi(N))$. This proves that (1.1.21) is false. □

Kanoveĭ [**18**] showed that $\exists N \in \omega$ can be replaced by $\forall N \in \omega$ in Proposition 1.1.16.

REMARK. If we fix some infinite N and take the class $\{u \mid u\,\mathrm{sst}\,N\}$ as the class of standard sets, then we obtain a model of nonstandard set theory in which the transfer principle holds along with the implication $\Longrightarrow$ in the idealization principle. In this model the set of standard natural numbers has 'holes', that is, there are nonstandard natural numbers between two standard natural numbers. This can be thought of intuitively as follows. If we regard standard numbers as numbers having a nice description, then we can imagine that there is a very large but nicely described number (for example, $10^{10^{10}}$) with the property that there are numbers smaller than it which do not have a nice description. It would be interesting to represent a system of axioms of nonstandard mathematics from which the existence of such holes would follow.

7°. There are different approaches to dealing with difficulties analogous to those described in the introduction to this section. The first work of this kind was based on the use of iterated nonstandard extensions, that is, the ultrapowers ${}^*V(\mathbb{R})$ [**70**]. Here the most advanced results are due to Péraire ([**25**], [**49**], [**50**]). A graded system of predicates $\mathrm{St}_p(x)$ ("x is standard of degree p^{-1}", $p \in \omega$) is introduced there, with $\mathrm{St}_p(x) \longrightarrow \mathrm{St}_{p+1}(x)$. For each p the corresponding transfer and idealization principles hold. Such a theory enables us to overcome some of the difficulties that arise in working with standard objects, but it is not possible to get a theorem analogous to Theorem 1.1.7 for arbitrary nonstandard objects, since it is false that $\forall x\ \exists p\ \mathrm{St}_p(x)$. This sentence cannot even be written in the theory under consideration. The approach closest to that described here is that of Benninghofen and Richter [**31**]. The concepts of Π-monads and superinfinitesimals introduced by them have natural definitions in terms of the predicate of relative standardness (whether st or sst).

We recall that the monad of a standard filter $\mathfrak{F}$ of sets is defined to be the external set $\mu(\mathfrak{F}) = \cap\{u \in \mathfrak{F} \mid \mathrm{st}(u)\}$. Similarly, the τ-monad of a τ-standard filter $\mathfrak{F}$ is defined to be the set $\mu^\tau(\mathfrak{F}) = \cap\{u \in \mathfrak{F} \mid u \,\mathrm{st}\, \tau\}$.

Suppose now that $\mathfrak{F} = \{\mathfrak{F}_j \mid j \in J\}$ is a standard family of filters (if $j \in J$ is standard, then $\mathfrak{F}_j$ is a filter on a set X_j). Denote by $\underline{\mathfrak{F}}$ the filter generated by this family on $\prod\{X_j \mid j \in J\}$. Now let $\tau \in J$ be an arbitrary (not necessarily standard) element. Benninghofen and Richter define the Π-monad of a family $\underline{\mathfrak{F}}$ to be the external set

$$\Pi_{\mu_\tau}(\underline{\mathfrak{F}}) = \{x \in X_\tau \mid \forall^{\mathrm{st}} F \in \underline{\mathfrak{F}}\ (x \in F(\tau))\},$$

where $F(\tau)$ is the projection of F on the component τ.

It was shown in [**10**] that $\Pi_{\mu_\tau}(\underline{\mathfrak{F}}) = \mu^\tau(\mathfrak{F}_\tau)$ ($\mathfrak{F}_\tau$ is a τ-standard filter on the τ-standard set X_τ). This implies that the concept of infinite closeness based on Π-monads (in the case of filters of neighborhoods in topological spaces) coincides with the concept of τ-infinite closeness. In [**52**] there are many applications of these concepts to convergence spaces and topological vector spaces. All these results can easily be obtained in terms of relative standardness. In the next section the concept of relative standardness is employed in a situation where the use of Π-monads is considerably less natural or even impossible.

In [**50**] consistency is proved for nonstandard set theory with a predicate of relative standardness that satisfies the relativized transfer, idealization, and standardization principles (the last with the same restrictions on C as in §1, 5°).

§2. Countably finite Loeb spaces

In this and all subsequent sections in the first part we use a countably saturated nonstandard universe ${}^*\mathbb{R}$. In some cases it is assumed that ${}^*\mathbb{R}$ satisfies the idealization principle of Nelson. Theorems, sections, or subsections in which these conditions are used will be labeled IST in accordance with the convention adopted earlier.

1°. In this subsection we consider a certain generalization of Theorem 0.4.17 to the case of Loeb spaces with an infinite measure ν_Δ.

Suppose that $(X, S_\Delta, \nu_\Delta)$ is a Loeb space, and $M \in S_\Delta$ is such that there exists an increasing sequence $\{M_n \mid n \in \omega\}$ of internal sets satisfying the two conditions:

$$(1.2.1) \qquad 1) \quad M = \bigcup_{n\in\omega} M_n; \qquad 2) \quad \Delta \cdot |M_n| \ll +\infty \forall n \in \omega$$

(see Theorem 1.2.2). In this case we denote by S_Δ^M the σ-algebra $\{A \cap M \mid A \in S_\Delta\}$ of subsets of M, and by ν_Δ^M the restriction of ν_Δ to S_Δ^M. The space $\Xi = (M, S_\Delta^M, \nu_\Delta^M)$ is called a σ-finite subspace of the Loeb space $(X, S_\Delta, \nu_\Delta)$.

DEFINITION 1.2.1. An internal function $F\colon X \longrightarrow {}^*\mathbb{R}$ is said to be $\mathcal{S}_M$-integrable if it satisfies the following conditions:
1) $\Delta \sum_{\xi\in X} |F(\xi)| \ll \infty$;
2) $A \in \mathfrak{A} \ \wedge \ \Delta \cdot |A| \approx 0 \Longrightarrow \Delta \cdot \sum_{\xi\in A} |F(\xi)| \approx 0$;
3) $B \in \mathfrak{A} \ \wedge \ B \subseteq X \setminus M \Longrightarrow \Delta \cdot \sum_{\xi \in B} |F(\xi)| \approx 0$.
(recall that $\mathfrak{A} = {}^*\mathcal{P}(X)$, that is, A and B are arbitrary internal subsets of X).

In the case when $X = M$ it follows from ω_1-saturation that $X = M_n$ for some $n \in \omega$ (since in this case M is an internal set), and $\nu_\Delta(X) < \infty$. Then the definition of $\mathcal{S}_M$-integrability coincides in view of Theorem 0.4.15 with the definition of $\mathcal{S}$-integrability.

An internal function $F\colon X \longrightarrow {}^*\mathbb{R}$ is said to be a lifting of a function $f\colon X \longrightarrow {}^*\overline{\mathbb{R}}$ if $f(\xi) = {}^\circ F(\xi)$ for ν_Δ^M-almost all $\xi \in M$.

THEOREM 1.2.2. *A function $f\colon X \longrightarrow {}^*\overline{\mathbb{R}}$ is ν_Δ^M-integrable if and only if it has an $\mathcal{S}_M$-integrable lifting F. In this case*

$$(1.2.2) \qquad \int_M f \, d\nu_\Delta^M = {}^\circ\Big(\Delta \sum_{\xi\in X} F(\xi)\Big).$$

The rest of this subsection is devoted to a proof of this theorem.

We introduce some notation. Let $\mathcal{L}$ be the internal hyperfinite-dimensional space of functions $F\colon X \longrightarrow {}^*\mathbb{R}$ with norm $\|F\| = \Delta \cdot \sum_{\xi\in X} |F(\xi)|$. If $A \in \mathfrak{A}$, and χ_A is the characteristic function of A, then $\|F\|_A = \|F \cdot \chi_A\| = \Delta \sum_{\xi\in A} |F(\xi)|$. Let $\operatorname{fin}(\mathcal{L})$ be the external subspace of $\mathcal{L}$ consisting of elements of finite norm, and let $\mathcal{L}_0 \subseteq \operatorname{fin}(\mathcal{L})$ be the elements of infinitesimal norm. Recall (Theorem 0.4.10) that $\mathcal{L}^\# = \operatorname{fin}(\mathcal{L})/\mathcal{L}_0$, with norm defined as the shadow of the norm in $\mathcal{L}$, is a nonseparable Banach space (in the case when $|X|$ is an infinite number). Denote by $\mathcal{S}(M)$ the subspace of $\operatorname{fin}(\mathcal{L})$ consisting of the $\mathcal{S}_M$-integrable functions.

In the course of this subsection the set M is fixed, so instead of $\mathcal{S}(M)$ and $\mathcal{S}_M$ we write simply $\mathcal{S}$, and instead of ν_Δ^M, simply ν_Δ. Finally, we write $F \sim G$ if $\|F - G\| \approx 0$.

LEMMA 1.2.3. *If $\|F\| \approx 0$, then $F(\xi) \approx 0$ for ν_Δ-almost all ξ.*

PROOF. Assume that there is an $A \in S_\Delta$ such that $\nu_\Delta(A) > 0$ and $\forall \xi \in A$ $({}^\circ F(\xi) \neq 0)$. We show that in this case there exists a $B \in \mathfrak{A}$ satisfying the same conditions.

Indeed, if $\nu_\Delta(A)$ is finite, then $\nu_\Delta(A) = \sup\{\nu_\Delta(B) \mid B \in \mathfrak{A},\ B \subseteq A\}$ by Theorem 0.4.11. If $\nu_\Delta(A) = +\infty$, then the same theorem gives us that either $\exists B \in \mathfrak{A}$ $(\nu_\Delta(B) = +\infty,\ B \subseteq A)$ or there is a sequence $\{A_n\}$ of internal sets with $A \subseteq \bigcup_{n=0}^\infty A_n$ and $\nu_\Delta(A_n) < +\infty \ \forall n \in \omega$. In the last case $\nu_\Delta(A \cap A_n) \longrightarrow +\infty$, that is, there exists an n such that $\nu_\Delta(A \cap A_n) > 0$, and again we can use Theorem 0.4.11.

Thus, suppose that B satisfies the conditions above. Since the set $T = \{|F(\xi)| \mid \xi \in B\}$ is internal, it follows that $^\circ(\inf T) > 0$, and $^\circ\|F\| \geq {}^\circ(\inf T\Delta|B|) = {}^\circ(\inf T)\nu_\Delta(B) > 0$. Contradiction.

LEMMA 1.2.4. *If $F \in \mathcal{S}$ and $G \sim F$, then $G \in \mathcal{S}$.*

PROOF. If A satisfies one of the conditions 2) or 3) in Definition 1.2.1, then $\|F\|_A \approx 0$, and since $\|F-G\| \approx 0$, it follows that $\|F-G\|_A \approx 0$, that is, $\|G\|_A \approx 0$.□

The next result follows from similar arguments.

LEMMA 1.2.5. *$\mathcal{S}^\#$ is a closed subspace of $\mathcal{L}^\#$.*

We proceed immediately to a proof of Theorem 1.2.2. It clearly suffices to prove the theorem for nonnegative f. Accordingly, suppose that $f \in L_1(\nu_\Delta)$ and $f \geq 0$. Let $f_n = f \cdot \chi_{M_n}$. The sequence f_n is monotonically increasing and converges pointwise to the integrable function f, so

$$\int_M f_n\, d\nu_\Delta \longrightarrow \int_M f\, d\nu_\Delta, \tag{1.2.3}$$

that is,

$$\int_M |f_n - f_m|\, d\nu_\Delta \longrightarrow 0, \quad n, m \longrightarrow +\infty. \tag{1.2.4}$$

Since the support of f_n lies in the internal set M_n of finite measure, there exists by Theorem 0.4.17 an $\mathcal{S}$-integrable lifting F_n of f_n that is concentrated on M_n (since $F_n(\xi) = 0$ for $\xi \in X \setminus M_n$, condition 3) in Definition 1.2.1 holds), and the equality (1.2.2) is valid. Then it follows from (1.2.4) that $^\circ\|F_n - F_m\| \longrightarrow 0$ as $m, n \longrightarrow +\infty$. By Lemma 1.2.5 it now follows that there is an $F \in \mathcal{S}$ such that $^\circ\|F_n - F\| \longrightarrow 0$. Then $^\circ\|F_n\| \longrightarrow {}^\circ\|F\|$, and since (1.2.2) holds for $^\circ\|F_n\|$, (1.2.3) implies that (1.2.2) holds also for F. It remains to show that $f(\xi) = {}^\circ F(\xi)$ for ν_Δ-almost all ξ. Fix an arbitrary natural number $k > 0$. Then $^\circ\|F_n \cdot \chi_{M_k} - F \cdot \chi_{M_k}\| \longrightarrow 0$ as $n \to \infty$. If $n > k$, then $M_n \supseteq M_k$, that is, $f_n \cdot \chi_{M_k} = f_k \cdot \chi_{M_k}$. Consequently, $F_n \cdot \chi_{M_k} \approx F_k \cdot \chi_{M_k}$ ν_Δ-almost everywhere, and then also $F_n \cdot \chi_{M_k} - F \cdot \chi_{M_k} \approx F_k \cdot \chi_{M_k} - F \cdot \chi_{M_k}$ ν_Δ-almost everywhere. Since the functions in the last equality are $\mathcal{S}$-integrable and concentrated on the set M_k of finite measure, it follows that $^\circ\|F_n \cdot \chi_{M_k} - F \cdot \chi_{M_k}\| = {}^\circ\|F_k \cdot \chi_{M_k} - F \cdot \chi_{M_k}\|$. Passing in the last equality to the limit as $n \to \infty$, we get that $^\circ\|F_k \cdot \chi_{M_k} - F \cdot \chi_{M_k}\| = 0$ for any k. By Lemma 1.2.3 we now conclude that $F_k \cdot \chi_{M_k} \approx F \cdot \chi_{M_k}$ ν_Δ-almost everywhere. Consequently, $f \cdot \chi_{M_k} \approx F \cdot \chi_{M_k}$ ν_Δ-almost everywhere, and then $f \approx F|M$ ν_Δ-almost everywhere because $M = \bigcup_{k=0}^\infty M_k$.

Suppose now that $F \in \mathcal{S}$ and $f \colon M \longrightarrow \mathbb{R}$ is such that $f(\xi) = {}^\circ F(\xi)$ for almost all $\xi \in M$. We show that $f \in L_1(\nu_\Delta)$ and that (1.2.2) holds. Let $F_n = F \cdot \chi_{M_k}$. Then $f_n = f \cdot \chi_{M_k} \approx F_n$ ν_Δ-almost everywhere, and Theorem 0.4.17 is applicable since $\nu_\Delta(M_n)$ is finite, that is,

$$\int_M |f_n|\, d\nu_\Delta = {}^\circ\|F_n\| \leq {}^\circ\|F\|.$$

Since the sequence $|f_n|$ is monotonically increasing and converges to $|f|$, it follows from the monotone convergence theorem that $f \in L_1(\nu_\Delta)$. Then by what was proved above, there exists an internal function $G\colon X \longrightarrow {}^*\mathbb{R}$ that is $\mathcal{S}$-integrable and satisfies $\int_M |f_n|\,d\nu_\Delta = {}^\circ\|G\|$.

To prove (1.2.2) it remains to show that $\|F - G\| \approx 0$. First of all, if $A \in \mathfrak{A}$ and $A \subseteq M$, then in view of the ω_1-saturation of the nonstandard universe there exists an n with $A \subseteq M_n$. We fix such an A and a corresponding n. Let $F_M = F \cdot \chi_M$. Then $f \approx F_M$ ν_Δ-almost everywhere and $f \approx G_M$ ν_Δ-almost everywhere, so $F_{M_n} \approx G_{M_n}$ ν_Δ-almost everywhere, and then $\|F - G\|_A \approx 0$ because F and G are $\mathcal{S}$-integrable.

We consider the family $\Gamma_{m,n}(A) = \{M_n \subseteq A \wedge \|F - G\|_A \le m^{-1}\}$ of formulas. Since each finite subfamily of it is realized, the whole family is also realized, and hence $\exists A \in \mathfrak{A}\ (M \subseteq A \wedge \|F - G\|_A \approx 0)$. Then in view of the fact that $F, G \in \mathcal{S}$ we have that $\|F - G\|_{X\setminus A} \approx 0 \Longrightarrow \|F - G\| \approx 0$.

Now let $p \in [1, +\infty)$, and let $\mathcal{L}_p$ be the internal space of functions $F\colon X \longrightarrow {}^*\mathbb{R}$ with norm $\|F\|_p = {}^\circ(\Delta \sum_{\xi\in X} |F(\xi)|^p)^{1/p}$ (sometimes this space will be denoted by $\mathcal{L}^X_{p,\Delta}$, and the norm in it by $\|\cdot\|_{p,\Delta}$). For any $F, G \in \mathcal{L}_p$ we write $F \underset{p}{\sim} G$ if $\|F - G\|_p \approx 0$. The space $\mathcal{L}^\#_p$ is defined just as above. For any $A \in \mathfrak{A}$ we have $\|F\|_{p,A} = \|F \cdot \chi_A\|_p$. Denote by $\mathcal{S}_p(M)$ the subspace of $\mathcal{L}_p$ consisting of the $F \in \mathcal{L}_p$ such that $|F|^p$ is $\mathcal{S}_M$-integrable. If it does not lead to confusion, we write simply $\mathcal{S}_p$, omitting the M. Since $\|F\|_{p,A} = \||F|^p\|_A$ for any internal function F and any $A \in \mathfrak{A}$, Lemmas 1.2.3–1.2.5 remain valid when $\mathcal{L}$, $\mathcal{S}$, and $\|\cdot\|$ are replaced in them by $\mathcal{L}_p$, $\mathcal{S}_p$, and $\|\cdot\|_p$, respectively.

We can also consider the complex spaces $\mathcal{L}_p$ and $\mathcal{S}_p$ in a completely analogous way. Furthermore, if $F\colon X \longrightarrow \mathbb{C}$ is an internal function, then $F = \operatorname{Re} F + i \operatorname{Im} F$, and for any $A \in \mathfrak{A}$

$$\|\operatorname{Re} F\|_{p,A}, \|\operatorname{Im} F\|_{p,A} \le \|F\|_{p,A} \le \|\operatorname{Re} F\|_{p,A} + \|\operatorname{Im} F\|_{p,A};$$

from these inequalities it follows that $F \in \mathcal{L}_p\ (\mathcal{S}_p) \Longleftrightarrow \operatorname{Re} F, \operatorname{Im} F \in \mathcal{L}_p\ (\mathcal{S}_p)$. It is now easy to prove

COROLLARY 1.2.6. *If $f\colon M \longrightarrow K$, where $K = \mathbb{R}$ or $\mathbb{C}$, then $f \in \mathcal{L}_p(\Xi)$ if and only if there exists a lifting $F\colon X \longrightarrow {}^*K$ such that $F \in \mathcal{S}_p(M)$. Further, $\|f\|_p = {}^\circ\|F\|_p$.*

2° (IST). In this subsection it is shown that from any measure space $(\mathfrak{X}, \Omega, \mu)$ with a σ-finite measure μ we can construct a Loeb space $(X, S_\Delta, \nu_\Delta)$ with $X \subseteq {}^*\mathfrak{X}$ and a σ-finite subspace $(M, S^M_\Delta, \nu^M_\Delta)$ of it such that for any $p \in [1, \infty)$ there exists a norm-preserving imbedding $j_p\colon L_p(\mu) \longrightarrow L_p(\nu^M_\Delta)$. Further, for any function $f \in L_p(\mu)$ the internal function $F = {}^*f|X$ is in $\mathcal{S}_p(M)$ and is a lifting of the function $j_p(f)$. This implies, in particular, that for any $f \in L_p(\mu)$ (more precisely, any representative of the class f)

$$\int_{\mathfrak{X}} f\,d\mu = {}^\circ\Big(\Delta \sum_{\xi\in X} {}^*f(\xi)\Big) \tag{1.2.5}$$

(see Theorem 1.2.2).

Let τ be an arbitrary element of ${}^*\mathbb{R}$ (recall that in this subsection ${}^*\mathbb{R}$ satisfies (IST), and let (Y, Σ, λ) be a τ-standard space with σ-additive probability measure λ.

The concept of a random element extends to τ-standard spaces (see Theorem 0.4.23 and the remark after it). An element $y \in Y$ is said to be τ-random if for any τ-standard set $A \in \Sigma$ with $\lambda(A) = 0$ we have that $y \notin A$. In particular, if the number

τ is standard, that is, (Y, Σ, λ) is a standard probability space, then an element $y \in {}^*Y$ is random in the previous sense. Of course, τ-random elements are defined also in any standard space (Y, Σ, λ) for nonstandard τ. For τ-standard measure spaces Theorem 0.4.23 remains true, because its proof uses only the concurrence principle, which is true also for τ-standard sets in view of the relativized idealization principle I_τ. Thus, we have

PROPOSITION 1.2.7. *There exists an internal set $B \in \Sigma$ of full measure whose elements are all τ-random.*

Suppose now that (Y, Σ, λ) is the product of the τ-standard probability spaces $(Y_1, \Sigma_1, \lambda_1)$ and $(Y_2, \Sigma_2, \lambda_2)$. If $y = \langle y_1, y_2 \rangle$ is a τ-random element of Y, then it is easy to see that y_i is a τ-standard element of Y_i. The converse is false, of course. For example, if $y_1 = y_2$ and the measure of the diagonal in Y is equal to zero, then the element $\langle y_1, y_2 \rangle$ is not τ-random, not even for τ-random y_1, because it belongs to a τ-standard set of measure zero (the diagonal).

PROPOSITION 1.2.8. *If y_1 is a τ-random element of Y_1 and y_2 a $\langle \tau, y_1 \rangle$-random element of Y_2, then $\langle y_1, y_2 \rangle$ is a τ-random element of $Y = Y_1 \times Y_2$.*

PROOF. Let $A \subseteq Y_1 \times Y_2$ be a τ-standard set with $\lambda(A) = 0$. Then $\forall z_1 \in Y_1$ $(A_{z_1} = \{z_2 \in Y_2 \mid \langle z_1, z_2 \rangle \in A\} \operatorname{st}\langle \tau, z_1 \rangle)$. Let $C = \{z_1 \in Y \mid \lambda_2(A_{z_1}) = 0\}$. Then C is τ-standard and $\lambda_1(C) = 1$ by Fubini's theorem. Consequently, $y_1 \in C$, so $\lambda_2(A_{y_1}) = 0$, and $y_2 \notin A_{y_1}$ because $A_{y_1} \operatorname{st}\langle \tau, y_1 \rangle$, that is, $\langle y_1, y_2 \rangle \notin A$. □

DEFINITION 1.2.9. If (Y, Σ, λ) is a τ-standard probability space, then an internal sequence $\langle y_n \mid n \in {}^*\mathbb{Z} \rangle$ is called an independent sequence of τ-random elements of Y if it is a τ-random element of the τ-standard space $Y^{{}^*\mathbb{Z}}$ with the product measure λ^∞.

THEOREM 1.2.10. *If $\langle y_n \mid n \in {}^*\mathbb{Z} \rangle$ is an independent sequence of τ-random elements of Y, then for any τ-standard function $f \colon Y \longrightarrow {}^*\mathbb{R}$ with $f \in L_1(\lambda)$*

$$\int_Y f \, d\lambda = \lim_{n \to \infty} \frac{1}{n} \sum_{k=0}^{n-1} f(y_k). \tag{1.2.6}$$

The proof of the theorem is based on the following 'standard' lemma relating to ergodic theory.

LEMMA 1.2.11. *Let (Y, Σ, λ) be a probability space, and λ^∞ the product measure on $Y^{\mathbb{Z}}$. For any integrable function $f \colon Y \longrightarrow \mathbb{R}$ let $\mathcal{A}_f$ be the set of $\langle y_n \mid n \in \mathbb{Z} \rangle \in Y^{\mathbb{Z}}$ for which* (1.2.6) *holds. Then $\lambda^\infty(\mathcal{A}_f) = 1$.*

PROOF OF THE LEMMA. Denote by $T \colon Y^{\mathbb{Z}} \longrightarrow Y^{\mathbb{Z}}$ the Bernoulli shift $T(\langle y_n \mid n \in \mathbb{Z} \rangle) = \langle y'_n \mid n \in \mathbb{Z} \rangle$, where $y'_n = y_{n+1}$. It is known that T is ergodic ([**21**], Chapter 8, §1, Theorem 1), that is, for any $\varphi \in L_1(\lambda^\infty)$

$$\int_{Y^{\mathbb{Z}}} \varphi \, d\lambda^\infty = \lim_{n \to \infty} \frac{1}{n} \sum_{n=0}^{n-1} \varphi(T^k \overline{y}) \tag{1.2.7}$$

holds for almost all $\overline{y} \in Y^{\mathbb{Z}}$. Let $\Pi_0 \colon Y^{\mathbb{Z}} \longrightarrow Y$ be the mapping $\Pi_0(\langle y_n \mid n \in \mathbb{Z} \rangle) = y_0$. By the definition of the measure λ^∞, Π_0 is measure-preserving, and hence if

$\varphi = f \circ \Pi_0$, then $\int_{Y^{\mathbb{Z}}} \varphi \, d\lambda^\infty = \int_Y f \, d\lambda$. Moreover, the right-hand side of (1.2.7) clearly coincides with the right-hand side of (1.2.6), and this proves the lemma. □

PROOF OF THEOREM 1.2.10. Applying the relativized transfer principle T_τ to Lemma 1.2.11, we conclude that the τ-standard set $\mathcal{A}_f$ has full measure, and hence any independent sequence $\langle y_n \mid n \in \mathbb{Z} \rangle$ of τ-standard elements is in $\mathcal{A}_f$, that is, (1.2.6) is satisfied. □

THEOREM 1.2.12. *If (Y, Σ, δ) is a τ-standard measure space with finite measure δ,*[1] *then there exists an internal hyperfinite set $Y_0 \subseteq Y$ such that for any τ-standard integrable function $f \colon Y \longrightarrow {}^*\mathbb{R}$*

$$\int_Y f \, d\delta \overset{\tau}{\approx} \frac{\delta(Y)}{|Y_0|} \sum_{y \in Y_0} f(y). \tag{1.2.8}$$

PROOF. We pass from the space (Y, Σ, δ) to the probability space (Y, Σ, λ) by setting $\lambda = \frac{1}{\delta(Y)} \delta$. Then

$$\int_Y f \, d\delta = \delta(Y) \int_Y f \, d\lambda. \tag{1.2.9}$$

If $\overline{y} = \langle y_n \mid n \in {}^*\mathbb{Z} \rangle$ is an independent sequence of τ-random elements of (Y, Σ, λ), then since the sequence on the right-hand side of (1.2.6) is $\langle \tau, \overline{y} \rangle$-standard, we get by using Theorem 1.1.7, assertion 3 and taking a $\langle \tau, \overline{y} \rangle$-infinite hypernatural number N that

$$\int_Y f \, d\lambda \overset{\langle \tau, \overline{y} \rangle}{\approx} \frac{1}{N} \sum_{k=0}^{N-1} f(y_k).$$

It now remains to let $Y_0 = \{y_0, \ldots, y_{N-1}\}$, and use (1.2.9) and the fact that $\alpha \overset{\langle \tau, \overline{y} \rangle}{\approx} \beta \Longrightarrow \alpha \overset{\tau}{\approx} \beta$ (Proposition 1.1.7). □

Let us now return to the standard space $(\mathfrak{X}, \Omega, \mu)$ with σ-finite measure μ. There exists an increasing sequence of sets $\mathfrak{X}_n \in \Omega$ such that $\forall n \in \omega \; (\mu(\mathfrak{X}_n) < \infty)$ and $\mathfrak{X} = \bigcup_{n \in \omega} \mathfrak{X}_n$. If Ω_n denotes the σ-algebra $\{A \cap \mathfrak{X}_n \mid A \in \Omega\}$, and μ_n the restriction of μ to Ω_n, then for any integrable function $f \colon \mathfrak{X} \longrightarrow \mathbb{R}$

$$\int_{\mathfrak{X}} f \, d\mu = \lim_{n \to \infty} \int_{\mathfrak{X}_n} f_n \, d\mu_n, \tag{1.2.10}$$

where $f_n = f | \mathfrak{X}_n$.

THEOREM 1.2.13. *There exists an internal hyperfinite set $X \subseteq {}^*\mathfrak{X}$ and a $\Delta \in {}^*\mathbb{R}$ such that* (1.2.5) *holds for any standard function $f \in L_1(\mu)$.*

---

[1]The condition that δ be finite means only that $\delta(Y) \neq \infty$, but is a τ-standard hyperreal number, possibly infinite.

PROOF. Let τ be an infinite hypernatural number. Then it follows from (1.2.10) that

$$\int_{\mathfrak{X}} f\,d\mu = \int_{{}^*\mathfrak{X}_n} {}^*f_\tau\,d\mu_\tau. \tag{1.2.11}$$

Let $(Y,\Sigma,\delta) = ({}^*\mathfrak{X}_\tau, {}^*\Omega_\tau, {}^*\mu_\tau)$. Then (Y,Σ,δ) satisfies the conditions of Theorem 1.2.12. The function ${}^*f_\tau$ is a τ-standard integrable function on ${}^*\mathfrak{X}_\tau$. Since $Y_0 \subseteq {}^*\mathfrak{X}_\tau$, it follows that ${}^*f_\tau|Y_0 = {}^*f|Y_0$. We now deduce (1.2.5) from (1.2.8) and (1.2.11) by setting $\Delta = \delta(Y)|Y_0|^{-1}$ and $X = Y_0$ and using the fact that $\int_{\mathfrak{X}} f\,d\mu$ is a standard number. □

It follows immediately from (1.2.5) that for any standard $A \in \Omega$

$$\mu(A) = {}^\circ(\Delta \cdot |X \cap {}^*A|) \tag{1.2.12}$$

(if $t \in {}^*\mathbb{R}$ and $t \sim +\infty$, then ${}^\circ t = +\infty$).

Now let $M_n = X \cap {}^*\mathfrak{X}_n$ $\forall n \in \omega$, and $M = \bigcup_{n\in\omega} M_n$. Then $\Xi = (M, S_\Delta^M, \nu_\Delta^M)$ is a σ-finite subspace of the Loeb space $(X, S_\Delta, \nu_\Delta)$ in view of (1.2.12) and the fact that $\mu({}^*\mathfrak{X}_n) < +\infty$.

THEOREM 1.2.14. *Let $(\mathfrak{X},\Omega,\mu)$ be a standard measure space with a σ-finite measure, where $\mathfrak{X} = \bigcup_{n\in\omega} \mathfrak{X}_n$ and $\mu(\mathfrak{X}_n) < +\infty$ $\forall n \in \omega$. Suppose that $X \subseteq {}^*\mathfrak{X}$ and $\Delta \in {}^*\mathbb{R}$ satisfy the conditions of Theorem 1.2.13, and let $M_n = X \cap {}^*\mathfrak{X}_n$ and $M = \bigcup_{n\in\omega} M_n$. Then for any $p \in [1,+\infty)$ and any $f \in L_p(\mu)$ the internal function $F(f) = {}^*f|X$ is in $\mathcal{S}_p(M)$, and if $j_p(f) = {}^\circ F(f)$, then $j_p\colon L_p(\mu) \longrightarrow L_p(\nu_\Delta^M)$ is a norm-preserving imbedding. In particular, if $f \in L_1(\mu)$, then*

$$\int_{\mathfrak{X}} f\,d\mu = \int_M j_1(f)\,d\nu_\Delta^M. \tag{1.2.13}$$

Before proving the theorem, we present some needed facts from the theory of complete normalized Boolean algebras. All these facts can be found, for example, in the book [**2**].

Let $\mathbb{B}$ be a Boolean algebra, and let m be a strictly positive finitely additive finite measure on $\mathbb{B}$. Then $\mathbb{B}$ satisfies the *countable chain condition* (CCC): each subset $E \subseteq \mathbb{B}$ consisting of disjunct elements is at most countable. It is easy to see that a σ-algebra satisfying the CCC is complete. A complete Boolean algebra with a strictly positive countably additive measure is said to be *normalized* (a CNBA). Let $(\mathbb{B}_1, m_1)$ and $(\mathbb{B}_2, m_2)$ be two CNBA's. It is known that any measure-preserving (and thus injective) homomorphism $\varphi\colon \mathbb{B}_1 \longrightarrow \mathbb{B}_2$ is completely additive, that is, $\varphi(\sup E) = \sup \varphi(E)$ for any $E \subseteq \mathbb{B}_1$. This implies that $\varphi(\mathbb{B}_1)$ is a true subalgebra of $\mathbb{B}_2$, that is, $\sup^{\varphi(\mathbb{B}_1)} E = \sup^{\mathbb{B}_2} E$ for all $E \subseteq \varphi(\mathbb{B}_1)$.

Suppose now that (Y,Σ,δ) is a measure space with finite measure, and let $n(\Sigma) = \{c \in \Sigma \mid \delta(c) = 0\}$. Then $\overline{\Sigma} = \Sigma/n(\Sigma)$ is a CNBA with strictly positive measure $\overline{\delta}$ such that $\overline{\delta}([c]) = \delta(c)$, where $[c]$ is the class of an element $c \in \Sigma$ in $\overline{\Sigma}$. The CNBA $\overline{\Sigma}$ will be called the Lebesgue algebra of the space (Y,Σ,δ). With each measurable function $f\colon Y \longrightarrow \mathbb{R}$ we can associate a monotonically increasing right-continuous family $\{e_f^t \mid t \in \mathbb{R}\}$ of elements in $\overline{\Sigma}$ called the resolution of the identity for f and defined by $e_f^t = [\{y \mid f(y) \le t\}]$. Note that $\sup\{e_f^t\} = \mathbf{1}_{\mathbb{B}}$ and

$\inf\{e_f^t\} = \mathbf{0}_{\mathbb{B}}$. It is known that $f \in L_1(\delta)$ if and only if the integral $\int_{-\infty}^{\infty} t\, d\bar{\delta}(e_f^t)$ converges, and

$$\int_Y f\, d\delta = \int_{-\infty}^{\infty} t\, d\bar{\delta}(e_f^t). \tag{1.2.14}$$

LEMMA 1.2.15. *If (Y, Σ, δ) is a standard measure space with finite measure, $Y_0 \subseteq {}^*Y$ satisfies the conditions of Theorem 1.2.12 (with standard τ), $\lambda = \delta(Y) \cdot |Y_0|^{-1}$, and $(Y_0, S_\lambda, \nu_\lambda)$ is the corresponding Loeb space, then the mapping $\psi\colon \overline{\Sigma} \longrightarrow \overline{S}_\lambda$ defined by $\psi([c]) = [{}^*c \cap Y_n]$ $\forall c \in \Sigma$ is a measure-preserving monomorphism.*

The proof of the lemma follows immediately from the formula (1.2.8), applied to the characteristic functions of sets in Σ.

The next lemma contains a proof of Theorem 1.2.14 for the case of finite measures. In this case the concept of $\mathcal{S}_M$-integrability coincides with the concept of $\mathcal{S}$-integrability (Definition 0.4.14).

LEMMA 1.2.16. *Under the conditions of Lemma 1.2.15 suppose that $h \in L_1(\delta)$, $H = {}^*h|Y_0$, and $\widetilde{h}\colon Y_0 \longrightarrow \overline{\mathbb{R}}$ is such that $\widetilde{h}(y) = {}^{\circ *}h(y)$ $\forall y \in Y_0$. Then: 1) $\widetilde{h} \in L_1(\nu_\lambda)$; 2) $\int_Y h\, d\delta = \int_{Y_0} \widetilde{h}\, d\nu_\lambda$; 3) H is $\mathcal{S}$-integrable.*

PROOF. For any standard $t \in \mathbb{R}$ let $C_t = \{y \in Y \mid h(y) \leq t\}$, $c^t = [C_t] \in \overline{\Sigma}$, $E_t = \{y \in Y_0 \mid \widetilde{h}(y) \leq t\}$, and $e^t = [E_t] \in \overline{S}_\lambda$. Then $\{c^t \mid t \in \mathbb{R}\}$ is the resolution of the identity of the function h in $\overline{\Sigma}$, and $\{e^t \mid t \in \mathbb{R}\}$ is the resolution of the identity of the element $\widetilde{h}$. Since the $\psi\colon \overline{\Sigma} \longrightarrow \overline{S}_\lambda$ defined in Lemma 1.2.15 preserves bounds, we get by setting $\widetilde{e}^t = \psi(c^t)$ that $\{\widetilde{e}^t\}$ is a resolution of the identity. It follows from the transfer principle that $\widetilde{e}^t = \{y \in Y_0 \mid {}^*h(y) \leq t\}$. Using the definition of $\widetilde{h}$, we get that $e^{t_1} < \widetilde{e}^{t_2}$ and $\widetilde{e}^{t_1} < e^{t_2}$ for any standard $t_1 < t_2$. This and the right-continuity of the families $\{e^t\}$ and $\{\widetilde{e}^t\}$ imply that $e^t = \widetilde{e}^t$ for any t. We now get the assertions 1) and 2) of the lemma from the equality (1.2.14) and the fact that ψ is measure-preserving. Assertion 3) follows from Theorem 0.4.15 (b). □

PROOF OF THEOREM 1.2.14. It suffices to prove the theorem for $p = 1$ and $f \geq 0$.

We consider the Loeb space $(M_n, S_{\Delta_n}, \nu_{\Delta_n})$, where $\Delta_n = \mu(\mathfrak{X}_n) \cdot |M_n|^{-1} \approx \Delta \cdot |M_n| \cdot |M_n|^{-1} = \Delta$. This is a measure space with finite measure, and the conditions of Lemma 1.2.16 hold if $(\mathfrak{X}_n, \Omega_n, \mu_n)$ is taken in place of (Y, Σ, δ) and M_n in place of Y_0. Let $f_n = f \cdot \chi_{\mathfrak{X}_n}\colon \mathfrak{X} \longrightarrow \overline{\mathbb{R}}$ and let $\overline{f}_n = f|\mathfrak{X}_n$, that is, $\overline{f}_n\colon \mathfrak{X}_n \longrightarrow \mathbb{R}$ and $\overline{f}_n = f|\mathfrak{X}_n$. By Lemma 1.2.16, ${}^*f|M_n$ is an $\mathcal{S}$-integrable lifting of the function ${}^{\circ *}f|M_n$ and

$$\int_{\mathfrak{X}_n} \overline{f}_n\, d\mu_n = \int_{M_n} {}^{\circ *}f|M_n\, d\nu_{\Delta_n} = {}^{\circ}\Big(\Delta_n \sum_{\xi \in M_n} {}^*\overline{f}_n(\xi)\Big) = {}^{\circ}\Big(\Delta_n \sum_{\xi \in X} {}^*f_n(\xi)\Big).$$

The last equality holds because $\frac{\Delta}{\Delta_n} \approx 1$ and ${}^*f_n(\xi) = 0$ for $\xi \in X \setminus M_n$. This implies also that ${}^*f_n|X$ is an $\mathcal{S}_M$-integrable lifting of $j_1(f_n)$, and (1.2.13) holds for f_n. Since $j_1(f_n) \longrightarrow j_1(f)$ is monotonically increasing and $\int_{\mathfrak{X}} f_n\, d\mu \longrightarrow \int_{\mathfrak{X}} f\, d\mu$, it follows from (1.2.13) for f_n that $j_1(f) \in L_1(\nu_\Delta)$ and f satisfies (1.2.13). By Theorem 1.2.13, ${}^{\circ}(\Delta \sum_{\xi \in X} |{}^*f_n(\xi) - {}^*f(\xi)|) = \int_{\mathfrak{X}} |f_n - f|\, d\mu \longrightarrow 0$. Using the fact that $\mathcal{S}(M)^{\#}$ is closed in $\mathcal{L}^{\#}$ (Lemma 1.2.5), we now conclude that ${}^*f|X$ is in $\mathcal{S}(M)$, and hence is an $\mathcal{S}_M$-integrable lifting of the function $j_1(f)$. □

3°. For many concrete spaces (X, Ω, μ) with a σ-finite measure there are other imbeddings of $L_p(\mu)$ in $L_p(\nu_\Delta^M)$ besides the one described in the preceding subsection for some σ-finite subspace $(M, S_\Delta^M, \nu_\Delta^M)$ of a suitable Loeb space $(X, S_\Delta, \nu_\Delta)$. Most of them are based on a construction of a measure-preserving mapping $\varphi\colon M \longrightarrow \mathfrak{X}$. In this case φ induces for any $p \in [1, \infty)$ a mapping $j_p^\varphi\colon L_p(\mu) \longrightarrow L_p(\nu_\Delta^M)$ defined by the formula $j_p^\varphi(f) = f \circ \varphi\ \forall f \in L_p(\mu)$. By Theorem 1.2.2 and its Corollary 1.2.6, $j_p^\varphi(f)$ has a lifting $F \in \mathcal{S}_p(M)$ which we call the lifting of f. In particular, if $f \in L_1(\mu)$ and F is an $\mathcal{S}_M$-integrable lifting of f, then

$$\int_{\mathfrak{X}} f\, d\mu = {}^\circ\Big(\Delta \sum_{\xi \in X} F(\xi)\Big). \tag{1.2.15}$$

Here the problem of constructing F from f is of interest. This problem was solved in an especially simple way in the preceding subsection, where $X \subseteq {}^*\mathfrak{X}$ and $\forall f \in L_1(\mu)$

$$F = {}^*f|X. \tag{1.2.16}$$

In the general case the equality (1.2.16) fails to hold even when $X \subseteq {}^*\mathfrak{X}$. In this subsection we consider a type of imbedding for which (1.2.16) holds for a sufficiently broad class of functions. We first consider a well-known example.

EXAMPLE 1. Let $X = [0, 1]$, and let μ be Lebesgue measure. We fix an arbitrary $\Delta \approx 0$, and let $N = [\Delta^{-1}]$ and $X = \{k\Delta \mid k = 1, \ldots, N\}$. In this case the Loeb space $(X, S_\Delta, \nu_\Delta)$ is a measure space with finite measure: $\nu_\Delta(X) = 1$, and therefore $M = X$. As a mapping $\varphi\colon X \longrightarrow \mathfrak{X}$ we take st (recall that $\mathrm{st}(k\Delta) = {}^\circ(k\Delta)$). It can be proved that, $\forall \mathcal{A} \subseteq [0, 1]$, $\mathcal{A}$ is Lebesgue-measurable if and only if $\mathrm{st}^{-1}(\mathcal{A})$ is Loeb-measurable, and $\mu(\mathcal{A}) = \nu_\Delta(\mathrm{st}^{-1}(\mathcal{A}))$. For an arbitrary Lebesgue-integrable function f the equality (1.2.16) does not hold here, as is easily seen by assuming that $\Delta \in {}^*\mathbb{Q}$ and considering the Dirichlet function. But if f is Riemann-integrable on $[0, 1]$ and F satisfies (1.2.16), then (1.2.15) holds in view of the corollary to Theorem 0.3.21. We show that in this case $F = {}^*f|X$ really is an $\mathcal{S}$-integrable lifting of f. Since f is bounded and the internal (in the sense of nonstandard analysis) uniform measure ν_Δ is finite, it follows that F satisfies the condition b) of Theorem 0.4.15, and hence is $\mathcal{S}$-integrable. If $\mathcal{A}$ is the set of points of discontinuity of f, then $\mu(\mathcal{A}) = 0$, because f is Riemann-integrable. If $k\Delta \in X \setminus \mathrm{st}^{-1}(\mathcal{A})$, then f is continuous at the point ${}^\circ(k\Delta)$, so that ${}^*f(k\Delta) \approx f({}^\circ(k\Delta))$. Thus, ${}^\circ F(n) = f(\mathrm{st}(n))$ for almost all $n \in X$, that is, F is a lifting of the function $f \circ \mathrm{st}$, and this means that F is a lifting of f.

Suppose now that $\mathfrak{X}$ is a separable locally compact Hausdorff topological space, μ is a Borel measure on $\mathfrak{X}$ that is finite on compact sets (μ is regular because $\mathfrak{X}$ is separable), and Ω is the completion of the σ-algebra of Borel sets with respect to μ. Assume that $\mathfrak{X} = \bigcup_{n \in \omega} \mathfrak{X}_n$, where $\mathfrak{X}_n$ is compact and $\mu(\mathfrak{X}_n) < \infty\ \forall n \in \omega$. It is then clear that $\mathrm{Ns}({}^*\mathfrak{X}) = \cup\, {}^*\mathfrak{X}_n$. Recall that the mapping $\mathrm{st}\colon \mathrm{Ns}({}^*\mathfrak{X}) \longrightarrow \mathfrak{X}$ is defined from the condition $\mathrm{st}(x) \approx x\ \forall x \in \mathrm{Ns}({}^*\mathfrak{X})$ (see §4, 2° of Chapter 0).

DEFINITION 1.2.17. Suppose that X is a hyperfinite set, $j\colon X \longrightarrow {}^*\mathfrak{X}$ is an internal mapping, $\Delta \in {}^*\mathbb{R}$, and $M = j^{-1}(\mathrm{Ns}({}^*\mathfrak{X}))$. The triple (X, j, Δ) is called a hyperfinite realization (HR) of the space $(\mathfrak{X}, \Omega, \mu)$ if the mapping $\varphi\colon (M, S_\Delta^M, \nu_\Delta^M) \longrightarrow (\mathfrak{X}, \Omega, \mu)$ defined by $\varphi = \mathrm{st} \circ j|M$ is measurable and measure-preserving (note that

$M = \bigcup_{n\in\omega} j^{-1}({}^*\mathfrak{X}_n)$, so that in this case $(M, S^M_\Delta, \nu^M_\Delta)$ is a σ-finite subspace of the Loeb space $(X, S_\Delta, \nu_\Delta)$).

PROPOSITION 1.2.18. *Let (X, j, Δ) be an HR of the space $(\mathfrak{X}, \Omega, \mu)$. Then any μ-integrable bounded function $f\colon \mathfrak{X} \longrightarrow \mathbb{R}$ that is continuous μ-almost everywhere satisfies the condition*

$$\forall B \in {}^*\mathcal{P}(X) \left(B \subseteq X \setminus M \Longrightarrow \Delta \sum_{x\in B} |{}^*f(j(x))| \approx 0\right), \tag{1.2.17}$$

the function $F = {}^*f \circ j$ is an $\mathcal{S}_M$-integrable lifting of f, and hence

$$\int_{\mathfrak{X}} f\, d\mu = {}^\circ\left(\Delta \sum_{x\in X} {}^*f(j(x))\right). \tag{1.2.18}$$

PROOF. First of all we show that F is $\mathcal{S}_M$-integrable. Condition 2) of Definition 1.2.1 holds because f is bounded, and condition 3) follows from the condition (1.2.17). To verify condition 1) we observe that ${}^*f \circ j | j^{-1}({}^*\mathfrak{X}_n)$ is an $\mathcal{S}$-integrable function (below, $j^{-1}({}^*\mathfrak{X}_n) = M_n$). This follows from the boundedness of f and the finiteness of $\nu_\Delta(M_n)$ (see condition b) of Theorem 0.4.15). An argument analogous to that at the end of Example 1 now shows that ${}^*f \circ j | M_n$ is a lifting of the function $f|\mathfrak{X}_n$, and thus by (1.2.15),

$$\int_{\mathfrak{X}_n} |f|\, d\mu = {}^\circ\left(\Delta \sum_{x\in M_n} |{}^*f \circ j|(x)\right) \leq \int_{\mathfrak{X}} |f|\, d\mu.$$

This implies that there is a standard constant C such that for any internal $\mathcal{D} \subseteq \bigcup_{n\in\omega} M_n$

$$\Delta \sum_{x\in\mathcal{D}} |F|(x) < C. \tag{1.2.19}$$

Using the countable saturation, we now find an internal $\mathcal{D} \supset M$ satisfying (1.2.19). Condition 1) of Definition 1.2.1 follows from the relation (1.2.17), applied to $B = X \setminus \mathcal{D}$. Since ${}^*f \circ j | M_n$ is a lifting of $f|\mathfrak{X}_n$ for any $n \in \omega$, ${}^*f \circ j$ is a lifting of f. $\square$

REMARK. Proposition 1.2.18 implies the analogous assertion for a bounded function $f \in L_p(\mu)$ that is μ-almost everywhere continuous, $p \in [1, \infty)$.

EXAMPLE 2. Let $\mathfrak{X} = \mathbb{R}$, let Ω be the σ-algebra of Lebesgue-measurable subsets of $\mathbb{R}$, and let μ be Lebesgue measure on $\mathbb{R}$. We choose a hypernatural $N \in {}^*\omega \setminus \omega$ and a $\Delta \approx 0$ such that $N\Delta \sim +\infty$. For convenience of notation we assume that $N = 2L + 1$ and consider the hyperfinite set $X = \{k\Delta \mid k = -L, \dots, L\}$. Here it can be assumed that $\mathfrak{X}_n = [-n, n]$, and $j\colon X \subseteq {}^*\mathbb{R}$ is an imbedding. Then $M_n = X \cap {}^*[-n, n]$, and $M = X \cap \bigcup_{n\in\omega} {}^*[-n, n]$. Here it is easy to see that the condition (1.2.17) can be written in the form

$$\forall k, l \left(|k| < |l| < L \ \wedge\ |k|\Delta \sim +\infty \Longrightarrow \Delta \sum_{n=k}^{l} |{}^*f(n\Delta)| \approx 0\right). \tag{1.2.20}$$

Since the last relation must be true for any L with $L\Delta \sim +\infty$, (1.2.20) is equivalent to the standard condition

$$\lim_{\substack{\Delta\to 0\\ A\to\infty}} \Delta \sum_{|k|>\frac{A}{\Delta}} |f(k\Delta)| = 0. \tag{1.2.21}$$

For a function f absolutely Riemann-integrable on $\mathbb{R}$ the last condition is equivalent to

$$\int_{-\infty}^{\infty} f(x)\,dx = \lim_{h\to 0} h \sum_{k=-\infty}^{\infty} f(kh). \tag{1.2.22}$$

It is known that the class of such functions is sufficiently broad.

The condition (1.2.17) in Proposition 1.2.18 is a condition of sufficiently rapid decrease of functions at infinity. It holds automatically for functions with compact support, and it is superfluous for spaces with finite measure. A deficiency of this condition is that it is formulated in terms depending on the HR, but the preceding example shows that in concrete cases it can be reformulated in standard terms independent of the HR. Moreover, it often happens that an HR can be chosen so that the condition (1.2.17) is superfluous. Let us return to Example 2 and choose L and Δ in the following way. We fix an infinite $\tau \in {}^*\mathbb{R}$, and take $\Delta \overset{\tau}{\approx} 0$ and $L = \left[\frac{\tau}{\Delta}\right]$. As above, the triple (X, j, Δ) is an HR of the space $(\mathfrak{X}, \Omega, \mu)$, where $\mathfrak{X} = \mathbb{R}$, Ω is the σ-algebra of Lebesgue-measurable sets, and μ is Lebesgue measure. The corollary to Theorem 1.1.13 shows that (1.2.18) holds now for any bounded function that is absolutely Riemann-integrable on $\mathbb{R}$ and continuous almost everywhere. If f is such a function, then it is easy to prove that ${}^*f \circ j$ is $\mathcal{S}_M$-integrable by using the $\mathcal{S}$-integrability of ${}^*f|{}^*[-n, n] \circ j$ $\forall n \in \omega$, the equality (1.2.18), and the closedness of $\mathcal{S}_M^{\#}$ in $\mathcal{L}^{\#}$ (see the proof of Theorem 1.2.14).

The main results of this section, including the concepts of a σ-finite subspace of a Loeb space and $\mathcal{S}_M$-integrability, are due to the author. The fact of the matter is that up to the present time the most substantial applications of Loeb measures have been in problems relating to probability theory, and in this connection most attention has been focused on the study of finite Loeb measures. Finite Radon measures induced by Loeb measures and the mapping st have been studied in [**30**] and other publications (see the survey [**35**] mentioned above and the book [**1**]). However, the question of when $j \circ {}^*f|X$ is a lifting of f was not treated there. A different construction of a lifting from a function in the case of the interval $[0, 1]$ is found in [**35**]. The treatment of σ-finite Loeb measures is essential for the present book, because in the next chapter Loeb measures are applied to the study of Haar measures on locally compact Abelian groups, and such measures are not finite.

Solovay introduced the concept of a random number as a number not lying in any set of measure zero having a constructive description in the Gödel sense, and he used it for proving that various propositions in measure theory are independent of the axioms of ZFC. In the Kolmogorov theory of complexity there is the analogous concept (due to Martin-Löf) of a random 0–1 sequence as a sequence not lying in any set of measure zero having a constructive description in the Markov sense. An analogous concept of a random element for $[0, 1]$ with Lebesgue measure and of an independent sequence of random elements in this case was introduced in [**54**], where Theorem 1.2.12 was proved for this case. The proof used the law of large numbers. For a standard space with finite measure Theorem 1.2.12 was proved in [**37**] using quite different arguments.

§3. Hyperfinite approximations of integral operators

1°. We recall that for hyperfinite X, standard $p \in [1, \infty]$, and $\Delta \in {}^*\mathbb{R}_+$ the internal space of functions $F \colon X \longrightarrow {}^*\mathbb{K}$ ($\mathbb{K} = \mathbb{R}$ or $\mathbb{C}$) with the norm $\|F\|_{p,\Delta} = {}^\circ(\Delta \sum_{\xi \in X} |F(\xi)|^p)^{1/p}$ is denoted by $\mathcal{L}^X_{p,\Delta}$. This space is hyperfinite-dimensional: $\dim \mathcal{L}^X_{p,\Delta} = |X|$. It determines the external Banach space $\mathcal{L}^{X\#}_{p,\Delta}$, which is nonseparable if $|X| \in {}^*\omega \setminus \omega$. For $p = 2$, $\|F\|_{2,\Delta}$ is defined in terms of the inner product $(F, G) = \Delta \sum_{\xi \in X} F(\xi)\overline{G}(\xi)$, and hence $\mathcal{L}^{X\#}_{2,\Delta}$ is a nonseparable Hilbert space. As above, we write simply $\mathcal{L}_p$ instead of $\mathcal{L}^X_{p,\Delta}$ if doing so does not cause confusion.

If $(M, S^M_\Delta, \nu^M_\Delta)$ is a σ-finite subspace of the Loeb space $(X, S_\Delta, \nu_\Delta)$, then $\mathcal{S}_p(M)^\#$ is a closed subspace of $\mathcal{L}^\#_p$ (Lemma 1.2.5). Corollary 1.2.6 shows that this subspace is isomorphic to $L_p(\nu^M_\Delta)$. An isomorphism is established by associating with each function $f \in L_p(\nu^M_\Delta)$ the class $F^\#$ of its lifting $F \in \mathcal{S}_p(M)$. Therefore, it is simply assumed below that $L_p(\nu^M_\Delta) \subseteq \mathcal{L}^\#_p$.

Suppose now that $(\mathfrak{X}_i, \Omega_i, \mu_i)$ $(i = 1, 2)$ are standard measure spaces with σ-finite measures, and suppose we have two Loeb spaces $(X_i, S_{\Delta_i}, \nu_{\Delta_i})$, σ-finite subspaces $(M_i, S^{M_i}_{\Delta_i}, \nu^{M_i}_{\Delta_i})$ of them, and imbeddings $j^{M_i}_{p_i} \colon L_{p_i}(\mu_i) \longrightarrow L_{p_i}(\nu^{M_i}_{\Delta_i}) \subseteq \mathcal{L}^\#_{p_i}$ for some $p_1, p_2 \in [1, +\infty]$.

DEFINITION 1.3.1. Let $\mathcal{A} \colon L_{p_1}(\mu_1) \longrightarrow L_{p_2}(\mu_2)$ be a bounded linear operator. An internal operator $A \colon \mathcal{L}_{p_1} \longrightarrow \mathcal{L}_{p_2}$ with finite norm is called a *hyperfinite-dimensional approximation* of the operator $\mathcal{A}$ if for any $f \in \mathcal{L}_{p_1}(\mu_1)$ we have that $\|G - A(F)\|_{p_2,\Delta_2} \approx 0$ when $F \in \mathcal{L}_{p_1}$ is a lifting of f and $G \in \mathcal{L}_{p_2}$ a lifting of $\mathcal{A}(f)$.

If, as in §2, 3°, the imbedding j^M_p is induced by some internal mapping $j \colon X \longrightarrow {}^*\mathfrak{X}$, and the lifting of f is ${}^*f \circ j$, that is, it can be regarded as the table of f formed at the nodes making up the hyperfinite set $j(X)$ (the equality (1.2.18) shows that this is a natural point of view), then the operator A approximates $\mathcal{A}$ if it carries the table of f into a vector infinitely close to the table of the image $\mathcal{A}(f)$ of f.

The assertion that $\mathcal{A}$ has a hyperfinite approximation can be written in the language $\mathcal{LRS}$ (or in the language of IST), and hence we can use Nelson's algorithm for it to get an equivalent proposition formulated in standard mathematical terms (an internal sentence of IST). In general form the corresponding formulation is fairly cumbersome, but its meaning is that there is a sequence of finite-dimensional normed spaces and operators acting in them for which there exist corresponding sequences of finite subsets (nodes of tables in the spaces $\mathfrak{X}_i$) and sequences of multipliers Δ such that the table of the function at each set of nodes is a vector in the corresponding space, the integral of the function is approximated by the sums of the values of the function at the nodes multiplied by Δ, and the values of the finite-dimensional operators of the tables of a function f converge to the tables of $\mathcal{A}(f)$. Below we give precise standard formulations in some concrete cases.

It is easy to see that for any internal normed spaces E_1 and E_2 every internal linear operator $T \colon E_1 \longrightarrow E_2$ with finite norm induces a bounded linear operator $T^\# \colon E_1^\# \longrightarrow E_2^\#$ such that $T^\#(e^\#) = (T(e))^\# \ \forall e \in E_1$.

The condition that $A \colon \mathcal{L}_{p_1} \longrightarrow \mathcal{L}_{p_2}$ be a hyperfinite-dimensional approximation of $\mathcal{A} \colon L_{p_1}(\mu_1) \longrightarrow L_{p_2}(\mu_2)$ is now equivalent to the condition that the following diagram be commutative:

$$
(1.3.1)\qquad
\begin{array}{ccc}
L_{p_1}(\mu_1) & \xrightarrow{\ \mathcal{A}\ } & L_{p_2}(\mu_2) \\
{\scriptstyle j_{p_1}^{M_1}}\downarrow & & \downarrow{\scriptstyle j_{p_2}^{M_2}} \\
\mathfrak{L}_{p_1}^{\#} & \xrightarrow{\ A^{\#}\ } & \mathfrak{L}_{p_2}^{\#}
\end{array}
$$

The following obvious lemma is often used in concrete cases to prove that A is a hyperfinite-dimensional approximation of $\mathcal{A}$.

LEMMA 1.3.2. *Let $\mathfrak{M} \subseteq L_{p_1}(\mu_1)$ be such that the linear span of $\mathfrak{M}$ is dense in $L_{p_1}(\mu_1)$ and the conditions of Definition* 1.3.1 *hold $\forall f \in \mathfrak{M}$. Then A is a hyperfinite-dimensional approximation of $\mathcal{A}$.*

PROOF. If $\mathfrak{L}(\mathfrak{M})$ stands for the linear span of $\mathfrak{M}$, then by the condition of the lemma, the diagram (1.3.1), restricted to $\mathfrak{L}(\mathfrak{M})$, is commutative, and hence it is also commutative on $L_{p_1}(\mu_1)$ because $\mathfrak{L}(\mathfrak{M})$ is dense in $L_{p_1}(\mu_1)$. □

2°. Here we give some properties of operators of the type $T^{\#}\colon E_1^{\#} \longrightarrow E_2^{\#}$ for internal hyperfinite-dimensional normed spaces E_1 and E_2.

PROPOSITION 1.3.3. *Let E_1, E_2, and E_3 be internal normed spaces over the field ${}^*\mathbb{K}$ $(= {}^*\mathbb{R}$ or ${}^*\mathbb{C})$, and $T_1, T_2\colon E_1 \longrightarrow E_2$ and $T_3\colon E_2 \longrightarrow E_3$ internal operators with finite norm. Then:*

a) $\|T_1^{\#}\| = {}^\circ\|T_1\|$;

b) $(T_1 + T_2)^{\#} = T_1^{\#} + T_2^{\#}$;

c) $\forall\lambda \in \mathrm{fin}({}^*\mathbb{K})\ ((\lambda T_1)^{\#} = ({}^\circ\lambda)\cdot T_1^{\#})$;

d) $(T_3 \circ T_2)^{\#} = T_3^{\#} \circ T_2^{\#}$. □

Suppose now that E is an internal inner product space. Then, as mentioned, $E^{\#}$ is a nonseparable Hilbert space. The inner product on $E^{\#}$ is defined by $(e_1^{\#}, e_2^{\#}) = {}^\circ(e_1, e_2)\ \forall e_1, e_2 \in E$.

If $\mathcal{D}\colon \mathcal{H} \longrightarrow \mathcal{H}$ is a linear operator acting in a Hilbert space $\mathcal{H}$, then $\mathcal{D}'$ denotes the operator adjoint to $\mathcal{D}$, $\Lambda(\mathcal{D})$ the spectrum of $\mathcal{D}$, and $\Pi(\mathcal{D})$ the point spectrum of $\mathcal{D}$ (the set of eigenvalues of $\mathcal{D}$).

PROPOSITION 1.3.4. *If E is an internal space with a positive-definite inner product, and $T\colon E \longrightarrow E$ an internal linear operator with finite norm, then:*

a) $(T')^{\#} = (T^{\#})'$;

b) *if T is a selfadjoint (normal, unitary) operator, then so is $T^{\#}$;*

c) *if E is hyperfinite-dimensional, then $\Lambda(T^{\#}) = \Pi(T^{\#})$.* □

The proofs of these two propositions are left to the reader as exercises (see also [**46**]).

PROPOSITION 1.3.5. *If under the conditions of the preceding proposition E is hyperfinite-dimensional and T a normal operator, then $\Lambda(T^{\#}) = \{{}^\circ\lambda \mid \lambda \in \Lambda(T)\}$.*

PROOF. Let $B = \{{}^\circ\lambda \mid \lambda \in \Lambda(T)\}$. Then B is a closed subset of $\mathbb{C}$, because $\Lambda(T)$ is internal ([**1**], Proposition 2.1.8). Obviously, $B \subseteq \Lambda(T^{\#})$. Suppose that $\mu \notin B$. Then there exists a standard $\delta > 0$ such that $\forall\zeta \in B\ (|\mu - \xi| \geq \delta)$. Consequently, $|\mu - \xi| \geq \delta\ \forall\lambda \in \Lambda(T)$. Let us consider the operator $(T - \mu I)^{-1}$, where I is the identity operator. Since $\mu \notin \Lambda(T)$, $(T - \mu I)^{-1}$ exists.

Obviously, $\Lambda((T-\mu I)^{-1}) = \{\frac{1}{\lambda-\mu},\ \lambda \in \Lambda(T)\}$. Since T is normal, it follows from the last equality that $\|(T-\mu I)^{-1}\| \le \delta^{-1}$, that is, $(T-\mu I)^{-1}$ has finite norm. Proposition 1.3.3, d) now shows that $(T-\mu I)^{-1\#} = (T^{\#} - \mu I^{\#})^{-1}$. Since $I^{\#}$ is the identity operator on $E^{\#}$, it follows that $\mu \notin \Lambda(T^{\#})$. □

3°. We return to hyperfinite-dimensional approximations of operators. Assume that $\mathcal{A}_k\colon L_{p_1}(\mu_1) \longrightarrow L_{p_2}(\mu_2)$ is an integral operator, that is, there exists a measurable function $k\colon \mathfrak{X}_1 \times \mathfrak{X}_2 \longrightarrow \mathbb{K}$ such that $\forall f \in L_{p_1}(\mu_1)$

$$\mathcal{A}_k(f)(s) = \int_{\mathfrak{X}_1} k(s,t)f(t)\,d\mu_1(t). \tag{1.3.2}$$

Then the question naturally arises of whether we can construct a hyperfinite-dimensional approximation of $\mathcal{A}$ by using a lifting of the kernel k, if it exists. Here it should be kept in mind that in general[2]

$$(M_1 \times M_2, S_{\Delta_1}^{M_1} \otimes S_{\Delta_2}^{M_2}, \nu_{\Delta_1}^{M_1} \otimes \nu_{\Delta_2}^{M_2}) \neq (M_1 \times M_2, S_{\Delta_1\Delta_2}^{M_1\times M_2}, \nu_{\Delta_1\Delta_2}^{M_1\times M_2}).$$

However, Anderson [**30**] showed that if M_1 and M_2 are spaces with finite Loeb measure (in this case M_1 and M_2 can be assumed to be internal sets), then $S_{\Delta_1}^{M_1} \otimes S_{\Delta_2}^{M_2} \subseteq S_{\Delta_1\Delta_2}^{M_1\times M_2}$, and the identity imbedding is measure-preserving. It is easy to see that this result remains true also for the case of σ-finite subspaces of Loeb spaces, because each σ-finite subspace is a disjoint union of a countable family of Loeb spaces with finite measure. Then $L_r(\nu_{\Delta_1}^{M_1} \otimes \nu_{\Delta_2}^{M_2})$ is imbedded with preservation of norm in $L_r(\nu_{\Delta_1\Delta_2}^{M_1\times M_2})$ for any $r \in [1,\infty)$. Since the imbedding $j_r^{M_i}\colon L_r(\mu) \longrightarrow L_r(\nu_{\Delta_i}^{M_i})$ $(i=1,2)$ induces an imbedding $j_r^{M_1\times M_2}\colon L_r(\mu_1 \otimes \mu_2) \longrightarrow L_r(\nu_{\Delta_1}^{M_1} \otimes \nu_{\Delta_2}^{M_2}) \subseteq L_r(\nu_{\Delta_1\Delta_2}^{M_1\times M_2})$, we can conclude that any function $k \in L_r(\mu_1 \otimes \mu_2)$ has a lifting $K \in \mathcal{S}_r(M_1 \times M_2)$. It is natural to investigate the question of conditions under which the matrix K determines a hyperfinite approximation A of $\mathcal{A}$ with kernel K. Here this question is answered for the case when $\mathcal{A}$ is a Hilbert–Schmidt operator. Below we consider a space $(\mathfrak{X}, \Omega, \mu)$ with a σ-finite measure, a σ-finite subspace $(M, S_\Delta^M, \nu_\Delta^M)$ of the Loeb space $(X, S_\Delta, \nu_\Delta)$, and an imbedding $j_2\colon L_r(\mu) \longrightarrow L_r(\nu_\Delta^M)$.

THEOREM 1.3.6. *If $\mathcal{A}_k\colon L_2(\mu) \longrightarrow L_2(\mu)$ is a Hilbert–Schmidt operator with kernel $k \in L_2(\mu \otimes \mu)$, and $K \in \mathcal{S}_2(M \times M)$ is a lifting of k (that is, K is a lifting of $j_2^{M\times M}(k) \in L_2(\nu_{\Delta^2}^{M\times M})$), then the operator $A_K\colon \mathfrak{L}_2 \longrightarrow \mathfrak{L}_2$ defined $\forall F \in \mathfrak{L}_2$ $\forall \xi \in X$ by*

$$A_K(F)(\xi) = \Delta \sum_{\eta \in X} K(\xi,\eta)F(\eta) \tag{1.3.3}$$

is a hyperfinite-dimensional approximation of the operator $\mathcal{A}_k$ (given by the formula (1.3.2)*).*

PROOF. It is well known that

$$\|\mathcal{A}_k\| \le \left(\int_{\mathfrak{X}\times\mathfrak{X}} |k|^2\,d\mu\otimes d\mu\right)^{1/2}. \tag{1.3.4}$$

[2]The right side of this is a σ-finite subspace of the Loeb space $(X_1 \times X_2, S_{\Delta_1\Delta_2}^{X_1\times X_2}, \nu_{\Delta_1\Delta_2}^{X_1\times X_2})$.

Similarly,

$$\|A_K\| \leq \left(\Delta^2 \sum_{\xi,\eta\in X} |K(\xi,\eta)|^2\right)^{1/2}. \tag{1.3.5}$$

We show first of all that the set of k for which the theorem holds is closed in $L_2(\mu\otimes\mu)$. Suppose that the theorem holds for a sequence $k_n \in L_2(\mu\otimes\mu)$ $(n\in\omega)$, and $\|k-k_n\|_{L_2(\mu\otimes\mu)} \longrightarrow 0$ as $n\to\infty$. Since $\mathcal{A}_k - \mathcal{A}_{k_n} = \mathcal{A}_{k-k_n}$ (1.3.2), it follows from (1.3.4) that $\|\mathcal{A}_k - \mathcal{A}_{k_n}\| \longrightarrow 0$ as $n\to\infty$. Let $K_n \in \mathcal{S}_2(M\times M)$ be a lifting of k_n, and let $K\in\mathcal{S}_2(M\times M)$ be a lifting of k. Then $\|A_K^\# - A_{K_n}^\#\| = {}^\circ\|A_K - A_{K_n}\| = {}^\circ\|A_{K-K_n}\| \leq {}^\circ(\Delta^2\sum_{\xi,\eta\in X}|K(\xi,\eta)-K_n(\xi,\eta)|^2)^{1/2} = \|k-k_n\|_{L_2(\mu\otimes\mu)} \longrightarrow 0$ as $n\to\infty$. We now use the fact that for $\mathcal{A}_{k_n}$ and A_{K_n} the diagram (1.3.1) commutes. Let $f\in L_2(\mu)$. Then $j_2(\mathcal{A}_k(f)) = \lim_{n\to\infty} j_2(\mathcal{A}_{k_n}(f)) = \lim_{n\to\infty} A_{K_n}^\#(j_2(f)) = A_K^\#(j_2(f))$.

To complete the proof of the theorem we show that it is true for functions k with the form $\varphi(s)\cdot\psi(t)$. Since linear combinations of such functions are dense in $L_2(\mu\otimes\mu)$, this will prove the theorem. Thus, let $k(s,t)=\varphi(s)\cdot\psi(t)$, where $\varphi,\psi\in L_2(\mu)$. Denote by $\Phi,\Psi\in\mathcal{S}_2(M)$ liftings of φ and ψ, respectively. Assume also that $F\in\varphi_2(M)$ is a lifting of a function $f\in L_2(\mu)$, and $G\in\mathcal{S}_2(M)$ a lifting of $\mathcal{A}_k(f)$. Obviously, $\Psi(\eta)\cdot F(\eta)$ is a lifting of the function $\psi(t)\cdot f(t)$. By using the Cauchy–Schwarz–Bunyakovskiĭ inequality and the fact that $\Psi, F\in\mathcal{S}_2(M)$ it is easy to show that $\Psi\cdot F\in\mathcal{S}(M)$, that is,

$$\alpha = \int_{\mathfrak{X}} \psi(t) f(t)\, d\mu(t) \approx \sum_{\eta\in X}\Psi(\eta)F(\eta) = \beta.$$

Since $\mathcal{A}_k(f)=\alpha\cdot\varphi$, the lifting G of $\mathcal{A}_k(f)$ is equal to $\alpha\cdot\Phi$. It follows from the formula (1.3.3) that $A_k(F)=\beta\cdot\Phi$, and then $\|G-A_K(F)\|_2 = |\alpha-\beta|\cdot\|\Phi\|_2\approx 0$.□

Using Theorem 1.2.14, we get

COROLLARY 1.3.7 (IST). *Every Hilbert–Schmidt operator acting in a space $L_2(\mu)$ with σ-finite measure μ has a hyperfinite-dimensional approximation.*

Suppose now that $\mathfrak{X}$ is a separable locally compact space, μ is a Borel measure on $\mathfrak{X}$, Ω is the completion of the σ-algebra of Borel sets with respect to μ, and (X,j,Δ) is a hyperfinite realization of the space $(\mathfrak{X},\Omega,\mu)$ (Definition 1.2.17). We consider the space $\mathfrak{X}\times\mathfrak{X}$ with the product topology. Obviously, $\mathrm{Ns}({}^*\mathfrak{X}\times{}^*\mathfrak{X}) = \mathrm{Ns}({}^*\mathfrak{X})\times\mathrm{Ns}({}^*\mathfrak{X})$ and $\mathrm{st}(\langle\xi,\eta\rangle) = \langle\mathrm{st}(\xi),\mathrm{st}(\eta)\rangle$ $\forall\xi,\eta\in\mathrm{Ns}({}^*\mathfrak{X})$. From this it follows easily that $(X\times X, j\otimes j,\Delta^2)$ is an HR of the space $(\mathfrak{X}\times\mathfrak{X},\Omega\otimes\Omega,\mu\otimes\mu)$. It is also easy to see that, for a bounded function $f\colon\mathfrak{X}\times\mathfrak{X}\longrightarrow\mathbb{R}$ that is continuous $\mu\otimes\mu$-almost everywhere, the condition (1.2.17) is equivalent to the condition

$$\forall B\in{}^*\mathcal{P}(X)\ \Bigg(B\subseteq X\setminus M \Longrightarrow \Delta^2\left(\sum_{x\in B}\sum_{y\in X}|{}^*f(j(x),j(y))| + \sum_{x\in X}\sum_{y\in B}|{}^*f(j(x),j(y))|\right)\approx 0\Bigg). \tag{1.3.6}$$

COROLLARY 1.3.8. *Let $\mathfrak{X}$ be a separable locally compact topological space with a Borel measure μ, and let (X, j, Δ) be an HR of the space $(\mathfrak{X}, \Omega, \mu)$. Then for any bounded function $k \in L_2(\mu \otimes \mu)$ that is continuous $\mu \otimes \mu$-almost everywhere and is such that $|k|^2$ satisfies* (1.3.6), *the operator A_K with $K = {}^*k|j(X) \times j(X)$ is a hyperfinite-dimensional approximation of the operator $\mathcal{A}_k$ ($\mathcal{A}_k$ and A_K are given by the formulas* (1.3.2) *and* (1.3.3), *respectively*).

We present a simple sufficient condition for $f \colon \mathfrak{X} \times \mathfrak{X} \longrightarrow \mathbb{R}$ to satisfy (1.3.6).

PROPOSITION 1.3.9. *If there exist bounded integrable functions φ_1 and φ_2 that are continuous μ-almost everywhere, satisfy the condition* (1.2.17), *and are such that the function $f \colon \mathfrak{X}^2 \longrightarrow \mathbb{R}$ satisfies the inequality*

$$\forall x, y \in \mathfrak{X}\ (|f(x,y)| \leq \varphi_1(x) \cdot \varphi_2(y)), \tag{1.3.7}$$

then f satisfies (1.3.6).

PROOF. By assumption, φ_i satisfies (1.2.17) and (1.2.18), that is, if $B \subseteq X \backslash M$, then $\Delta^2 \sum_{x \in B} \sum_{y \in X} |{}^*f(j(x), j(y))| \leq \Delta \sum_{x \in B} {}^*\varphi_1(j(x)) \cdot \Delta \sum_{y \in X} {}^*\varphi_2(j(y)) \approx \int_{\mathfrak{X}} \varphi_2 \, d\mu \cdot \Delta \sum_{x \in B} {}^*\varphi_1(j(x)) \approx 0$. □

Corollary 1.3.8 is equivalent to the following assertion.

COROLLARY 1.3.8′. *Let $k \in L_2(\mu \otimes \mu)$ be a bounded function that is continuous almost everywhere and such that $|k|^2$ satisfies* (1.3.6) (*or the inequality* (1.3.7)). *Then for any bounded function $f \in L_2(\mu)$ that is continuous almost everywhere and satisfies* (1.2.17),

$$\Delta \sum_{x \in X} \left| \int_{^*\mathfrak{X}} {}^*k(j(x), \eta) {}^*f(\eta) \, d^*\mu(\eta) - \Delta \sum_{y \in X} {}^*k(j(x), j(y)) {}^*f(j(y)) \right|^2 \approx 0. \tag{1.3.8}$$

If we now consider the particular case analyzed in Example 2 of the preceding section, we obtain

COROLLARY 1.3.10. *Let $k \in L_2(\mathbb{R}^2)$ be a bounded function that is continuous almost everywhere and is such that $|k|^2$ satisfies the inequality* (1.3.7) *for some absolutely integrable functions φ_1 and φ_2 on $\mathbb{R}$ that satisfy the condition* (1.2.22). *Assume that $\Delta \approx 0$ and $L \in {}^*\omega$ are such that $L\Delta \sim +\infty$. Then*

$$\Delta \sum_{\alpha=-L}^{L} \left| \int_{-\infty}^{\infty} {}^*k(\alpha\Delta, y) {}^*f(y) \, dy - \Delta \sum_{\beta=-L}^{L} {}^*k(\alpha\Delta, \beta\Delta) {}^*f(\beta\Delta) \right|^2 \approx 0 \tag{1.3.9}$$

for any bounded function $f \in L_2(\mathbb{R})$ that is continuous almost everywhere and satisfies (1.2.22).

If we set $X = \{k \in {}^*\omega \mid |k| \leq L\}$, define $j \colon X \longrightarrow {}^*\mathbb{R}$ by $j(k) = k\Delta$, and consider $K \colon X^2 \longrightarrow {}^*\mathbb{R}$ such that $K = {}^*k|j(X) \times j(X)$, then the relation (1.3.9) is equivalent to the assertion that A_K is a hyperfinite-dimensional approximation of the operator $\mathcal{A}_k$ (Corollary 1.3.8′).

If, as in the preceding section, we choose L and Δ in a special way, namely, if we consider a $\tau \sim +\infty$ and take $\Delta \overset{\tau}{\approx} 0$ and $L = \left[\frac{\tau}{\Delta}\right]$, then by analogy with the corollary to Theorem 1.1.13 we get that

$$\int_{-\infty}^{\infty}\int_{-\infty}^{\infty} |k(x,y)|^2\,dx\,dy = {}^\circ\Big(\Delta^2 \sum_{\alpha,\beta=-L}^{L} |{}^*k(\alpha\Delta,\beta\Delta)|^2\Big),$$

and then it is easy to show as above that K is in $\mathcal{S}_2(M\times M)$ and is a lifting of k, even if $|k|^2$ does not satisfy (1.3.6), that is, we have

COROLLARY 1.3.11. *If $\tau \sim +\infty$ and $L = \left[\frac{\tau}{\Delta}\right]$, then the relation (1.3.9) holds for any bounded almost-everywhere continuous functions $k \in L_2(\mathbb{R}^2)$ and $f \in L_2(\mathbb{R})$.*

Obviously, Corollaries 1.3.10 and 1.3.11 remain true if instead of $\mathbb{R}$ we consider $\mathbb{R}^n$ for arbitrary $n \geq 1$ (in this case $k \in L_2(\mathbb{R}^{2n})$).

By using various conditions on k under which the integral operator with kernel k is a bounded operator acting from L_p to L_q (see, for example, [**17**]), we can get results analogous to Theorem 1.3.6 and its corollaries also for certain other integral operators.

4°. In this subsection we consider standard versions of some of the results obtained above. One possible approach to getting such versions is to write the needed result in the language $\mathcal{LRS}$ (or IST) and then to use Nelson's algorithm.

As an example we consider Corollary 1.3.7 to Theorem 1.3.6. By Theorem 1.2.14, it can be assumed that $X \subseteq {}^*\mathfrak{X}$, and the lifting in $\mathcal{S}_2(M)$ of a function $f \in L_2(\mu)$ is ${}^*f|X$. We remark that for an $\mathcal{S}_2(M)$-lifting $K\colon X^2 \longrightarrow {}^*\mathbb{R}$ of a function $k \in L_2(\mu\otimes\mu)$ the analogous equality does not hold in general, since $\langle X^2,\Delta\rangle$ can fail to satisfy the conditions of Theorem 1.2.13 with respect to $(\mathfrak{X}^2, \Omega\otimes\Omega, \mu\otimes\mu)$ (in any case the corresponding theorem has not been proved). Nevertheless, such a lifting K exists in view of the remarks preceding Theorem 1.3.6. Corollary 1.3.7 can be written as follows in the language of $\mathcal{LRS}$.

Let $(\mathfrak{X},\Omega,\mu)$ be a standard measure space with σ-finite measure, and let $k \in L_2(\mu\otimes\mu)$ be standard. Then

$$\begin{aligned}&\exists^{\mathrm{fin}} X \subseteq \mathfrak{X}\ \exists\Delta\in\mathbb{R}_+\ \exists K\colon X^2 \longrightarrow \mathbb{R}\ \forall^{\mathrm{st}} f\in L_2(\mu)\ \forall^{\mathrm{st}}\varepsilon\in\mathbb{R}_+\\ &\quad\Big(\Big|\iint_{\mathfrak{X}^2} |k|^2\,d\mu\otimes\mu - \Delta^2\sum_{x,y\in X}|K(x,y)|^2\Big| < \varepsilon\\ &\qquad\wedge\ \Delta\cdot\sum_{x\in X}\Big|\int_{\mathfrak{X}} k(x,\eta)f(\eta)\,d\mu(\eta) - \Delta\sum_{y\in X}K(x,y)f(y)\Big|^2 < \varepsilon\Big).\end{aligned}$$

We denote the quantifier-free part of this sentence by $\mathcal{W}$ and apply the idealization principle (I) to it, obtaining the equivalent sentence

$$\begin{aligned}&\forall^{\mathrm{st}}\Phi\in\mathcal{P}^{\mathrm{fin}}(L_2(\mu))\ \forall^{\mathrm{st}}\Xi\in\mathcal{P}^{\mathrm{fin}}(\mathbb{R}_+)\ \exists^{\mathrm{fin}}X\subseteq\mathfrak{X}\ \exists\Delta\in\mathbb{R}_+\\ &\qquad\qquad \exists K\colon X^2\longrightarrow\mathbb{R}\ \forall f\in\Phi\ \forall\varepsilon\in\Xi\ (\mathcal{W}).\end{aligned}$$

Since $(\mathfrak{X},\Omega,\mu)$ and k are standard, we can use the transfer principle (T) and remove the st from the first two quantifiers in the last sentence. By passing from Ξ to $\varepsilon = \min\Xi$ it is easy to see that we can replace the second quantifier by $\forall\varepsilon\in\mathbb{R}_+$, and we can remove the last quantifier entirely. We arrive at last at the following proposition.

PROPOSITION 1.3.12. *Let $(\mathfrak{X}, \Omega, \mu)$ be a measure space with σ-finite measure, and let $k \in L_2(\mu \otimes \mu)$. Then for any finite set $\Phi \subseteq L_2(\mu)$ and any $\varepsilon > 0$ there exist $X \subseteq \mathfrak{X}$, $\Delta \in \mathbb{R}$, and $K\colon X^2 \longrightarrow \mathbb{R}$ such that*

$$\left| \iint_{\mathfrak{X}^2} |k|^2 \, d\mu - \Delta^2 \sum_{x,y} |K(x,y)|^2 \right| < \varepsilon, \tag{1.3.10}$$

and for any $f \in \Phi$

$$\Delta \sum_{x \in X} \left| \int_{\mathfrak{X}} k(x,\eta) f(\eta)\, d\mu(\eta) - \Delta \sum_{y \in X} K(x,y) f(y) \right|^2 < \varepsilon.$$

A more careful analysis of the translation shows that while the nonstandard X, Δ, and K in Corollary 1.3.7 depend only on the elements of $L_2(\mu)$ and $L_2(\mu \otimes \mu)$ and not on their representatives, and while the $k \in L_2(\mu \otimes \mu)$ in Proposition 1.3.12 can be regarded as a class (that is, X, Δ, and K do not depend on the concrete function in the class k), Φ must be regarded as a finite set of square-integrable functions.

In fact, Proposition 1.3.12 is not a complete analogue of Corollary 1.3.7, because the inequality (1.3.10) is only one of the consequences of the fact that K is an $\mathcal{S}_2(M)$-lifting of the function k. Essentially external objects actually appear in the statement "K is an $\mathcal{S}_2(M)$-lifting of the function k," and it cannot be written in $\mathcal{LRS}$, but it is still possible to give many internal consequences of this statement and thereby strengthen Proposition 1.3.12.

We now consider a standard version of Corollary 1.3.11. Denote by $\mathcal{H}_n$ the space of bounded functions $h \in L_2(\mathbb{R}^n)$ that are continuous almost everywhere with respect to Lebesgue measure on $\mathbb{R}^n$. Then the indicated corollary can be written as follows in $\mathcal{LRS}$:

$$\exists L \in \omega\ \exists \Delta \in \mathbb{R}_+ \Bigg(L\Delta \sim +\infty \ \wedge\ \Delta \approx 0 \ \wedge\ \forall^{\mathrm{st}} f \in \mathcal{H}_1 \ \wedge\ \forall^{\mathrm{st}} k \in \mathcal{H}_2$$
$$\left(\Delta \sum_{\alpha=-L}^{L} \left| \int_{-\infty}^{\infty} {}^*k(\alpha\Delta, y)\, {}^*f(y)\, dy - \Delta \sum_{\beta=-L}^{L} {}^*k(\alpha\Delta, \beta\Delta)\, {}^*f(\beta\Delta) \right|^2 \approx 0 \right)\Bigg).$$

Let us rewrite this without $\sim$ and $\approx$. We get

$$\exists L \in \omega\ \exists \Delta \in \mathbb{R}_+\ \forall^{\mathrm{st}} n \in \omega\ \forall^{\mathrm{st}} f \in \mathcal{H}_1\ \forall^{\mathrm{st}} k \in \mathcal{H}_2 \Bigg(L\Delta > n \ \wedge\ \Delta < n^{-1}$$
$$\wedge \left(\Delta \sum_{\alpha=-L}^{L} \left| \int_{-\infty}^{\infty} k(\alpha\Delta, \eta) f(\eta)\, d\eta - \Delta \sum_{\beta=-L}^{L} k(\alpha\Delta, \beta\Delta) f(\beta\Delta) \right|^2 < \frac{1}{n} \right)\Bigg).$$

As above, this sentence is easily transformed to standard form. The result is

PROPOSITION 1.3.13. *For any finite sets of functions $\Phi \subseteq \mathcal{H}_1$ and $\mathcal{K} \subseteq \mathcal{H}_2$ and any $n \in \omega$ there exist numbers $L \in \omega$ and $\Delta \in \mathbb{R}_+$ such that $L\Delta > n$, $\Delta < n^{-1}$, and*

$$\Delta \sum_{\alpha=-L}^{L} \left| \int_{-\infty}^{\infty} k(\alpha\Delta, \eta) f(\eta)\, d\eta - \Delta \sum_{\beta=-L}^{L} k(\alpha\Delta, \beta\Delta) f(\beta\Delta) \right|^2 < \frac{1}{n} \tag{1.3.11}$$

for any functions $f \in \Phi$ and $k \in \mathcal{K}$.

It is even simpler to formulate a standard version of Corollary 1.3.10. Denote by $\widetilde{\mathcal{H}}_1$ the set of functions $f \in \mathcal{H}_1$ for which $|f|^2$ satisfies the condition (1.2.21), and by $\widetilde{\mathcal{H}}_2$ the set of $k \in \mathcal{H}_2$ such that $|k|^2$ satisfies the inequality (1.3.7) for some φ_1 and φ_2 for which (1.2.21) holds ($\widetilde{\mathcal{H}}_n$ is defined similarly). The consequence (1.3.10) can now be written in IST as

$$\forall L \in \omega\ \forall \Delta \in \mathbb{R}_+ \left(L\Delta \sim +\infty\ \wedge\ \Delta \approx 0 \Longrightarrow \forall^{\mathrm{st}} f \in \widetilde{\mathcal{H}}_1\ \forall^{\mathrm{st}} k \in \widetilde{\mathcal{H}}_2 \right.$$
$$\left. \left(\Delta \sum_{\alpha=-L}^{L} \left| \int_{-\infty}^{\infty} k(\alpha\Delta, \eta) f(\eta)\, d\eta - \Delta \sum_{\beta=-L}^{L} k(\alpha\Delta, \beta\Delta) f(\beta\Delta) \right|^2 \approx 0 \right) \right).$$

Using the fact that $a \approx 0 \Longleftrightarrow \forall^{\mathrm{st}} n\ (|a| < 1/n)$ and the rules of predicate logic, we arrive at the equivalent sentence

$$\forall^{\mathrm{st}} f \in \widetilde{\mathcal{H}}_1\ \forall^{\mathrm{st}} k \in \widetilde{\mathcal{H}}_2\ \forall^{\mathrm{st}} n \in \omega\ \forall L \in \omega\ \forall \Delta \in \mathbb{R}_+\ (L\Delta \sim +\infty\ \wedge\ \Delta \approx 0 \Longrightarrow \mathcal{W}),$$

where $\mathcal{W}$ denotes the inequality (1.3.11). Writing the predicates $L\Delta \sim +\infty$ and $\Delta \approx 0$ in $\mathcal{LRS}$ and using the idealization and transfer principles, we now come to the sentence

$$\forall f \in \widetilde{\mathcal{H}}_1\ \forall k \in \widetilde{\mathcal{H}}_2\ \forall n \in \omega\ \exists^{\mathrm{fin}} \mathcal{N} \subseteq \omega\ \forall L \in \omega\ \forall \Delta \in \mathbb{R}_+$$
$$(\forall m \in \mathcal{N}\ (\Delta < 1/m\ \wedge\ L\Delta > m) \Longrightarrow \mathcal{W}),$$

which is obviously equivalent to the following proposition.

PROPOSITION 1.3.14. *For any $f \in \widetilde{\mathcal{H}}_1$ and $k \in \widetilde{\mathcal{H}}_2$*

$$\lim_{\substack{\Delta \to 0 \\ L\Delta \to \infty}} \Delta \sum_{\alpha=-L}^{L} \left| \int_{-\infty}^{\infty} k(\alpha\Delta, \eta) f(\eta)\, d\eta - \Delta \sum_{\beta=-L}^{L} k(\alpha\Delta, \beta\Delta) f(\beta\Delta) \right|^2 = 0.$$

We consider a standard version of the definition of a hyperfinite-dimensional approximation of an arbitrary bounded operator $\mathcal{A}\colon L_p(\mathbb{R}) \longrightarrow L_q(\mathbb{R})$ $(p, q \geq 1)$. Denote by $\widetilde{\mathcal{H}}^{(r)}$ the space of bounded functions $f \in L_r(\mathbb{R})$ that are continuous almost everywhere on $\mathbb{R}$ and such that $|f|^2$ satisfies the condition (1.2.21). Note that $\widetilde{\mathcal{H}}^{(r)}$ is dense in $L_r(\mathbb{R})$ $(r \geq 1)$. We also assume that $\mathcal{A}$ satisfies the condition

$$\{f \in \widetilde{\mathcal{H}}^{(p)} \mid \mathcal{A}(f) \in \widetilde{\mathcal{H}}^{(q)}\} \text{ is dense in } L_p(\mathbb{R}). \tag{1.3.12}$$

DEFINITION 1.3.15. Let $\mathbb{T} = \{T_{L,\Delta} \mid L \in \omega,\ \Delta \in \mathbb{R}_+\}$ be a family of matrices $T_{L,\Delta} = (t_{\alpha,\beta})_{\alpha,\beta=-L}^{L}$ (that is, each is an $N \times N$ matrix, where $N = 2L + 1$) depending on two parameters and with norms uniformly bounded. We say that the family $\mathbb{T}$ approximates a bounded operator $\mathcal{A}\colon L_p(\mathbb{R}) \longrightarrow L_q(\mathbb{R})$ satisfying the condition (1.3.12) if for any function $f \in \widetilde{\mathcal{H}}^{(p)} \cap \mathcal{A}^{-1}(\widetilde{\mathcal{H}}^{(q)})$

$$\lim_{\substack{\Delta \to 0 \\ L\Delta \to \infty}} \Delta \sum_{\alpha=-L}^{L} \left| \mathcal{A}(f)(\alpha\Delta) - \sum_{\beta=-L}^{L} t_{\alpha\beta} f(\beta\Delta) \right|^q = 0. \tag{1.3.13}$$

An immediate consequence of this definition and Theorem 0.3.21, i) is

PROPOSITION 1.3.16. *If the family* $\mathbb{T}$ *of matrices satisfies the conditions of Definition* 1.3.15, *then it approximates the bounded operator* $\mathcal{A}\colon L_p(\mathbb{R}) \longrightarrow L_q(\mathbb{R})$ *if and only if for any* $L \in {}^*\omega$ *and* $\Delta \in {}^*\mathbb{R}_+$ *with* $\Delta \approx 0$ *and* $L\Delta \sim +\infty$ *the operator* ${}^*T_{L,\Delta}\colon \mathcal{L}^X_{\Delta,p} \longrightarrow \mathcal{L}^X_{\Delta,q}$, *where* $X = \{-L, \dots, L\}$, *is a hyperfinite-dimensional approximation of* $\mathcal{A}$.

The following standard proposition is now a consequence of Lemma 1.3.2.

PROPOSITION 1.3.17. *Under the conditions of Definition* 1.3.15 *suppose that the equality* (1.3.13) *holds for any function* f *in some set* $\mathfrak{M} \subseteq \widetilde{\mathcal{H}}^{(p)} \cap \mathcal{A}^{-1}(\widetilde{\mathcal{H}}^{(q)})$ *whose linear span is dense in* $L_p(\mathbb{R})$. *Then it holds also for any function* $f \in \widetilde{\mathcal{H}}^{(p)} \cap \mathcal{A}^{-1}(\widetilde{\mathcal{H}}^{(q)})$.

REMARKS. 1) The nonstandard Definition 1.3.1 is essentially more general than Definition 1.3.15 even in the case of the spaces $L_p(\mathbb{R})$, because it does not in general assume the existence of a standard family $\mathbb{T}$ of matrices satisfying the conditions of Definition 1.3.15. The difference between these definitions is very clear upon comparison of Propositions 1.3.13 and 1.3.14. The latter can be refined by a complete translation of Corollary 1.3.11 into standard language that takes into account the relation $\Delta \overset{T}{\approx} 0$. However, such a refinement leads to an essential complication of the formulation of the proposition.

2) Below we need the case when in the approximation of $\mathcal{A}$ the table of f is made up with step Δ and the table of $\mathcal{A}(f)$ with step Δ_1. Of course, $\Delta_1 \to 0$ and $L\Delta_1 \to \infty$ (for example, $\Delta_1 = ((2L+1)\Delta)^{-1}$). The general definition of a hyperfinite-dimensional approximation (Definition 1.3.1) encompasses also this case. Therefore, after obvious modifications in the formulations of Definition 1.3.15 and Propositions 1.3.16 and 1.3.17 the latter two remain valid.

3) Definition 1.3.15 and Propositions 1.3.16 and 1.3.17 can be generalized to the case when a hyperfinite-dimensional approximation of the operator $\mathcal{A}\colon L_p(\mu) \longrightarrow L_q(\mu)$, where μ is a σ-finite measure on a separable locally compact topological space $\mathfrak{X}$, is constructed on the basis of a hyperfinite realization of this space (see Corollary 1.3.8). Here it is necessary to give a standard version of the definition of a hyperfinite realization as a certain family X of finite sets, mappings $j\colon X \longrightarrow \mathfrak{X}$, and numbers Δ satisfying definite conditions. As already noted, the main difficulties here arise in the translation of the conditions (1.2.17) and (1.3.6) to standard language. We shall return to these questions when we treat hyperfinite approximations of locally compact Abelian groups.

5°. In this subsection we present some auxiliary results that will be needed in the next chapter.

PROPOSITION 1.3.18. *Let* E *be an internal hyperfinite-dimensional unitary (Euclidean, if* $\mathbb{K} = \mathbb{R}$*) space with* $\dim E = N \in {}^*\omega$, *and let* $\mathcal{A}\colon E \longrightarrow E$ *be an internal selfadjoint operator whose matrix* $A = (a_{ij})_{i,j=1}^N$ *in some orthonormal basis satisfies* $\sum_{i,j=1}^N |a_{ij}|^2 < +\infty$. *Then* 1) *all the eigenvalues of* A *are finite, and* 2) *the multiplicity of an eigenvalue that is not infinitesimal is a standard natural number.*

PROOF. This follows immediately from the elementary equality

$$\sum_{i=1}^{s} n_i|\lambda_i|^2 = \sum_{i,j=1}^{N} |a_{ij}|^2,$$

where $\lambda_1, \dots, \lambda_s$ is a complete set of distinct eigenvalues, and $n_1, \dots, n_s$ are their multiplicities. □

PROPOSITION 1.3.19. *Under the conditions of Proposition* 1.3.17 *if* T *is an internal subspace of* E, *then* $(T^{\perp})^{\#} = (T^{\#})^{\perp}$.

PROOF. Let P_T be the orthogonal projection onto the subspace T of E, and let $P_{T^\#}$ be the orthogonal projection onto the subspace $T^{\#}$ of $E^{\#}$. We show that $P_T^{\#} = P_{T^\#}$. It follows immediately from Propositions 1.3.3 and 1.3.4 that $P_T^{\#}$ is an orthogonal projection. It must be shown that, for any $\xi \in E^{\#}$, $P_T^{\#}\xi = \xi$ if and only if $\xi \in T^{\#}$. If $\xi = x^{\#}$ and $x \in T$, then $P_T^{\#}x^{\#} = (P_T x)^{\#} = x^{\#} = \xi$. Conversely, suppose that $P_T^{\#}\xi = \xi$. If $\xi = y^{\#}$, then $(P_T y)^{\#} = y^{\#}$, that is, $P_T y - y \approx 0$. Setting $x = P_T y$, we now get that $\xi = x^{\#}$, but $x = P_T y \in T$, that is, $\xi \in T^{\#}$. To conclude the proof we note that $H = T^{\perp}$ if and only if $P_H + P_T = I$ and $P_H P_T = P_T P_H = 0$. In this case, again using Proposition 1.3.3, we get that $P_{H^\#} + P_{T^\#} = I^{\#}$, and $P_{H^\#} P_{T^\#} = P_{T^\#} P_{H^\#} = 0^{\#}$, that is, $H^{\#} = T^{\#\perp}$. □

PROPOSITION 1.3.20. *Let* E *be an internal hyperfinite-dimensional unitary space*, $A\colon E \longrightarrow E$ *a bounded selfadjoint operator* (*that is*, $\|A\| < +\infty$), *and* $\lambda \in \mathbb{R}$ *a standard number. Assume that the set* $\mathcal{M}$ *of eigenvalues of* A *infinitely close to* λ *is internal and standardly finite, that is*, $\mathcal{M} = \{\lambda_1, \dots, \lambda_n\}$, *where* $n \in \omega$. *Assume also that the multiplicity of each* $\lambda_i \in \mathcal{M}$ *is standard, and denote by* $\{\varphi_1, \dots, \varphi_m\}$ *a complete orthonormal system of eigenvectors associated with the eigenvalues in* $\mathcal{M}$. *In this case if* $f \in E^{\#}$ *is an eigenvector of* $A^{\#}$ *belonging to the eigenvalue* λ, *then* f *can be expressed linearly in terms of* $\varphi_1^{\#}, \dots, \varphi_m^{\#}$.

PROOF. Let H be the internal linear span of $\varphi_1, \dots, \varphi_m$. Since m is standard, it is easy to show that $H^{\#} \subseteq E^{\#}$ is the linear span of $\varphi_1^{\#}, \dots, \varphi_m^{\#}$. Denote by $E_\lambda^{\#}$ the set of eigenvectors of $A^{\#}$ belonging to the eigenvalue λ. Then it is clear that $H^{\#} \subseteq E_\lambda^{\#}$. If $H^{\#} \neq E_\lambda^{\#}$, then $(H^{\#})^{\perp} \cap E_\lambda^{\#} \neq \{0\}$. By Proposition 1.3.19, $H^{\#\perp} = (H^{\perp})^{\#}$. Since $H^{\perp}$ is an invariant subspace for A (because H is an invariant subspace for A by definition), $(H^{\perp})^{\#}$ is an invariant subspace for $A^{\#}$. Suppose that $0 \neq f \in (H^{\perp})^{\#} \cap E_\lambda^{\#}$. Then λ is an eigenvalue for $A^{\#}|(H^{\perp})^{\#}$, and there exists an eigenvalue γ of the operator $A|H^{\perp}$ by Proposition 1.3.5. An eigenvector belonging to it is orthogonal to all the φ_i, and this contradicts the conditions of the proposition. □

CHAPTER 2

Nonstandard Analysis on Locally Compact Abelian Groups

§1. Hyperfinite approximation of the Fourier transformation in $L_2(\mathbb{R})$

Here we study approximation of the Fourier transformation (FT) $\mathcal{F}\colon L_2(\mathbb{R}) \longrightarrow L_2(\mathbb{R})$ by a discrete Fourier transformation (DFT). The same idealogy is used as in §3 of Chapter 1, that is, the matrix of the DFT is applied to the table of values of a function f formed at the nodes $-L\Delta_1, \dots, L\Delta_1$. The vector obtained is compared with the table of $\mathcal{F}(f)$ formed at the nodes $-L\Delta, \dots, L\Delta$. We investigate the question of conditions under which the norm of the difference of these two vectors tends to zero as $\Delta, \Delta_1 \to 0$ and $L\Delta, L\Delta_1 \to +\infty$ (or, what is the same, is infinitesimal for $\Delta, \Delta_1 \approx 0$ and $L\Delta, L\Delta_1 \sim +\infty$). While this always holds for Hilbert–Schmidt operators (see Remark 2) after Proposition 1.3.17), the answer to the question posed for the FT depends essentially on the relation among L, Δ, and Δ_1. The sufficient condition obtained here arises in a well-known theorem of Kotel′nikov in a somewhat different situation.

In fact, the results of this section have a group nature, enabling us to generalize them to the case of arbitrary separable locally compact Abelian groups in the sections to follow. However, the case studied here of the additive group of the field $\mathbb{R}$ nevertheless requires a separate treatment.

1°. In the course of this section $X = \{k \in {}^*\omega \mid -L \le k \le L\}$, where $L \in {}^*\omega \setminus \omega$, $N = 2L+1$, $\Delta, \Delta' \approx 0$, $N\Delta, N\Delta' \sim +\infty$, and $j\colon X \longrightarrow {}^*\mathbb{R}$ and $j'\colon X \longrightarrow {}^*\mathbb{R}$ are defined by $j(k) = k\Delta$ and $j'(k) = k\Delta'$. Thus, (X, j, Δ) and (X, j', Δ') are hyperfinite realizations of the space $\langle \mathbb{R}, \Omega, dx \rangle$, where Ω is the σ-algebra of Lebesgue-measurable sets, and dx is Lebesgue measure on $\mathbb{R}$ (see Definition 1.2.17). With slightly different notation this HR was studied in detail in Example 2 of §2 in Chapter 1.

In the internal hyperfinite-dimensional space $\mathbb{C}^X$ we fix the basis $\{E_k \mid k \in X\}$ with $E_k(m) = \delta_{km}$. Below, all internal linear operators acting in $\mathbb{C}^X$ are given by matrices in this basis. In correspondence with the notation adopted earlier (see the beginning of §3 in Chapter 1) the space $\mathbb{C}^X$ with the inner product $(\cdot,\cdot)_\Delta$ ($(\cdot,\cdot)_{\Delta'}$) determined from the condition $(E_k, E_m)_\Delta = \Delta\delta_{km}$ ($(E_k, E_m)_{\Delta'} = \Delta'\delta_{km}$) is denoted by $\mathcal{L}_{2\Delta}$ ($\mathcal{L}_{2\Delta'}$).

A DFT is an operator $\Phi\colon \mathbb{C}^X \longrightarrow \mathbb{C}^X$ with matrix $(\exp(-2\pi i\alpha\beta/N))_{\alpha,\beta=-L}^{L}$. Below we consider the DFT $\Phi_\Delta = \Delta\Phi$, $\Phi_\Delta\colon \mathcal{L}_{2\Delta} \longrightarrow \mathcal{L}_{2\Delta'}$. Then it is easy to verify that

$$(\Phi_\Delta(F), \Phi_\Delta(G))_{\Delta'} = N\Delta\Delta'(F, G)_\Delta. \tag{2.1.1}$$

Thus, $\|\Phi_\Delta\| = N\Delta\Delta'$, that is, a nonzero operator $\Phi_\Delta^\#\colon \mathcal{L}_{2\Delta}^\# \longrightarrow \mathcal{L}_{2\Delta'}^\#$ is determined only when $0 < {}^\circ(N\Delta\Delta') < +\infty$.

Unless a statement to the contrary is made, it is assumed in this section that

$$N\Delta\Delta' \approx 1, \tag{2.1.2}$$

with the case of precise equality singled out specially, and the notation

$$\widehat{\Delta} = (N\Delta)^{-1} \tag{2.1.3}$$

is used. The HR $(X, \widehat{j}, \widehat{\Delta})$ of the space $\langle \mathbb{R}, \Omega, dx \rangle$ corresponds to this case.

In correspondence with Definition 1.2.17 let $M = j^{-1}(\mathrm{Ns}({}^*\mathbb{R}))$, and $\varphi = \mathrm{st} \circ j|M \colon M \longrightarrow \mathbb{R}$. Then φ induces a mapping $j_2 \colon L_2(\mathbb{R}) \longrightarrow L_2(\nu_\Delta^M) \subseteq \mathcal{L}_{2,\Delta}^{\#}$. Recall that if $f \in L_2(\mathbb{R})$, then $j_2(f) = F^{\#}$, where $F \in \mathcal{S}_2(M)$ is a lifting of the function $f \circ \varphi$ (which is simply called a lifting of f). Furthermore, if f is continuous almost everywhere and bounded, and if $|f|^2$ satisfies condition (1.2.21), then as F we can take $X_\Delta(f) \in \mathbb{C}^X$, where $X_\Delta(f)_k = {}^*f(k\Delta)$ $\forall k \in X$ (see Example 2 in §2 of Chapter 1).

The quantities M', φ', j_2', and $X_{\Delta'}(f)$ ($\widehat{M}$, $\widehat{\varphi}$, $\widehat{j}_2$, and $X_{\widehat{\Delta}}(f)$) are defined from Δ' ($\widehat{\Delta}$) similarly.

Finally, let $\mathcal{F} \colon L_2(\mathbb{R}) \longrightarrow L_2(\mathbb{R})$ denote the Fourier transformation: $\mathcal{F}(f)(y) = \int_{-\infty}^{\infty} f(x)\exp(-2\pi ixy)\,dx$.

THEOREM 2.1.1. *If condition* (2.1.2) *holds, then the DFT* $\Phi_\Delta \colon \mathcal{L}_{2,\Delta} \longrightarrow \mathcal{L}_{2,\Delta'}$ *is a hyperfinite-dimensional approximation of the operator* $\mathcal{F} \colon L_2(\mathbb{R}) \longrightarrow L_2(\mathbb{R})$.

PROOF. As a set $\mathfrak{M}$ satisfying the conditions of Lemma 1.3.2 we consider the set of characteristic functions of intervals of the form $[0, a]$ and $[-a, 0]$, where $a > 0$. Let $f = \chi_{[0,a]}$. Let T be an infinite natural number such that $(T-1)\Delta \le a < T\Delta$.

To verify the conditions of Definition 1.3.1 we must prove that

$$\Delta' \sum_{k=-L}^{L} \left| \Delta \sum_{n=0}^{T-1} \exp(-2\pi ink/N) - \int_0^a \exp(-2\pi ixk\Delta')\,dx \right|^2 \approx 0. \tag{2.1.4}$$

Obviously, it suffices to consider only $\sum_{k=1}^{L}$ in (2.1.4). Moreover, in view of (2.1.2)

$$\Delta' \sum_{k=-L}^{L} \left| \int_a^{T\Delta} \exp(-2\pi ixk\Delta')\,dx \right|^2 \le N\Delta^2\Delta' \approx \Delta \approx 0,$$

and hence $\int_0^a$ can be replaced by $\int_0^{T\Delta}$ in (2.1.4). After this and the necessary computations we get that it suffices to prove the relation

$$\Delta' \sum_{k=-L}^{L} |\Delta(1 - \exp(-2\pi ikT/N))(1 - \exp(-2\pi ik/N))^{-1} - (2\pi ik\Delta')^{-1}(1 - \exp(-2\pi ikT\Delta\Delta'))|^2 \approx 0. \tag{2.1.5}$$

If we replace Δ' in the last relation by $\widehat{\Delta} = (N\Delta)^{-1}$ (2.1.3), then we arrive at the relation

$$\frac{\Delta}{N} \sum_{k=1}^{L} |1 - \exp(-2\pi ikT/N)|^2 |(1 - \exp(-2\pi ik/N))^{-1} - N/(2\pi ik)|^2 \approx 0. \tag{2.1.6}$$

For any $k \in [1, L]$ we have $0 < 2\pi k/N < \pi$, and the function $(1-\exp(-i\varphi))^{-1} - (i\varphi)^{-1}$ has a finite limit as $\varphi \to 0$, so it is bounded on $[0,\pi]$. This makes (2.1.6) obvious. By (2.1.6) and (2.1.2), it is easy to see that (2.1.5) follows from the relation

$$\Delta' \sum_{k=1}^{L} |(2\pi i k\Delta')^{-1}(1-\exp(-2\pi i kT\Delta\Delta')) \\ - N\Delta(2\pi i k)^{-1}(1-\exp(-2\pi i kT/N))|^2 \approx 0,$$

which in turn follows from the two relations

$$\frac{1}{\Delta'}\sum_{k=1}^{L} k^{-2}|(N\Delta\Delta'-1)(1-\exp(2\pi i kT/N))|^2 \approx 0 \tag{2.1.7}$$

and

$$\frac{1}{\Delta'}\sum_{k=1}^{L} k^{-2}|\exp(-2\pi i kT\Delta\Delta') - \exp(-2\pi i kT/N)|^2 \approx 0.$$

We prove the first of these, since the second is proved similarly.

Let $\alpha = N\Delta\Delta' - 1 \approx 0$ and let $\overline{a} = T\Delta \approx a$. Then (2.1.7) is equivalent to

$$\frac{\alpha^2}{\Delta'}\sum_{k=1}^{L} \frac{1}{k^2}\sin^2\frac{\pi k\overline{a}}{N\Delta} \approx 0. \tag{2.1.8}$$

If $\alpha^2 M\Delta' \approx 0$, then (2.1.8) follows immediately from the fact that $\overline{a}$ is finite and from the inequality $\sin^2 x \le x^2$. But if $\alpha^2 M\Delta' \not\approx 0$, then we set $S = \left[\frac{1}{\alpha\Delta'}\right]$. In this case $S\Delta'$ is infinite and $\alpha^2 S\Delta' \approx 0$, and thus $S < M$. We now have that

$$\frac{\alpha^2}{\Delta'}\sum_{k=1}^{L} \frac{1}{k^2}\sin^2\frac{\pi k\overline{a}}{N\Delta} = \frac{\alpha^2}{\Delta'}\sum_{k=1}^{S} \frac{1}{k^2}\sin^2\frac{\pi k\overline{a}}{N\Delta} + \frac{\alpha^2}{\Delta'}\sum_{k=S+1}^{L} \frac{1}{k^2}\sin^2\frac{\pi k\overline{a}}{N\Delta}.$$

The first term on the right-hand side is infinitesimal because $\alpha^2 S\Delta' \approx 0$, and the second because

$$\frac{\alpha^2}{\Delta'}\sum_{k=S+1}^{L}\frac{1}{k^2} \le \frac{\alpha^2}{\Delta'}(S^{-1} - M^{-1}),$$

and $S\Delta'$ and $M\Delta'$ are infinite. □

COROLLARY 2.1.2. *If $N\Delta\Delta' \approx 2\pi h$, where $h > 0$ is a standard constant, then the DFT $\Phi_\Delta : \mathfrak{L}_{2,\Delta} \longrightarrow \mathfrak{L}_{2,\Delta'}$ is a hyperfinite-dimensional approximation of the FT $\mathfrak{F}_h : L_2(\mathbb{R}) \longrightarrow L_2(\mathbb{R})$ with $\mathfrak{F}_h(f)(y) = \int_{-\infty}^{\infty} f(x)\exp(-ixy/h)\,dx$.*

This is proved by passing from a function $f(t) \in L_2(\mathbb{R})$ to the function $\varphi(t) = f(2\pi h t)$ and replacing Δ by $\Delta/2\pi h$. □

COROLLARY 2.1.3. *Suppose that f and $\mathfrak{F}(f)$ are bounded and continuous almost everywhere, and $|f|^2$ and $|\mathfrak{F}(f)|^2$ satisfy the condition (1.2.21). Then*

$$\lim_{n\to\infty} \Delta'_n \sum_{k=-n}^{n} \bigg| \int_{-\infty}^{\infty} f(x)\exp(-2\pi i xk\Delta'_n)\,dx \\ - \Delta_n \sum_{m=-n}^{n} f(m\Delta_n)\exp(-2\pi i km/n)\bigg|^2 = 0$$

for any sequences $\Delta_n, \Delta'_n \to 0$ $(n \to \infty)$ such that $n\Delta_n \cdot \Delta'_n \to 1/2$.

This is a standard version of Theorem 2.1.1 that follows immediately from Proposition 1.3.16 (see also Remark 2 after Proposition 1.3.17 and example b) in the Introduction).

The relation (2.1.2) arising here is encountered in a somewhat different situation in the well-known theorem of Kotel′nikov asserting that if the spectrum of a bounded function f is concentrated in the interval $[-a, a]$, then f is completely determined by its values on the set $\{n\lambda \mid -\infty < n < +\infty\}$, where $\lambda \leq 1/(2a)$, according to the formula

$$f(t) = \sum_{k=-\infty}^{\infty} f(k\lambda)\frac{\sin 2\pi a(t-k\lambda)}{2\pi a(t-k\lambda)}.$$

In our case the values of the function are computed at the points $k\Delta$, where $\Delta \approx 1/(N\Delta')$, and it is not difficult to see that $N\Delta'$ is precisely the length of the interval on which $\mathcal{F}(f)$ is being considered.

The proposition to follow also points to the close connection between the relation $N\Delta\Delta' \approx 2\pi h$ given in Corollary 2.1.2 and uncertainty relations.

We consider the one-parameter groups of unitary operators $U(u) = \exp(-iuP)$ and $V(v) = \exp(-ivQ)$, where Q and P are the coordinate and momentum operators, respectively, that is, Q is the operator of multiplication by the independent variable, and $P = \frac{h}{i}\frac{d}{dx}$, where $h > 0$ is a standard constant (Planck's constant). We recall that $U(u)\varphi(x) = \varphi(x - uh)$, $V(v)\varphi(x) = \exp(-ivx)\varphi(x)$, and the commutation relations

$$U(u)V(v) = \exp(ihuv)V(v)U(u) \tag{2.1.9}$$

hold, which can be regarded as one form of the uncertainty relation.

We introduce the hyperfinite-dimensional operators $U_d, V_d \colon \mathbb{C}^X \longrightarrow \mathbb{C}^X$ with $(U_dF)(k) = F(k\dot{-}1)$ and V_d the diagonal matrix $(\exp(-2\pi ink/N)\delta_{nk})_{n,k=-L}^{L}$. In this section $\dot{+}$ and $\dot{-}$ are the addition and subtraction operations in the additive group of the ring ${}^*\mathbb{Z}/N{}^*\mathbb{Z} = \{-L, \dots, L\}$. It is easy to verify that the operators U_d^r and V_d^m satisfy the following commutation relations for any $r, m \in {}^*\mathbb{Z}/N{}^*\mathbb{Z}$:

$$U_d^r V_d^m = \exp(2\pi irm/N)V_d^m U_d^r. \tag{2.1.10}$$

PROPOSITION 2.1.4. *If $r, m \in {}^*\mathbb{Z}/N{}^*\mathbb{Z}$ are such that $r\Delta \approx uh$, and $2\pi m\widehat{\Delta} \approx v$ (2.1.3), where u and v are standard numbers, then U_d^r and $V_d^m \colon \mathfrak{L}_{2,\Delta} \longrightarrow \mathfrak{L}_{2,\Delta}$ are hyperfinite-dimensional approximations of the respective operators $U(u)$ and $V(v)$.*

PROOF. As a set $\mathfrak{M}$ satisfying the conditions of Lemma 1.3.2 we consider the set of characteristic functions of closed intervals. Then

$$U(u)(\chi_{[a,b]}) = \chi_{[a+uh,b+uh]}. \tag{2.1.11}$$

We remark that if $k\Delta$ and $m\Delta$ are near-standard, then $|k|, |m|, |k-m| < L$, so that $k\dot{-}m = k - m$, and then $U_d^r(F)(k) = F(k-r)$. If $f = \chi_{[a,b]}$, then

$$X_\Delta(f)(n) = \begin{cases} 0 & \text{if } n\Delta \notin {}^*[a,b], \\ 1 & \text{if } n\Delta \in {}^*[a,b], \end{cases}$$

and in view of the remark just made,

$$U_d^r(X_\Delta(f))(n) = \begin{cases} 0 & \text{if } n\Delta \notin {}^*[a+r\Delta, b+r\Delta], \\ 1 & \text{if } n\Delta \in {}^*[a+r\Delta, b+r\Delta]. \end{cases}$$

Similarly,

$$X_\Delta(U(u)(f))(n) = \begin{cases} 0 & \text{if } n\Delta \notin {}^*[a+uh, b+uh], \\ 1 & \text{if } n\Delta \in {}^*[a+uh, b+uh]. \end{cases}$$

It is now obvious that

$$\Delta \sum_{k=-L}^{L} |U_d^r(X_\Delta(f))(n) - X_\Delta(U(u)(f))(n)|^2 = 2\Delta\big[|(uh - r\Delta)/\Delta|\big] \approx 0,$$

since $uh \approx r\Delta$.

It can be directly verified that $V(v) = \mathcal{F}_h^{-1}U^{-1}(v)\mathcal{F}_h$ (see Corollary 2.1.2), and $V_d = \Phi_\Delta^{-1}U_d^{-1}\Phi_\Delta$.

We let $\Delta_1 = 2\pi p\widehat{\Delta}$ and regard U_d^{-1} as an operator acting from the space $\mathcal{L}_{2,\Delta_1}$ into $\mathcal{L}_{2,\Delta_1}$. By the conditions of the proposition, $m\Delta_1 \approx vh$, and by what was proved above, U_d^{-m} is a hyperfinite-dimensional approximation of $U^{-1}(v)$. By Corollary 2.1.2, $\Phi_\Delta\colon \mathcal{L}_{2,\Delta} \longrightarrow \mathcal{L}_{2,\Delta_1}$ is a hyperfinite-dimensional approximation of $\mathcal{F}_h$, that is, $V_d^m = \Phi_\Delta^{-1}U_d^{-m}\Phi_\Delta$ is a hyperfinite-dimensional approximation of $\mathcal{F}_h^{-1}U^{-1}(v)\mathcal{F}_h = V(v)$. □

Under the conditions of the proposition $\exp(2\pi irm/N) \approx \exp(\pi ihuv)$, that is, the relations (2.1.10) pass into the relations (2.1.9).

Let us compare the result on hyperfinite-dimensional approximation of the FT obtained here with the results on hyperfinite approximation of Hilbert–Schmidt operators (§3 of Chapter 1). The Fourier transformation $\mathcal{F}$ is an operator with bounded analytic kernel $k_{\mathcal{F}}(x,y) = \exp(-2\pi ixy)$. In the case of Hilbert–Schmidt operators with bounded kernels $k \in L_2(\mathbb{R}^2)$ that are continuous almost everywhere and have $|k|^2$ satisfying condition (1.3.7) (that is, $k \in \widetilde{\mathcal{H}}_2$) Corollary 1.3.10 holds, and this shows that the operator $A_k\colon \mathcal{L}_{2,\Delta} \longrightarrow \mathcal{L}_{2,\Delta}$ with matrix $(\Delta^* k(\alpha\Delta, \beta\Delta))_{\alpha,\beta=-L}^L$ is a hyperfinite-dimensional approximation of the operator with kernel k. If Φ_Δ is regarded as an operator acting from $\mathcal{L}_{2,\Delta}$ to $\mathcal{L}_{2,\Delta}$, then for it to be a hyperfinite-dimensional approximation of $\mathcal{F}$ it suffices according to Theorem 2.1.1 to require that $N\Delta^2 = 1$ (or $N\Delta^2 \approx 1$). Under this relation the matrix of Φ_Δ is obviously

$$(\Delta \cdot k_{\mathcal{F}}(\alpha\Delta, \beta\Delta))_{\alpha,\beta=-L}^L. \tag{2.1.12}$$

The next proposition shows that under different relations between N and Δ the operator with matrix (2.1.12) can fail to be a hyperfinite-dimensional approximation of $\mathcal{F}$.

PROPOSITION 2.1.5. *If $N\Delta^2 = 2$, then the operator $B\colon \mathcal{L}_{2,\Delta} \longrightarrow \mathcal{L}_{2,\Delta}$ with matrix* (2.1.12) *is not a hyperfinite-dimensional approximation of the FT $\mathcal{F}$.*

PROOF. The operator B is given by the matrix

$$(\Delta \exp(-4\pi inm/N))_{n,m=-L}^L.$$

We set $f = \chi_{[0;\sqrt{3/2}]}$ and show that

$$^\circ(\|B(X_\Delta(f)) - X_\Delta(\mathcal{F}(f))\|_\Delta) > 0. \tag{2.1.13}$$

We choose an infinite natural number T such that $(T-1)\Delta \leq \sqrt{3/2} < T\Delta$. Elementary computations show that (2.1.13) follows from the inequality

$$^\circ\left(\Delta^3 \sum_{m=L-T}^{L} |1-\exp(-4\pi imT/N)|^2|(1-\exp(-4\pi im/N))^{-1} - N/(4\pi im)|^2\right) > 0.$$

Since $T/N \approx 0$, it is easy to show that the last sum is greater than or equal to

$$\Delta^3 \sum_{m=L-T}^{L} \sin^2 \frac{2\pi mT}{N} \Big/ \sin^2 \frac{2\pi m}{N}. \tag{2.1.14}$$

Obviously, $\sin^2(2\pi m/N)$ is decreasing for $L-T \leq m \leq L$. Let $S = [2T/3]$. Then $\pi T - 0.75\pi - \gamma \leq 2\pi mT/N \leq \pi T - \pi/2 - \delta$ for $M-T \leq m \leq M-S$, where $\gamma, \delta \approx 0$, and hence $\sin^2(2\pi mT/N)$ is increasing in this interval. It follows that the terms with such indices in the sum (2.1.4) are increasing. Further, the term corresponding to $m = L-T$ is greater than or equal to $\mathfrak{D}\cdot\Delta^{-2}$ for some standard constant $\mathfrak{D} > 0$. It is now obvious that the whole sum (2.1.4) is greater than or equal to $\mathfrak{D}(T-S)\Delta$. It is easy to see that this number is not infinitesimal. □

2°. In this subsection Theorem 2.1.1 will be extended to a certain class of generalized functions of moderate growth (tempered distributions). Here it is assumed that $N = 2L+1$ and Δ satisfy the additional condition

$$0 < {}^\circ(N\Delta^2) < +\infty. \tag{2.1.15}$$

Note that if $0 < {}^\circ(N\Delta\Delta') < +\infty$, then the pair $\langle N, \Delta'\rangle$ also satisfies (2.1.15).

First of all we study how to associate elements of ${}^*\mathbb{C}^X$ with generalized functions. To do this we begin by considering the operator $\mathfrak{D}_d\colon \mathfrak{L}_{2,\Delta} \longrightarrow \mathfrak{L}_{2,\Delta}$ given by the matrix $((2\Delta)^{-1}d_{nk})_{n,k=-L}^{L}$, where

$$d_{nk} = \begin{cases} 1 & \text{if } k = n \dotplus 1, \\ -1 & \text{if } k = n \dot{-} 1, \\ 0 & \text{otherwise,} \end{cases}$$

that is, if $\mathbb{G} = \mathfrak{D}_d F$, then $\mathbb{G}(k) = (F(k+1) - F(k-1))(2\Delta)^{-1}$. It is clear that $\|\mathfrak{D}_d\| \sim +\infty$, that is, we cannot speak of the operator $\mathfrak{D}_d^{\#}$. Nevertheless, we have

PROPOSITION 2.1.6. *If f belongs to the Schwartz space $S(\mathbb{R})$ and n is a standard natural number, then*

$$\|\mathfrak{D}_d^n X_\Delta(f) - X_\Delta(f^{(n)})\|_{2,\Delta} \approx 0. \tag{2.1.16}$$

We first prove some lemmas.

LEMMA 2.1.7. *If $G^{(n)} = \mathfrak{D}_d^n F$, then*

$$G^{(n)}(k) = \frac{1}{(2\Delta)^n} \sum_{r=0}^{n} (-1)^r \binom{n}{r} F(k \dotplus n \dot{-} 2r). \tag{2.1.17}$$

Furthermore, if $|k| \leq L-n$, then in (2.1.17) $\dotplus$ and $\dot{-}$ can be replaced by $+$ and $-$, respectively.

LEMMA 2.1.8. *If in the notation of the preceding lemma*

$$\Delta^{-s}F(\pm(L-t)) \approx 0$$

for any standard s *and* t, *then* $G^{(n)}(\pm(L-m)) \approx 0$ *for any standard* n *and* m.

LEMMA 2.1.9. *If* $f \in S(\mathbb{R})$, *then* $\Delta^{-s*}f(\pm(L-t)\Delta) \approx 0$ *for any standard* s *and* t.

PROOF. Since $L\Delta \sim +\infty$ and t is standard, $(L-t)\Delta \sim +\infty$, and since $f \in S(\mathbb{R})$, we have $[(L-t)\Delta]^{-s}\,{}^*f(\pm(L-t)\Delta) \approx 0$. Thus,

$$0 \approx \Delta^{-s}[(L-t)\Delta^2]^s\,{}^*f(\pm(L-t)\Delta).$$

It now remains to use condition (2.1.15). □

PROOF OF PROPOSITION 2.1.6. By Lemmas 2.1.8 and 2.1.9,

$$\Delta \sum_{|k|>L-n} |(\mathcal{D}_d^n X_\Delta(f))(k) - X_\Delta(f^{(n)})(k)|^2 \approx 0,$$

that is, it suffices by (2.1.17) to show that

$$0 \approx S = \Delta \sum_{k=-L+n}^{L-n} \left| \frac{1}{(2\Delta)^n} \sum_{r=0}^{n} (-1)^r \binom{n}{r} {}^*f((k+n-2r)\Delta) - {}^*f^{(n)}(k\Delta) \right|^2.$$

By Taylor's formula,

$$f(k\Delta + (n-2r)\Delta) = \sum_{s=0}^{n} \frac{f^{(s)}(k\Delta)}{s!}(n-2r)^s\Delta^s + \frac{f^{(n+1)}(\xi r)}{(n+1)!}(n-2r)^{n+1}\Delta^{n+1}.$$

From this,

(2.1.18)

$$\begin{aligned} &\frac{1}{(2\Delta)^n} \sum_{r=0}^{n} (-1)^r \binom{n}{r} f((k+n-2r)\Delta) \\ &\qquad = \frac{1}{(2\Delta)^n} \sum_{s=0}^{n} \frac{f^{(s)}(k\Delta)}{s!} \Delta^s \sum_{r=0}^{n} (-1)^r \binom{n}{r} (n-2r)^s \\ &\qquad + \frac{\Delta}{2^n} \sum_{r=0}^{n} \frac{f^{(n+1)}(\xi r)}{(n+1)!} (n-2r)^{n+1}. \end{aligned}$$

Using the easily verified formula

$$\sum_{r=0}^{n} (-1)^r r^s \binom{n}{r} = \begin{cases} 0 & \text{if } s < n, \\ (-1)^n \cdot n! & \text{if } s = n, \end{cases}$$

we get that the first term on the right-hand side in (2.1.18) is equal to $f^{(n)}(k\Delta)$. Since $f^{(n+1)}$ is bounded on $\mathbb{R}$, the modulus of the second term on the right-hand side in (2.1.18) is bounded above by the number $B\Delta$ for some standard constant B. Then $S \le \Delta \cdot 2(L-n)B^2\Delta^2$, and it remains to employ the relation (2.1.15). □

We define the sequence $\langle \mathfrak{L}^{(n)} \mid n \in \omega \rangle$ of external subspaces of the space $\mathfrak{L}_{2,\Delta}$ as follows:

$$\mathfrak{L}^{(0)} = \left\{ F \in \mathfrak{L}_{2,\Delta} \;\middle|\; \forall^{\mathrm{st}} a\, \exists^{\mathrm{st}} C \left(\Delta \sum_{k=-[a/\Delta]}^{[a/\Delta]} |F(k)|^2 < C \right) \right\};$$

$$\mathfrak{L}^{(n+1)} = \mathfrak{D}_d(\mathfrak{L}^{(n)});$$

$$\mathfrak{L}^{(\sigma)} = \bigoplus_{n=0}^{\infty} \mathfrak{L}^{(n)}.$$

Let $F \in {}^*\mathbb{C}^X$ be such that ${}^\circ(\Delta \sum_{k=-L}^{L} F(k)^* f(k\Delta)) = {}^\circ(F, \overline{X_\Delta(f)}) < \infty$ for any standard $f \in C_0^\infty(\mathbb{R})$. Then a linear (not necessarily continuous) functional $\psi_F^\Delta \colon C_0^\infty(\mathbb{R}) \longrightarrow \mathbb{C}$ is defined by

$$\psi_F^\Delta(f) = {}^\circ(F, X_\Delta(f))_\Delta. \tag{2.1.19}$$

THEOREM 2.1.10. 1. *$\psi_F^\Delta \in (C_0^\infty(\mathbb{R}))'$ for any $F \in \mathfrak{L}^{(\sigma)}$.*

2. *If $f \in (C_0^\infty(\mathbb{R}))'$ and $f = \varphi^{(k)}$ for some regular generalized function φ $(k \geq 0)$, then there exists an $F \in \mathfrak{L}^{(\sigma)}$ such that $\psi_F^\Delta = f$.*

3. *$\psi_{\mathfrak{D}_d F}^\Delta = (\psi_F^\Delta)'$ for any $F \in \mathfrak{L}^{(\sigma)}$.*

PROOF. Let $(F,G)_\Delta^A = \Delta \sum_{k=-A}^{A} F(k)\overline{G(k)}$ for any $A \leq L$. Suppose that $f, f_n \in C_0^\infty(\mathbb{R})$ and $f_n \to f$ in $C_0^\infty(\mathbb{R})$ as $n \to \infty$. Then there exists a standard $a > 0$ such that $\operatorname{supp} f_n, \operatorname{supp} f \subseteq [-a,a]$ and f_n converges to f uniformly. It follows from the compactness of the supports that $\|f_n\|_{L_2}$ and $\|f\|_{L_2}$ are finite, and $\|f - f_n\|_{L_2} \to 0$ as $n \to \infty$. Let $A = [a/2]$; then

$$\|f\|_{L_2} = {}^\circ\|X_\Delta(f)\|_{2,\Delta}^A, \qquad \|f_n\|_{L_2} = {}^\circ\|X_\Delta(f_n)\|_{2,\Delta}^A$$
$${}^\circ\|X_\Delta(f - f_n)\|_{2,\Delta}^A \underset{n\to\infty}{\longrightarrow} 0.$$

If now $F \in \mathfrak{L}^{(0)}$, then $|\psi_F(f)| = |{}^\circ(F, \overline{X_\Delta(f)})_\Delta| = |{}^\circ(F, \overline{X_\Delta(f)})_\Delta^A| \leq {}^\circ\|F\|_{2,\Delta}^A \cdot {}^\circ\|X_\Delta(f)\|_{2,\Delta}^A$. It follows from the definition of $\mathfrak{L}^{(0)}$ that ${}^\circ\|F\|_{2,\Delta}^A$ is finite, and it was just proved that ${}^\circ\|X_\Delta(f)\|_{2,\Delta}^A$ is finite. Thus, $\psi_F^\Delta(f)$ is finite for any $f \in C_0^\infty(\mathbb{R})$. It is proved similarly that $\psi_F^\Delta(f_n - f) \to 0$ as $n \to \infty$, that is, $\psi_F^\Delta \in (C_0^\infty(\mathbb{R}))'$.

Further, $(\mathfrak{D}_d^n F, \overline{X_\Delta(f)}) = (-1)^n (F, \overline{\mathfrak{D}_d^n X_\Delta(f)})$. It follows from Proposition 2.1.6 that $\mathfrak{D}_d^n X_\Delta(f) = X_\Delta(f^{(n)}) + T$, where $\|T\|_{2,\Delta} \approx 0$. Moreover, since $\operatorname{supp} f^{(n)} \subseteq [-a,a]$, it follows that $X_\Delta(f^{(n)})(k) = 0$ for $|k| > A$. The formula (2.1.17) shows that $\mathfrak{D}_d^n X_\Delta(f)(k) = 0$ for $|k| > A + n$. Since n is standard, we arrive at the conclusion that $T(k) = 0$ for $|k| > [b/\Delta]$ for any standard $b > a$. Thus, if $B = [b/\Delta]$, then $(F,T)_\Delta = (F,T)_\Delta^B \approx 0$ because $\|F\|_{2,\Delta}^B$ is finite, and $\|T\|_{2,\Delta}^B = \|T\|_{2,\Delta} \approx 0$. This implies that $(\mathfrak{D}_d^n F, \overline{X_\Delta(f)}) \approx (-1)^n (F, \overline{X_\Delta(f^{(n)})})$. But this means that $\psi_{\mathfrak{D}_d^n F}^\Delta(f) = \psi_F^{\Delta(n)}(f)$. This at once proves assertions 1 and 3 of the theorem. To prove assertion 2 we observe that it can be assumed without loss of generality that φ is continuous, and then $X_\Delta(\varphi) \in \mathfrak{L}^{(0)}$. □

Now let $\mathfrak{C}^{(0)} = \operatorname{fin}(\mathfrak{L}_{2,\Delta})$, $\mathfrak{C}^{(n+1)} = \mathfrak{D}_d \mathfrak{C}^{(n)}$, and $\mathfrak{C}^{(\sigma)} = \bigoplus_{n=0}^{\infty} \mathfrak{C}^{(n)}$. Obviously, $\mathfrak{C}^{(0)} \subseteq \mathfrak{L}^{(0)}$, and hence $\mathfrak{C}^{(n)} \subseteq \mathfrak{L}^{(n)}$ for any n, therefore, $\mathfrak{C}^{(\sigma)} \subseteq \mathfrak{L}^{(\sigma)}$.

PROPOSITION 2.1.11. 1. *If $F \in \mathfrak{C}^{(\sigma)}$, then ψ_F^Δ is a generalized function of moderate growth.*

2. *If $f = \varphi^{(k)}$ and $\varphi \in L_2(\mathbb{R})$, then there exists an $F \in \mathfrak{C}^{(\sigma)}$ such that $f = \psi_F^\Delta$.*

The proof of the proposition is analogous to that of the preceding theorem. □

Suppose now that Δ' satisfies (2.1.2) and, as above, $\Phi_\Delta \colon \mathcal{L}_{2,\Delta} \longrightarrow \mathcal{L}_{2,\Delta'}$ is the DFT. It follows from the boundedness of the norm of Φ_Δ that $\Phi_\Delta(\mathfrak{C}^{(0)}) = \operatorname{fin}(\mathcal{L}_{2,\Delta'}) = \widehat{\mathfrak{C}}^{(0)}$, and $\Phi_\Delta^{-1}(\widehat{\mathfrak{C}}^{(0)}) = \mathfrak{C}^{(0)}$. We let $\mathcal{M}_d = \Phi_\Delta \mathcal{D}_d \Phi_\Delta^{-1} \colon \mathcal{L}_{2,\Delta'} \longrightarrow \mathcal{L}_{2,\Delta'}$ and define the sequence $\langle \widehat{\mathfrak{C}}^{(n)} \mid n \in \omega \rangle$ of external subspaces of $\mathcal{L}_{2,\Delta'}$ by setting $\widehat{\mathfrak{C}}^{(n+1)} = \mathcal{M}_d(\widehat{\mathfrak{C}}^{(n)})$. It is now obvious that $\Phi_\Delta(\mathfrak{C}^{(n)}) = \widehat{\mathfrak{C}}^{(n)}$ and $\widehat{\mathfrak{C}}^{(\sigma)} = \bigoplus_{n=0}^\infty \widehat{\mathfrak{C}}^{(n)} = \Phi_\Delta \mathfrak{C}^{(\sigma)}$.

A direct computation shows that the operator $\mathcal{M}_d$ is given by the matrix

$$\left(\frac{i}{\Delta} \sin \frac{2\pi\beta}{N} \delta_{\alpha\beta} \right)_{\alpha,\beta=-L}^{L}. \tag{2.1.20}$$

For any $G \in \mathbb{C}^X$ and any $f \in C_0^\infty(\mathbb{R})$ the quantity $\psi_G^{\Delta'}(f)$ is given by the formula (2.1.19) with Δ replaced by Δ' and F by G.

THEOREM 2.1.12. 1. $\psi_G^{\Delta'} \in (S(\mathbb{R}))'$ $\forall G \in \widehat{\mathfrak{C}}^{(\sigma)}$.

2. $\mathcal{F}(\psi_F^\Delta) = \psi_{\Phi_\Delta(F)}^{\Delta'}$ $\forall F \in \mathfrak{C}^{(\sigma)}$.

PROOF. Denote by $\mathcal{D} \colon S(\mathbb{R}) \longrightarrow S(\mathbb{R})$ the differentiation operator, and let $\mathcal{M} = \mathcal{F}\mathcal{D}\mathcal{F}^{-1}$. Then $\mathcal{M}(f)(x) = 2\pi i x f(x)$. We need

LEMMA 2.1.13. *For any $f \in S(\mathbb{R})$*

$$^\circ\|\mathcal{M}_d^n(X_{\Delta'}(f)) - X_{\Delta'}(\mathcal{M}^n(f))\| = 0. \tag{2.1.21}$$

PROOF OF LEMMA 2.1.13. In view of (2.1.20) it is required to show that

$$\Delta' \sum_{\beta=-L}^{L} \left| \left(\frac{1}{\Delta} \sin \frac{2\pi\beta}{N} \right)^n f(\beta\Delta') - (2\pi\beta\Delta')^n f(\beta\Delta') \right|^2 \approx 0. \tag{2.1.22}$$

We show first that if $T < L$ but $T\Delta' \sim +\infty$, then

$$\mathcal{W} = \Delta' \sum_{|\beta|>T} |f(\beta\Delta')|^2 \cdot \frac{1}{\Delta^{2n}} \left| \left(\sin \frac{2\pi\beta}{N} \right)^n - (2\pi\beta\Delta\Delta')^n \right|^2 \approx 0.$$

We have that

$$\left| \left(\sin \frac{2\pi\beta}{N} \right)^n - (2\pi\beta\Delta\Delta')^n \right|^2 = \left| \frac{\beta}{N} \right|^{2n} \left| \left(\frac{\beta}{N} \right)^{-n} \left(\sin \frac{2\pi\beta}{N} \right)^n - (2\pi N\Delta\Delta')^n \right|^2 \le C \left| \frac{\beta}{N} \right|^{2n}$$

for some standard constant C, by (2.1.2).

From this,

$$\mathcal{W} \leq C_1\Delta' \sum_{|\beta|>T} |f(\beta\Delta')|^2|\beta\Delta'|^{2n}. \tag{2.1.23}$$

If $\varphi(x) = x^{2n}f(x)$, then since $f(x) \in S(\mathbb{R})$, it follows that $\varphi(x) \in L_2(\mathbb{R})$. Moreover, φ is continuous and bounded, and it satisfies (1.2.21), that is, the internal function $G\colon X \longrightarrow \mathbb{C}$ given by $G(\beta) = {}^*f(\beta\Delta')(\beta\Delta')^n$ is a lifting of φ, and $G \in \mathcal{S}_2(M)$. This implies that the right-hand side of the inequality (2.1.23) is an infinitesimal, that is, $\mathcal{W} \approx 0$.

We take an $a \sim +\infty$ such that $N\Delta' \overset{a}{\sim} +\infty$, and $N\Delta\Delta' - 1 \overset{a}{\approx} 0$. This can be done on the basis of Proposition 1.1.12. Let $T = [a/\Delta']$. Then T satisfies the preceding conditions, that is, for a proof of (2.1.22) it suffices in view of the boundedness of f to show that

$$\mathcal{W}_1 = \Delta' \sum_{\beta=1}^{T} \left| \left(\frac{1}{\Delta}\sin\frac{2\pi\beta}{N}\right)^n - (2\pi\beta\Delta')^n \right|^2 \approx 0.$$

Let $1 \leq \beta \leq T$. Then $0 < \beta/N < a/(N\Delta') \overset{a}{\approx} 0$, that is,

$$\left(\sin\frac{2\pi\beta}{N}\right)\left(\frac{2\pi\beta}{N}\right)^{-1} = 1 - \alpha_\beta, \quad \text{where } \alpha_\beta \overset{a}{\approx} 0.$$

In this case $\Delta^{-n}\left(\sin\frac{2\pi\beta}{N}\right)^n = (1-\alpha_\beta)(2\pi\beta\Delta')^n(N\Delta\Delta')^{-n}$. By the choice of a, $(N\Delta\Delta')^{-n} = 1+\delta$, where $\delta \overset{a}{\approx} 0$. Finally, $\Delta^{-n}\left(\sin\frac{2\pi\beta}{N}\right)^n = (1+\gamma_\beta)(2\pi\beta\Delta')^n$, where $\gamma_\beta \overset{a}{\approx} 0$. If $\gamma = \max\{|\gamma_\beta| \mid 1 \leq \beta \leq T\}$, then $\gamma \overset{a}{\approx} 0$, so that

$$\mathcal{W}_1 \leq \Delta'\gamma^2 \sum_{\beta=1}^{T} (2\pi\beta\Delta')^{2n} \leq \Delta'\gamma^2(2\pi)^n(T\Delta')^{2n} \leq (2\pi)^n\gamma^2 a^{2n}\Delta' \approx 0,$$

because $\gamma \overset{a}{\approx} 0$. □

We continue the proof of the theorem. Its first assertion can now be proved just like Theorem 2.1.10 and Proposition 2.1.11. It obviously suffices to prove the second assertion for $F = \mathcal{D}_d^n G$, where $G \in \mathfrak{C}^{(\sigma)}$. We have that

$$\begin{aligned}
\psi^{\Delta'}_{\Phi_\Delta(F)}(f) &= {}^\circ(\Phi_\Delta(\mathcal{D}_d^n G), \overline{X_{\Delta'}(f)}) = {}^\circ(\mathcal{D}_d^n G, \Phi_\Delta^{-1}(X_{\Delta'}(\overline{f})))\\
&= (-1)^{n\circ}(G, \mathcal{D}_d^n\Phi_\Delta^{-1}X_{\Delta'}(\overline{f})) = (-1)^{n\circ}(G, \Phi_\Delta^{-1}\mathcal{M}_d X_{\Delta'}(\overline{f}))\\
&= (-1)^{n\circ}(G, \Phi_\Delta^{-1}X_{\Delta'}(\mathcal{M}^n(\overline{f}))) = (-1)^{n\circ}(G, X_{\Delta'}(\mathcal{F}^{-1}\mathcal{M}^n(\overline{f})))\\
&= (-1)^{n\circ}(G, X_\Delta(\mathcal{D}^n\mathcal{F}^{-1}(\overline{f}))) = (-1)^{n\circ}(G, \mathcal{D}_d^n X_\Delta(\mathcal{F}^{-1}(\overline{f})))\\
&= {}^\circ(\mathcal{D}_d^n G, X_\Delta(\mathcal{F}^{-1}(\overline{f}))) = {}^\circ(F, X_\Delta(\mathcal{F}^{-1}(\overline{f})))\\
&= \psi_F^\Delta(\mathcal{F}^{-1}(\overline{f})) = (\mathcal{F}\psi_F^\Delta)(f). \ \square
\end{aligned}$$

The theorem proved can serve as a basis for approximating the Fourier transformation of generalized functions of moderate growth by the discrete Fourier transformation. Unfortunately, such a basis has been obtained only for those functions that are derivatives of functions in $L_2(\mathbb{R})$.

The next example shows that this class can be enlarged.

EXAMPLE. Let $\varphi = 1$, so that φ is a regular generalized function, and by the proof of Theorem 2.1.10, $\varphi = \psi_F$, where $F = X_\Delta(f)$, that is, $F(k) = 1\ \forall k \in X$. We have that $\mathcal{F}(1) = \delta$. Further, $\Phi_\Delta(F)(k) = N\Delta\delta_{k0}$, that is, if $G = \Phi_\Delta(F)$, then $\psi_G^{\Delta'}(f) = {}^\circ(G, X_\Delta(f))_{\Delta'} = {}^\circ(N\Delta\Delta'{}^*f(0)) = f(0) = \delta(f)$, so, $\mathcal{F}(\psi_F) = \psi^{\Delta'}_{\Phi_\Delta(F)}$.

§2. Construction of LCA groups from hyperfinite Abelian groups

1°. We consider the additive group G of the ring ${}^*\mathbb{Z}/N{}^*\mathbb{Z} = \{-L, \dots, L\}$ (see §1, 1° in this chapter). It is clear that G is an internal hyperfinite Abelian group. In it we distinguish the two natural external subgroups $G_0 = \{k \in G \mid k\Delta \approx 0\}$ and $G_f = \{k \in G \mid {}^\circ|k\Delta| < +\infty\}$. Obviously, G_0 is the intersection of a countable family of internal sets ($G_0 = \bigcap_{n\in\omega} j^{-1}({}^*(-n^{-1}, n^{-1}))$), and G_f is the union of a countable family of internal sets ($G_f = \bigcup_{n\in\omega} j^{-1}({}^*[-n, n])$). Further, the mapping $\mathrm{st}\colon G_f \longrightarrow \mathbb{R}$ is an epimorphism, and $\ker \mathrm{st} = G_0$, that is, $\mathbb{R} \cong G_f/G_0$. Suppose that the internal set A is equal to $j^{-1}({}^*(a,b))$ for some $a, b \in \mathbb{R}$. It is easy to see that $\mathrm{st}(A) = [a,b]$. We define the external set $A^\circ = \{c \in A \mid c + G_0 \subseteq A\}$. Then it is easy to verify that $\mathrm{st}(A^\circ) = (a,b)$. A similar definition of A° applies for any internal set A. Here it is easy to prove that if A is internal, then $\mathrm{st}(A)$ is closed, and $\mathrm{st}(A^\circ)$ open. Thus, $\{\mathrm{st}(A^\circ) \mid A$ an internal subset of $G_f\}$ is a base for the topology of $\mathbb{R}$. It is also obvious that ${}^\circ(\Delta \cdot |j^{-1}({}^*[a,b])|) = b - a$. From this it follows at once that $\langle G_f, S_\Delta^{G_f}, \nu_\Delta^{G_f}\rangle$ is a σ-finite subspace of the Loeb space $\langle G, S_\Delta, \nu_\Delta\rangle$ (see §2, 2° of Chapter 1). In this case $\nu_\Delta^{G_f}$ is an invariant measure on G_f, and the preceding equality shows that the mapping $\mathrm{st}\colon G_f \longrightarrow \mathbb{R}$ is measure-preserving if Lebesgue measure is fixed as the Haar measure on $\mathbb{R}$.

Let $\widehat{G}$ be the group of characters of G. Then the internal mapping $n \longmapsto \chi_n$ with $\chi_n(m) = \exp(2\pi imn/N)\ \forall n, m \in G$ is an isomorphism of G and $\widehat{G}$. This assertion follows from the transfer principle, since it is true for any standard N.

In order for the character $\chi_n\colon G \longrightarrow {}^*\mathbb{C}$ to induce a character $\varkappa\colon \mathbb{R} \longrightarrow \mathbb{C}$ with the help of the homomorphism st it is necessary that $\chi_n|G_0 \approx 1$ (it is easy to see that this is a necessary and sufficient condition). It is thereby natural to single out in $\widehat{G}$ the external subgroup $H_f = \{\chi \in \widehat{G} \mid \chi|G_0 \approx 1\}$ and to define the monomorphism $\widehat{\mathrm{st}}\colon H_f \longrightarrow \widehat{\mathbb{R}}$ with $\widehat{\mathrm{st}}(\chi)(n) = {}^\circ\chi({}^\circ(n\Delta))\ \forall n \in \omega\ \forall \chi \in \widehat{G}$. Further, $\ker \widehat{\mathrm{st}} = H_0 = \{\chi \in H_f \mid \chi|G_f \approx 1\}$. It is now obvious that $H_f/H_0 \subseteq \widehat{\mathbb{R}}$. We prove the following equivalences: $\forall n \in G$

$$
\begin{aligned}
&\chi_n \in H_f \Longleftrightarrow {}^\circ(|n \cdot \widehat{\Delta}|) < +\infty,\\
&\chi_n \in H_0 \Longleftrightarrow n \cdot \widehat{\Delta} \approx 0,
\end{aligned}
\tag{2.2.1}
$$

where $\widehat{\Delta} = (N\Delta)^{-1}$.

Indeed, if $n\widehat{\Delta}$ is finite, then $\chi_n(m) = \exp(2\pi imn/N) = \exp(2\pi im\Delta \cdot n\widehat{\Delta}) \approx 1$ because $m\Delta \approx 0$. Conversely, suppose now that $n\widehat{\Delta}$ is infinite. Let $m = [N/(2\pi)]$. Then it is obvious that $m\widehat{\Delta} \approx 0$. In this case $\chi_n(m) = \exp 2\pi(\frac{N}{2\pi})\Delta \cdot \frac{n}{N\Delta} \approx -1$ because $0 \le \alpha < 1$. The proof of the second of the equivalences in (2.2.1) is completely analogous. Thus, the groups $\widehat{G}_f$ and $\widehat{G}_0$ are constructed just like the groups G_f and G_0, so that $\widehat{G}_f/\widehat{G}_0 \cong \widehat{\mathbb{R}} \cong \mathbb{R}$.

For an arbitrary LCA group $\mathfrak{G}$ the Fourier transformation $\mathcal{F}\colon L_2(\mathfrak{G}) \longrightarrow L_2(\widehat{\mathfrak{G}})$ is defined by $\mathcal{F}(f)(\chi) = (f, \widehat{\chi})$, therefore, it is easy to see that Theorem 2.1.1 has a group interpretation. We proceed to the general situation.

2°. Let G be an internal hyperfinite Abelian group, and let $G_0 \subseteq G_f$ be subgroups of it satisfying the following two conditions.

(A) There exists a sequence $\{A_n \mid n \in \mathbb{N}\}$ of internal sets such that $A_n \subseteq G_f$ $\forall n \in \mathbb{N}$ and $G_0 = \cap\{A_n \mid n \in \mathbb{N}\}$.

(B) There exists a sequence $\{B_n \mid n \in \mathbb{N}\}$ of internal sets such that $B_n \supseteq G_0$ $\forall n \in \mathbb{N}$ and $G_f = \cup\{B_n \mid n \in \mathbb{N}\}$.

Thus, the subgroups G_0 and G_f can be either internal or external.

LEMMA 2.2.1. *If the subgroups $G_0 \subseteq G_f$ of the group G satisfy conditions* (A) *and* (B), *then there exists a sequence $\{C_n \mid n \in \mathbb{Z}\}$ of symmetric internal subsets of G such that:*

a) $\bigcap_{n\in\mathbb{Z}} C_n = G_0$;
b) $\bigcup_{n\in\mathbb{Z}} C_n = G_f$;
c) $C_n + C_n \subseteq C_{n+1}$ $\forall n \in \mathbb{Z}$.

Further, if $F \subseteq G$ is an internal set, then:

d) $F \subseteq G_f \iff \exists n \in \mathbb{Z}\ (F \subseteq C_n)$;
e) $F \supseteq G_0 \iff \exists n \in \mathbb{Z}\ (F \supseteq C_n)$.

PROOF. The lemma follows easily from the countable saturation of the nonstandard universe. □

Below we work with a fixed sequence $\{C_n \mid n \in \mathbb{Z}\}$ satisfying the conditions of the lemma.

If $F \subseteq G_f$, then we write

$$\overset{\circ}{F} = \{g \in G_f \mid g + G_0 \subseteq F\}$$

The following is a direct consequence of Lemma 2.2.1.

LEMMA 2.2.2. *If $F \subseteq G_f$ is an internal set, then $\forall g \in G\ (g \in \overset{\circ}{F} \iff \exists m \in \mathbb{Z}$ $(g + C_m \subseteq F))$.* □

We denote the family of all internal subsets of G_f by $\mathrm{In}(G_f)$, and the family $\{F \in \mathrm{In}(G_f) \mid G_0 \subseteq F\}$ by $\mathrm{In}_0(G_f)$.

Let $G^\# = G_f/G_0$, and let $j\colon G_f \longrightarrow G^\#$ be the natural projection. If $g \in G_f$ and $A \subseteq G_f$, then instead of $j(g)$ and $j(A)$ we write $g^\#$ and $A^\#$, respectively.

THEOREM 2.2.3. 1. *The family $\mathfrak{U} = \{\overset{\circ}{F}{}^\# \mid F \in \mathrm{In}(G_f)\}$ forms a base of neighborhoods of zero for a topology on $G^\#$.* (*This topology will be called canonical below.*)

2. *If $F \in \mathrm{In}(G_f)$, then $F^\#$ is closed in the canonical topology.*

3. *The topological group $G^\#$ is complete with respect to the uniformity.*

PROOF. 1. In view of the criterion in §18 of the book [**26**] and in view of the fact that $G^\#$ is Abelian, to see that $\mathfrak{U}$ is a system of neighborhoods of zero it is necessary to check the following conditions:

a) $\cap\{U \mid U \in \mathfrak{U}\} = \{e\}$;
b) $\forall U, V \in \mathfrak{U}\ \exists W \in \mathfrak{U}\ (W \subseteq U \cap V)$;
c) $\forall U \in \mathfrak{U}\ \exists V \in \mathfrak{U}\ (V - V \subset U)$;
d) $\forall U \in \mathfrak{U}\ \forall \xi \in U\ \exists V \in \mathfrak{U}\ (V + \xi \subseteq U)$.

Let us verify condition a). If $g^\# \in \cap\{\overset{\circ}{F}{}^\# \mid F \in \mathrm{In}_0(G_f)\}$, then $\forall F \in \mathrm{In}_0(G_f)$ $((g+G_0)\cap \overset{\circ}{F} \neq \emptyset)$. Here we use the fact that $j^{-1}(g^\#) = g + G_0$ and $j^{-1}(\overset{\circ}{F}{}^\#) = \overset{\circ}{F} + G_0 = \overset{\circ}{F}$ (see the definition of $\overset{\circ}{F}$). Consequently, $\exists g_1 \in G_0$ $(g + g_1 + G_0 \subseteq F)$, that is, $g \in F$. Thus, $g \in \cap\,\mathrm{In}_0(G_f) = G_0$.

Condition b) follows from the inclusion $(F_1 \cap F_2)^{\circ\#} \subseteq \overset{\circ}{F}{}_1^\# \cap \overset{\circ}{F}{}_2^\#$, which is verified similarly.

We check condition c). If $F \in \mathrm{In}_0(G_f)$, then by Lemma 2.2.1, $\exists n \in \mathbb{Z}$ $(C_n + C_n + C_n \subseteq F)$, that is, $C_n + C_n + G_0 \subseteq F$. Therefore, $\overset{\circ}{C}_n + \overset{\circ}{C}_n \subseteq \overset{\circ}{F}$, and hence $(\overset{\circ}{C}_n + \overset{\circ}{C}_n)^\# = \overset{\circ}{C}{}_n^\# + \overset{\circ}{C}{}_n^\# \subseteq \overset{\circ}{F}{}^\#$. Since C_n is symmetric, $\overset{\circ}{C}{}_n^\# = -\overset{\circ}{C}{}_n^\#$. Condition d) is verified similarly.

2. Let $F \in \mathrm{In}(G_f)$. We show that $F^\#$ is closed. If $g^\# \notin F^\#$, then $(g+G_0)\cap F = \emptyset$, that is, $\exists n \in \mathbb{Z}$ $((g+C_n)\cap F = \emptyset)$ (ω^+-saturation). Thus, $(g+C_{n-1}+G_0)\cap F = \emptyset$ (Lemma 2.2.1), hence $(g^\# + C_{n-1}^\#)\cap F^\# = \emptyset$, so that $(g^\# + \overset{\circ}{C}{}_{n-1}^\#)\cap F^\# = \emptyset$ because $\overset{\circ}{C}_{n-1} \subseteq C_{n-1}$. This proves that $F^\#$ is closed, since $\overset{\circ}{C}{}_{n-1}^\# \in \mathfrak{U}$.

3. It follows from Lemma 2.2.1 that the canonical topology on $G^\#$ satisfies the first axiom of countability, so it suffices to show that every Cauchy sequence of elements of $G^\#$ has a limit. Let $g_n^\#$ be a Cauchy sequence. This means that there exists a function $\nu\colon \mathbb{Z} \longrightarrow \mathbb{N}$ such that $\forall m \in \mathbb{Z}\ \forall n_1, n_2 > \nu(m)$ $(g_{n_1}^\# - g_{n_2}^\# \in \overset{\circ}{C}{}_m^\#)$. We consider the countable family of sets

$$\Gamma = \{A_{m,n} \mid n > \nu(m),\ n, m \in \mathbb{N}\},$$

where

$$A_{m,n} = \{g \mid g_n - g \in C_{m+1}\},$$

and we show that each finite subset S of Γ has a nonempty intersection. Let $S = \{A_{m_1,n_1}, \ldots, A_{m_k,n_k}\}$, and let $n > \max\{\nu(m_1), \ldots, \nu(m_k)\}$. Then $g_{n_i}^\# - g_n^\# \in \overset{\circ}{C}{}_{m_i}^\#$, therefore, $\exists g \in G_0$ $(g_{n_i} - g_n + g \in \overset{\circ}{C}_{m_i})$, so that $g_{n_i} - g_n + g + G_0 \subseteq C_{m_i}$. Since $g + G_0 = G_0$, it follows that $g_{n_i} - g_n \in C_{m_i}$ $(i = 1, \ldots, k)$, that is, $g_n \in \cap S$. The ω^+-saturation now gives us the existence of a $g \in \cap\Gamma$. By using the fact that $C_{m+1} \subseteq \overset{\circ}{C}_{m+2}$ (which follows immediately from Lemma 2.2.1) it is easy to get that $g_n^\# \to g^\#$ as $n \to \infty$. □

We recall that if the relation $|A| \in \mathbb{N}$ (and not ${}^*\mathbb{N}$) holds for an internal set A, then A is said to be standardly finite. According to Proposition 0.3.16 and the remark after it, any finite subset or subclass of the universe ${}^*\mathbb{R}$ is internal and standardly finite. In what follows, subclasses of ${}^*\mathbb{R}$ will also be called sets, in correspondence with the considerations in §6 of Chapter 0.

THEOREM 2.2.4. 1. *The canonical topology on $G^\#$ is locally compact and separable (that is, has a countable base) if and only if*

$$\forall F_1, F_2 \in \mathrm{In}_0(G_f)\ (F_1 \subseteq F_2 \longrightarrow \exists B \subseteq F_2\ (|B| \in \mathbb{N}\ \wedge\ F_1 + B \supseteq F_2)). \tag{2.2.2}$$

2. *If condition (2.2.2) holds, then $G^\#$ is compact (discrete) if and only if G_f (G_0) is an internal subgroup of G.*

PROOF. We assume that (2.2.2) holds and show that $G^\#$ is locally compact. It suffices to prove that $F^\#$ is compact for any $F \in \mathrm{In}_0(G_f)$. The closedness of $F^\#$ was shown in Theorem 2.2.3. We prove that for any neighborhood U of zero in $G^\#$ there is a finite set $\{v_1, \dots, v_k\} \subseteq F^\#$ such that

$$\bigcup_{i=1}^{k} v_i + U \supseteq F^\#.$$

Let $n \in \mathbb{Z}$ be such that $C_n \subseteq F$ and $\overset{\circ}{C}{}_n^\# \subseteq U$ (Lemma 2.2.1). Then $C_{n-1} \subseteq \overset{\circ}{C}_n$ ($C_{n-1} + C_{n-1} \subseteq C_n$; $G_0 \subseteq C_{n-1}$). It follows from (2.2.2) that there is a finite $B \subseteq F$ such that $C_{n-1} + B \supseteq F$. Then $\overset{\circ}{C}_n + B \supseteq F$ and $\overset{\circ}{C}{}_n^\# + B^\# \supseteq F^\#$. Of course, $B^\#$ is finite because B is standardly finite. The separability of $G^\#$ follows from its metrizability and the fact that $G^\# = \cup\{C_n^\# \mid n \in \mathbb{Z}\}$ and each $C_n^\#$ is compact.

Suppose, conversely, that $G^\#$ is locally compact and separable. It is easy to see that $\exists n_0 \in \mathbb{Z}\ \forall n \leq n_0$ ($C_n^\#$ is compact). We show that $F^\#$ is compact $\forall F \in \mathrm{In}(G_f)$. Let $V = \overset{\circ}{C}{}_{n_0-1}^\#$. Then $\cup\{g^\# + V \mid g \in G_f\} = G^\#$. By separability, there is a sequence $\{g_n\}$ such that $G^\# = \bigcup_n g_n^\# + V$, that is, $G_f = \bigcup_n g_n + \overset{\circ}{C}_{n_0-1} + G_0 \subseteq \bigcup_n g_n + C_{n_0} \subseteq G_f$. Consequently, $F \subseteq \bigcup_n g_n + C_{n_0}$, and in view of ω_1-saturation there exists a finite set $\{n_1, \dots, n_k\}$ such that $F \subseteq \bigcup_{i=1}^k g_{n_i} + C_{n_0}$. Thus, $F^\# \subseteq \bigcup_{i=1}^k g_{n_i} + C_{n_0}$, which also implies that $F^\#$ is compact. Suppose now that $G_0 \subseteq F_1 \subseteq F_2$, and let $n \leq n_0$ be such that $C_n \subseteq F_1$. There exist $g_1 \dots, g_k$ such that $g_1^\#, \dots, g_k^\# \in F_2^\#$ and $\bigcup_{i=1}^k g_i + \overset{\circ}{C}_{n-1} \supseteq F_2$. Let $h_1, \dots, h_k \in F_2$ and $h_i - g_i \in G_0$, that is, $h_i + G_0 = g_i + G_0$. Then $F_2 + G_0 \subseteq \bigcup_{i=1}^k g_i + \overset{\circ}{C}_{n-1} + G_0 \subseteq \bigcup_{i=1}^k h_i + C_n \subseteq \{h_1, \dots, h_k\} + F_1$. Setting $B = \{h_1, \dots, h_k\}$, we get (2.2.2).

Suppose now that $G^\#$ is compact. We take an arbitrary $F \in \mathrm{In}_0(G_f)$ and find $g_1, \dots, g_k \in G_f$ such that

$$G^\# = \bigcup_{i=1}^{k} g_i^\# + \overset{\circ}{F}{}^\# = \left(\bigcup_{i=1}^{k} g_i + \overset{\circ}{F}\right)^\# = \left(\bigcup_{i=1}^{k} F + g_i\right)^\# \qquad (\overset{\circ}{F} \subseteq F \subseteq G_f).$$

Obviously, $K = \bigcup_{i=1}^k F + g_i \subseteq G_f$ is internal, and $K^\# = G^\#$, that is, $G_f = K + G_0 \subseteq K + F \subseteq G_f$. Thus, $G_f = K + F$ is an internal set.

Suppose, conversely, that G_f is an internal subgroup of G. We show that $G^\#$ is compact (condition (2.2.2) is assumed, of course). It suffices to show that $\forall n \in \mathbb{Z}$ $\exists B \subseteq G_f$ ($|B| \in \mathbb{N} \wedge G^\# = \bigcup_{g \in B} g^\# + \overset{\circ}{C}{}_n^\#$). In view of Lemma 2.2.1, $C_{n-1} \subseteq \overset{\circ}{C}_n$, and by (2.2.2) there is a standardly finite B such that $G_f = B + C_{n-1} = B + \overset{\circ}{C}_n = \bigcup_{g \in B} g + \overset{\circ}{C}_n$. Now $G^\# = \bigcup_{g \in B} g^\# + \overset{\circ}{C}{}_n^\#$. The simple proof of the last assertion of the theorem is omitted. □

COROLLARY 2.2.5. *Under the conditions of Theorem 2.2.4, $F^\#$ is compact for any $F \in \mathrm{In}(G_f)$.*

We present without proof some simple properties, needed in what follows, of the concepts introduced.

LEMMA 2.2.6. 1. *If* $F \subseteq G_f$, *then* $g+\overset{\circ}{F} = (g+F)^\circ$, $j^{-1}(j(\overset{\circ}{F})) = \overset{\circ}{F}$, $(g+\overset{\circ}{F})^\# = g^\# + \overset{\circ}{F}{}^\#$, *and* $\overset{\circ}{F}{}^\# = C(G_f \setminus F)^\#$ (CA *is the complement of the set* A).

2. *If* $F \in \mathrm{In}(G_f)$, *then* $\overset{\circ}{F}{}^\#$ *is open, and the family* $\{\overset{\circ}{F}{}^\# \mid F \in \mathrm{In}(G_f)\}$ *forms a base for the canonical topology of* $G^\#$. □

Everywhere below it is assumed that the conditions of Theorem 2.2.4 hold.

LEMMA 2.2.7. $\forall F_1, F_2 \in \mathrm{In}_0(G_f)$ $(0 < {}^\circ(|F_1|/|F_2|) < +\infty)$.

PROOF. By Theorem 2.2.4, there exists a finite B (that is, $|B| \in \mathbb{N}$) such that $F_1+B \supseteq F_1 \cup F_2 \supseteq F_2$. Then $|F_2| \leq |F_1+B| \leq |F_1| \cdot |B|$, that is, ${}^\circ(|F_2|/|F_1|) < +\infty$. Similarly, ${}^\circ(|F_1|/|F_2|) < +\infty$, which proves what is required. □

DEFINITION 2.2.8. A hyperreal number $\Delta \in {}^*\mathbb{R}_+$ is called a normalizing multiplier (NM) of the triple $\langle G; G_0, G_f\rangle$ if $0 < {}^\circ(\Delta \cdot |F|) < +\infty$ for any $F \in \mathrm{In}_0(G_f)$.

It follows immediately from Lemma 2.2.7 that for any $F \in \mathrm{In}_0(G_f)$ the number $\Delta = |F|^{-1} \in {}^*\mathbb{R}_+$ is an NM, that is, an NM exists for any triple satisfying the conditions of Theorem 2.2.4. It also gives us that if Δ is an NM, then Δ' is an NM if and only if $0 < {}^\circ\left(\frac{\Delta}{\Delta'}\right) < +\infty$.

Lemma 2.2.7 shows that if Δ is an NM for $\langle G; G_0, G_f\rangle$, then $\langle G_f, S_\Delta^{G_f}, \nu_\Delta^{G_f}\rangle$ is a σ-finite subspace of the Loeb space $\langle G, S_\Delta, \nu_\Delta\rangle$. Everywhere below we write simply S instead of $S_\Delta^{G_f}$, and ν_Δ instead of $\nu_\Delta^{G_f}$.

LEMMA 2.2.9. $\forall A \in S\ \forall g \in G_f\ ((g + A \in S) \wedge (\nu_\Delta(g + A) = \nu_\Delta(A)))$.

PROOF. Obvious. □

Denote by $\mathcal{B}$ the σ-algebra of Borel subsets of $G^\#$.

LEMMA 2.2.10. $\forall B \in \mathcal{B}\ (j^{-1}(B) \in S)$.

PROOF. In view of Lemma 2.2.6 it suffices to show that $\forall F \in \mathrm{In}(G_f)$ $(j^{-1}((G_f \setminus F)^\#) \in S)$. This follows from the equalities

$$j^{-1}((G_f \setminus F)^\#) = G_f \setminus F + G_0 = \bigcup_{n\in\mathbb{Z}} (C_n \setminus F) + \bigcap_{m\in\mathbb{Z}} C_m,$$

$$\bigcup_{n\in\mathbb{Z}} (C_n \setminus F) + \bigcap_{m\in\mathbb{Z}} C_m = \bigcup_{n\in\mathbb{Z}} (C_n \setminus F + \bigcap_{m\in\mathbb{Z}} C_m),$$

$$C_n \setminus F + \bigcap_{m\in\mathbb{Z}} C_m = \bigcap_{m\in\mathbb{Z}} (C_n \setminus F + C_m).$$

In the proof of the last equality we must use the ω^+-saturation of the nonstandard universe. □

We now define a measure μ_Δ on $\mathcal{B}$ by setting

$$\mu_\Delta(B) = \nu_\Delta(j^{-1}(B)). \tag{2.2.3}$$

It follows immediately from Lemma 2.2.9 that μ_Δ is invariant, and it is regular because $G^\#$ is separable. Thus, μ_Δ is a Haar measure on $G^\#$. We denote by L the completion of the σ-algebra $\mathcal{B}$ with respect to μ_Δ. The extension of μ_Δ to L will be denoted again by μ_Δ.

THEOREM 2.2.10. 1. $\forall A \subseteq G^{\#}$ $(A \in L \Longleftrightarrow j^{-1}(A) \in S)$.
2. *The equality* (2.2.3) *is valid for any* $B \in L$.

PROOF. The fact that $\forall A \in L$ $(j^{-1}(A) \in S)$ and that (2.2.3) holds follows immediately from the completeness of the Loeb measure ν_Δ. It obviously suffices to prove the converse assertion for $A \subseteq G^{\#}$ such that $j^{-1}(A) \subseteq F \in \mathrm{In}_0(G_f)$. Suppose that this holds and $j^{-1}(A) \in S$. To show that $A \in L$ and that (2.2.3) holds for A it suffices to show that $\mu_{\mathrm{in}}(A) = \mu_{\mathrm{out}}(A) = \nu_\Delta(j^{-1}(A))$, where $\mu_{\mathrm{in}}(A)$ ($\mu_{\mathrm{out}}(A)$) is the inner (outer) Haar measure of the set A.

Since $\nu_\Delta(F)$ is finite, and hence also $\nu_\Delta(j^{-1}(A))$, there exists for any standard $\varepsilon > 0$ an internal $\mathcal{D} \subseteq j^{-1}(A)$ such that $\nu_\Delta(\mathcal{D}) \geq \nu_\Delta(j^{-1}(A)) - \varepsilon$ (Proposition 0.4.13). From the facts that $\mathcal{D}^{\#} \subseteq A$ and $\mathcal{D}^{\#}$ is closed it is then easy to get that $\mu_{\mathrm{in}}(A) \geq \nu_\Delta(j^{-1}(A))$.

Suppose now that $H \in \mathrm{In}_0(G_f)$ is such that $A \subseteq \overset{\circ}{H}{}^{\#}$ (for example, $F + F$ can be taken as H), and let $B = \overset{\circ}{H}{}^{\#} \setminus A$. Then $j^{-1}(B) = \overset{\circ}{H} \setminus j^{-1}(A)$ and $\nu_\Delta(j^{-1}(A)) + \nu_\Delta(j^{-1}(B)) = \nu_\Delta(\overset{\circ}{H}) = \mu_\Delta(\overset{\circ}{H}{}^{\#})$ $(j^{-1}(\overset{\circ}{H}{}^{\#}) = \overset{\circ}{H}$ by Lemma 2.2.6). This implies that $\mu_{\mathrm{in}}(A) + \mu_{\mathrm{in}}(B) \geq \mu_\Delta(\overset{\circ}{H}{}^{\#})$. On the other hand, it follows from the regularity of μ_Δ that $\mu_\Delta(\overset{\circ}{H}{}^{\#}) = \sup\{\mu_\Delta(E) \mid E \subseteq \overset{\circ}{H}{}^{\#},\ E \text{ is closed}\} \geq \sup\{\mu_\Delta(C) + \mu_\Delta(\mathcal{D}) \mid C \subseteq A,\ \mathcal{D} \subseteq B,\ C,\ \mathcal{D} \text{ are closed}\} = \sup\{\mu_\Delta(C) \mid C \subseteq A,\ C \text{ is closed}\} + \sup\{\mu_\Delta(\mathcal{D}) \mid \mathcal{D} \subseteq B,\ \mathcal{D} \text{ is closed}\} = \mu_{\mathrm{in}}(A) + \mu_{\mathrm{in}}(B)$. Thus, $\mu_{\mathrm{in}}(A) + \mu_{\mathrm{in}}(B) = \mu_\Delta(\overset{\circ}{H}{}^{\#})$. Now if $\varepsilon > 0$, then there exists a closed $C \subseteq B$ such that $\mu_\Delta(C) \geq \mu_{\mathrm{in}}(B) - \varepsilon$, and since A is contained in the open set $\overset{\circ}{H}{}^{\#} \setminus C$, it follows that $\mu_{\mathrm{out}}(A) \leq \mu_\Delta(\overset{\circ}{H}{}^{\#} \setminus C) = \mu_\Delta(\overset{\circ}{H}{}^{\#}) - \mu_\Delta(C) \leq \mu_\Delta(\overset{\circ}{H}{}^{\#}) - \mu_{\mathrm{in}}(B) + \varepsilon = \mu_{\mathrm{in}}(A) + \varepsilon$. Thus, $\mu_{\mathrm{out}}(A) = \mu_{\mathrm{in}}(A)$. It is now easy to get the equality (2.2.3) for A. □

Theorem 2.2.10 shows that the mapping $j\colon G_f \longrightarrow G^{\#}$ is measure-preserving, so that the considerations in §2, 3° of Chapter 1 apply.

If $f\colon G^{\#} \longrightarrow \mathbb{R}$ is a μ_Δ-measurable function, then $f \circ j\colon G_f \longrightarrow \mathbb{R}$ is a ν_Δ-measurable function. A lifting φ of it will be called a lifting of f. Thus, the internal function $\varphi\colon G \longrightarrow {}^*\mathbb{R}$ is a lifting of f if $f(g^{\#}) = {}^\circ\varphi(g)$ for ν_Δ-almost all $g \in G_f$. If $\varphi \in \mathcal{S}_p(G_f)$, then φ is called an $\mathcal{S}_p$-integrable lifting of f. It will sometimes be indicated that φ is an $\mathcal{S}_{p,\Delta}$-lifting of f.

COROLLARY 2.2.11. *Let* $p \in [1, \infty]$. *A function* f *is in* $L_p(\mu_\Delta)$ *if and only if it has an* $\mathcal{S}_{p,\Delta}$-*integrable lifting* $\varphi\colon G \longrightarrow {}^*\mathbb{R}$. *Furthermore, for* $p = 1$

$$\int_{G^{\#}} f\, d\mu_\Delta = {}^\circ\Big(\Delta \sum_{g \in G} \varphi(g)\Big), \tag{2.2.4}$$

and for arbitrary $p \in [1, \infty]$

$$\int_{G^{\#}} |f|^p\, d\mu_\Delta = {}^\circ\Big(\Delta \sum_{g \in G} |\varphi(g)|^p\Big).$$

3°. Let $G^{\frown}$ be the internal group of characters of G. It follows from the hyperfiniteness of G and the transfer principle that $G^{\frown}$ is isomorphic to G, so that $G^{\frown}$ is an internal hyperfinite Abelian group.

Following [**26**], we represent the group S^1 (the unit circle) as the interval $[-\frac{1}{2}, \frac{1}{2}]$ with addition modulo 1. We take in S^1 the countable system $\{\Lambda_k \mid k = 1, 2 \dots\}$ of neighborhoods of zero with $\Lambda_k = (-\frac{1}{3k}, \frac{1}{3k})$.

The next lemma [**26**] is obvious.

LEMMA 2.2.12. *If $\gamma \in \Lambda_1, 2\gamma \in \Lambda_1, \dots, k\gamma \in \Lambda_1$, then $\gamma \in \Lambda_k$.*

We consider the two external subgroups $H_0 \subseteq H_f$ of $G^\frown$ such that

$$\alpha \in H_0 \Longleftrightarrow \forall g \in G_f \ (\alpha(g) \approx 0),$$
$$\alpha \in H_f \Longleftrightarrow \forall g \in G_0 \ (\alpha(g) \approx 0), \tag{2.2.5}$$

along with the countable family $\{W(C_n, \Lambda_k) \mid n \in \mathbb{Z},\ k \in \mathbb{N}\}$ of internal subsets of $G^\frown$ such that $\alpha \in W(C_n, \Lambda_k) \Longleftrightarrow \forall g \in C_n\ (\alpha(g) \in \Lambda_k)$; $W(F, \Lambda_k)$ is defined similarly for any $F \in \mathrm{In}(G_f)$.

LEMMA 2.2.13. *$H_0 = \bigcap_{n,k} W(C_n, \Lambda_k)$ and $H_f = \bigcup_n W(C_n, \Lambda_1)$.*

PROOF. The first equality is almost obvious. We prove the second. Suppose that $\alpha \in W(C_n, \Lambda_1)$ and $m \in \mathbb{Z}$ is such that $k \cdot C_m \subseteq C_n$. In this case if $g \in C_m$, then $\alpha(g), 2\alpha(g), \dots, k\alpha(g) \in \Lambda_1$, and hence $\alpha(g) \in \Lambda_k$ in view of Lemma 2.2.12. Consequently, $\forall k\ \forall g \in G_0\ (\alpha(g) \in \Lambda_k)$, so that $\alpha(g) \approx 0$, hence $\alpha \in H_f$. Conversely, let $\alpha \in H_f$, and assume that $\forall n\ (\alpha \notin W(C_n, \Lambda_1))$. Then $\forall n\ \exists g \in C_n\ (|\alpha(g)| \geq \frac{1}{3})$. It follows from ω_1-saturation that $\exists g \in G_0\ (|\alpha(g)| \geq \frac{1}{3})$, which contradicts the fact that $\alpha \in H_f$. □

Thus, the triple $\langle G^\frown, H_0, H_f \rangle$ satisfies the same conditions as $\langle G, G_0, G_f \rangle$, and thus we can define the canonical topology on the group $G^{\frown\#} = H_f/H_0$. It follows from Theorem 2.2.3 that $G^{\frown\#}$ is complete in this topology. If $\alpha \in H_f$, then the image of α in $G^{\frown\#}$ is denoted by $\alpha^\#$.

LEMMA 2.2.14. *Let the internal function $f\colon G \longrightarrow {}^*\mathbb{C}$ be such that $\forall g_1, g_2$ $(g_1 - g_2 \in G_0 \Longrightarrow f(g_1) \approx f(g_2))$ and $\forall g \in G_f\ ({}^\circ|f(g)| < \infty)$. Then the function $\widetilde{f}\colon G^\# \longrightarrow \mathbb{C}$ with $\widetilde{f}(g^\#) = {}^\circ f(g)\ \forall g \in G_f$ is uniformly continuous on $G^\#$.*

The simple proof of this lemma is omitted.

We now define a mapping $\psi\colon G^{\frown\#} \longrightarrow G^{\#\frown}$ by setting

$$\psi(\alpha^\#)(g^\#) = {}^\circ(\alpha(g))$$

for any $\alpha \in H_f$ and for any $g \in G_f$. The fact that $\psi(\alpha^\#)$ is in $G^{\#\frown}$ follows from Lemma 2.2.14. It can be verified directly that ψ is well defined, and it is also easy to verify that it is an injective homomorphism.

THEOREM 2.2.15. *The mapping $\psi\colon G^{\frown\#} \longrightarrow \psi(G^{\frown\#}) \subseteq G^{\#\frown}$ is a topological isomorphism.*

PROOF. Recall [**26**] that the topology on the group $G^{\#\frown}$ dual to $G^\#$ is determined by a base of neighborhoods of zero of the form $\{\mathcal{W}(\mathcal{F}, \Lambda_k) \mid \mathcal{F} \subseteq G^\#$ is compact, $k = 1, 2, \dots\}$, where $\mathcal{W}(\mathcal{F}, \Lambda_k) = \{h \in G^{\#\frown} \mid \forall \xi \in \mathcal{F},\ h(\xi) \in \Lambda_k\}$. It is now easy to see that $\{\mathcal{W}(C_n^\#, \Lambda_k) \mid n \in \mathbb{Z},\ k \in \mathbb{N}\}$ is also a base of neighborhoods of zero in $G^{\#\frown}$. The continuity of ψ and ψ^{-1} follows from the easily checked relations

$${}^\circ\psi(\overset{\circ}{W}(C_n, \Lambda_{k+1})) \subseteq \mathcal{W}(C_n^\#, \Lambda_k), \qquad \psi^{-1}(\mathcal{W}(C_n^\#, \Lambda_{k+1})) \subseteq W(C_n, \Lambda_k)^\#. \ \square$$

COROLLARY 2.2.16. *$\psi(G^{\wedge\#})$ is a closed subgroup of the group $G^{\#\wedge}$.*

The proof follows from the completeness of $G^{\wedge\#}$ and the uniform continuity of ψ.

COROLLARY 2.2.17. *$G^{\#\wedge}$ is a locally compact separable group.*

CONJECTURE. *If $\langle G, G_0, G_f\rangle$ satisfies the conditions of Theorem 2.2.4, then $\psi\colon G^{\wedge\#} \longrightarrow G^{\#\wedge}$ is a topological isomorphism, that is,*

$$\psi(G^{\wedge\#}) = G^{\#\wedge}. \tag{2.2.6}$$

Below we present a necessary and sufficient condition for the equality (2.2.6) to hold and consider several cases in which it does hold.

For brevity we denote $G^{\wedge}$ by H, that is, $G^{\#\wedge} = H^{\#}$, and we identify $H^{\#}$ with $\psi(H^{\#})$ by setting $h^{\#}(g^{\#}) = {}^{\circ}(h(g))\ \forall h \in H_f\ \forall g \in G_f$. Thus, $H^{\#}$ is a closed subgroup of $G^{\#\wedge}$.

Let $G_0' = \{g \in G \mid \forall \alpha \in H_f\ (\alpha(g) \approx 0)\}$ and $G_f' = \{g \in G \mid \forall \alpha \in H_0\ (\alpha(g) \approx 0)\}$. Obviously, $G_0' \supseteq G_0$ and $G_f' \supseteq G_f$. Let $G^{\#\prime} = G_f'/G_0'$.

THEOREM 2.2.18. *$H^{\#} = G^{\#\wedge} \Longleftrightarrow G_0' \cap G_f = G_0$.*

PROOF. Since $G = H^{\wedge}$, we can apply Theorem 2.2.15 and its corollary to the pair $\langle G_0', G_f'\rangle$ and get that $G^{\#\prime}$ is a closed subgroup of $H^{\#\wedge}$. Since $H^{\#}$ is a closed subgroup of $G^{\#\wedge}$, the Pontryagin duality theory [26] yields the equality

$$H^{\#\wedge} = G^{\#} / \operatorname{Ann} H^{\#}, \tag{2.2.7}$$

where $\operatorname{Ann} H^{\#} = \{\xi \in G^{\#} \mid \forall h \in H^{\#}\ (h(\xi) = 0)\}$.

The canonical image of an element $g \in G_f'$ in $G^{\#\prime}$ is denoted by $g^{\#\prime}$. Since $g^{\#\prime}$ is a character of the group $H^{\#}$, there exists by (2.2.7) an element $g_1 \in G_f$ such that $\forall h \in H_f\ (g^{\#\prime}(h^{\#}) = g_1^{\#}(h^{\#}))$, and thus $\forall h \in H_f\ (h(g - g_1) \approx 0)$, so that $g - g_1 \in G_0'$. Hence, $\forall g \in G_f'\ \exists g_1 \in G_f\ (g - g_1 \in G_0')$, and therefore $G^{\#\prime} = G_f/G_f \cap G_0'$.

Suppose now that $G_f \cap G_0' = G_0$. Then $G^{\#\prime} = G^{\#}$, and thus $\operatorname{Ann} H^{\#} = \{0\}$ in view of (2.2.7), so $H^{\#} = G^{\#\wedge}$. Conversely, suppose that $H^{\#} = G^{\#\wedge}$, that is, $H^{\#\wedge} = G^{\#\wedge\wedge}$, so that $H^{\#\wedge} = G^{\#}$. Then $\operatorname{Ann} H^{\#} = \{0\}$. However, if $g \in G_0' \cap G_f \setminus G_0$, then $g^{\#} \in \operatorname{Ann} H^{\#}$ and $g^{\#} \neq 0$. Contradiction. □

COROLLARY 2.2.19. *$G^{\#\prime} = H^{\#\wedge}$.*

PROOF. Since $G_f \subseteq G_f'$, the group $H_0' = \{h \in H \mid \forall g \in G_f'\ (h(g) \approx 0)\} \subseteq H_0$. We now use Theorem 2.2.18. □

LEMMA 2.2.20. *If K is an internal hyperfinite Abelian group, and $X\colon K \longrightarrow S^1$ is an internal character of K with $X(g) \approx 1\ \forall g \in K$, then $X \equiv 1$.*

PROOF. Here $S^1 = \{z \in \mathbb{C} \mid |z| = 1\}$. Let $|K| = N$. Then $X(g)^N = 1$, that is, $X(g) = \exp(2\pi i m_X(g)/N)$. Obviously, $m_X\colon G \longrightarrow \mathbb{Z}/N\mathbb{Z}$ is a homomorphism of Abelian groups. Therefore, $m_X(G)$ is a cyclic subgroup of $\mathbb{Z}/N\mathbb{Z}$, and hence $\exists d \mid N\ (m_X(G) = \{kd \mid 0 \le k < \frac{N}{d}\})$. If $\frac{N}{d}$ is even, then, setting $k = \frac{N}{2d}$, we get that $\exp(2\pi i k d/N) = \exp \pi i = -1$, which contradicts the condition of the lemma. If $\frac{N}{d}$ is odd, then we take $k = (N/d - 1)/2$. In this case $\exp(2\pi i k d/N) =$

$-\exp(-\pi id/N) \approx -1$ if $\frac{d}{N} \approx 0$. If $\frac{d}{N} \not\approx 0$, then $\left(\frac{d}{N}\right)^{-1} = m$ is a standard number, and $\exp(2\pi ikd/N) = -\exp(-\pi i/m) \not\approx 1$ for $m \neq 1$. Thus, $N = d$, and hence $m_X(G) = \{0\}$, that is, $X \equiv 1$. □

THEOREM 2.2.21. *If $G^\#$ is discrete or compact, then $G^{\#\frown} = G^{\frown\#}$.*

PROOF. Let $G^\#$ be a discrete group. Then by Theorem 2.2.4, 2, G_0 is an internal subgroup of G. According to Lemma 2.2.20, $H_f = \{h \in G^\frown \mid \forall g \in G_0\ (h(g) = 0)\}$, and thus H_f is an internal subgroup of $H = G^\frown$. But then $G_0' = \{g \in G \mid \forall h \in H_f\ (h(g) = 0)\}$ is an internal subgroup of G. Further, $H_f = \operatorname{Ann} G_0$ and $G_0' = \operatorname{Ann} H_f$. By the theorem on duality of annihilators, $G_0 = G_0'$ in view of the transfer principle. The case of a compact group $G^\#$, that is, of an internal group G_f, is treated in the next section. □

REMARK. We note that the equality $G^{\#\frown} = G^{\frown\#}$ means that every character $h\colon G^\# \longrightarrow S^1$ has the form $\widetilde{\chi}$, where $\chi\colon G \longrightarrow {}^*S^1$ is an internal character of G, $\chi|G_0 \approx 0$, and $\forall g \in G_f\ (\widetilde{\chi}(g^\#) = {}^\circ\chi(g))$.

THEOREM 2.2.22. *If there exists a subgroup $K \in \operatorname{In}_0(G_f)$, then $G^{\#\frown} = G^{\frown\#}$.*

PROOF. First of all we note that the triples $\langle G_0, K, G\rangle$ and $\langle K, G_f, G\rangle$ satisfy all the conditions of Theorem 2.2.4. By this theorem, $K^\# = K/G_0$ is a compact subgroup of $G^\#$, and G_f/K is a discrete group, that is, the conditions of Theorem 2.2.21 hold for these groups.

We show now that $G_0' \cap G_f = G_0$. If not, then $\exists g_0 \in G_0' \cap G_f \setminus G_0$, and two cases are possible.

1. $g_0 \in K$. Then by Theorem 2.2.21 and the remark after it, there exists an internal character $X\colon K \longrightarrow {}^*S^1$ such that $X|G_0 \approx 0$ and $X(g_0) \not\approx 0$. In view of the transfer principle X can be extended to an internal character $\varkappa \in H_f$. Further, $\varkappa(g_0) \not\approx 0$, contradicting the fact that $g_0 \in G_0'$.

2. $g_0 \notin K$. Then by Theorem 2.2.21 there exists an internal character $h\colon G \longrightarrow {}^*S^1$ such that $h|K \approx 1$, and $h(g_0) \not\approx 1$. Thus, $h \in H_f$, and we again contradict the fact that $g_0 \in G_0'$.

Note that both cases use the fact that the characters of an LCA group separate its points. □

THEOREM 2.2.23. *Let $\langle G, G_0, G_f\rangle$ be a triple of groups satisfying the conditions of Theorem 2.2.4. The group $G^\#$ has a compact open subgroup if and only if there exists an internal subgroup $K \in \operatorname{In}_0(G_f)$. Furthermore, $K^\#$ is an open compact subgroup of $G^\#$.*

PROOF. If $K \in \operatorname{In}_0(G_f)$, then $K + G_0 = K$, which implies that $K^\#$ is an open set. The compactness of $K^\#$ was established in Corollary 2.2.5. Obviously, if K is a subgroup of G, then $K^\#$ is a subgroup of $G^\#$.

Conversely, suppose that $U \subseteq G^\#$ is a compact open subgroup. We show that $j^{-1}(U)$ is an internal set. Let $F \in \operatorname{In}_0(G_f)$ be such that $F^\# \subseteq U$. For example, we can take C_{n-1} to be F if $\overset{\circ}{C}{}_n^\# \subseteq U$ (Lemma 2.2.1). Such a C_n exists because U is open. Since U is compact, there exist $g_1^\#, \dots, g_n^\# \in U$ such that $U = \bigcup_{i=1}^n g_i^\# + \overset{\circ}{F}{}^\#$, and since U is a subgroup and $F^\# \subseteq U$, it follows that $U = \bigcup_{i=1}^n g_i^\# + F^\#$. Thus, $j^{-1}(U) = (\bigcup_{i=1}^n g_i + \overset{\circ}{F}) \subseteq \bigcup_{i=1}^n g_i + F \subseteq j^{-1}(U)$. □

4°. Let Δ be an NM of the triple $\langle G, G_0, G_f\rangle$ satisfying Theorem 2.2.4 (Definition 2.2.8). As above in (2.1.3), $\widehat{\Delta} = (\Delta \cdot |G|)^{-1}$. We recall that $\mathfrak{L}_{2,\Delta}(G)$ is an internal hyperfinite-dimensional space in ${}^*\mathbb{C}^G$ with the inner product $(\varphi, \psi)_\Delta = \Delta \sum_{g\in G} \varphi(g)\overline{\psi(g)}$ for any internal $\varphi, \psi \in {}^*\mathbb{C}^G$; $\mathfrak{L}_{2,\widehat{\Delta}}(\widehat{G})$ is defined similarly.

The discrete Fourier transformation $\Phi_\Delta^G \colon \mathfrak{L}_{2,\Delta}(G) \longrightarrow \mathfrak{L}_{2,\widehat{\Delta}}(\widehat{G})$ is defined by

$$\forall \varphi \in \mathfrak{L}_{2,\Delta}(G)\ \forall \chi \in \widehat{G}\ (\Phi_\Delta^G(\varphi)(\chi) = (\varphi, \chi)_\Delta).$$

It is known that the DFT Φ_Δ^G preserves the inner product.

DEFINITION 2.2.24. A triple $\langle G, G_0, G_f\rangle$ of groups satisfying the conditions of Theorem 2.2.4 is said to be admissible if the following conditions hold:

1) $G^{\#\wedge} = G^{\wedge\#}$;

2) $\widehat{\Delta}$ is an NM for the triple $\langle \widehat{G}, H_0, H_f\rangle$ in (2.2.5);

3) Φ_Δ^G is a hyperfinite-dimensional approximation of the FT $\mathfrak{F}_\Delta^{G^\#} : L_2(\mu_\Delta) \longrightarrow L_2(\mu_{\widehat{\Delta}})$ defined by

$$\forall \varkappa \in G^{\#\wedge} \left(\mathfrak{F}_\Delta^{G^\#}(f)(\varkappa) = \int f(g)\overline{\varkappa(g)}\, d\mu(g) \right).$$

Theorem 2.1.1 shows that the triple $\langle G, G_0, G_f\rangle$ considered in the first subsection of this section is admissible.

THEOREM 2.2.25. *If there exists a subgroup* $K \in \mathrm{In}_0(G_f)$, *then* $\langle G, G_0, G_f\rangle$ *is admissible.*

PROOF. We first show that $\widehat{\Delta}$ is an NM for $\widehat{G}$. To do this we consider the subgroup $K^\perp = \{\chi \in \widehat{G} \mid \chi|K = 1\}$. It follows from Lemma 2.2.20 that $H_0 \subseteq K^\perp \subseteq H_f$. It therefore suffices to show that $0 < {}^\circ(\widehat{\Delta} \cdot |K^\perp|) < +\infty$. For this we note that $\widehat{K} = \widehat{G}/K^\perp$ (2.2.7), hence, $|K^\perp| = |\widehat{G}|/|\widehat{K}| = |G|/|K|$, so that $\widehat{\Delta} \cdot |K^\perp| = (\Delta \cdot |K|)^{-1}$, but $0 < {}^\circ(\Delta \cdot |K|) < +\infty$ because Δ is an NM for $\langle G, G_0, G_f\rangle$ and $K \in \mathrm{In}_0(G_f)$.

We know that $K^\#$ is a compact open subgroup of $G^\#$ (Theorem 2.2.23), therefore $G^\#/K^\#$ is discrete and countable in view of the separability of $G^\#$. Let $\{\xi_i \mid i \in \mathbb{N}\}$ be a complete system of representatives of the cosets of $G^\#/K^\#$, and let $\varkappa \in K^{\#\wedge}$. We define the function $f_{i\varkappa} \colon G^\# \longrightarrow \mathbb{C}$ by setting $\forall \eta \in G^\#$

$$f_{i\varkappa}(\eta) = \begin{cases} 0 & \text{if } \eta \notin \xi_i + K^\#, \\ \varkappa(\eta - \xi_i) & \text{if } \eta \in \xi_i + K^\#. \end{cases}$$

Let $\mathfrak{M} = \{f_{i\varkappa} \mid i \in \mathbb{N},\ \varkappa \in K^{\#\wedge}\}$. Since linear combinations of characters are dense in $L_2(K^\#)$ and $L_2(G^\#) = \bigoplus_{i=0}^\infty L_2(\xi_i + K^\#)$, the linear span of $\mathfrak{M}$ is dense in $L_2(G^\#)$, and therefore Lemma 1.3.2 can be used to prove the theorem.

Let $\{x_i \mid i \in \mathbb{N}\} \subseteq G_f$ be such that $\forall i \in \mathbb{N}\ (x_i^\# = \xi_i)$. We fix a $\varkappa_0 \in K^{\#\wedge}$ and choose a character $\chi_0 \in \widehat{K}$ such that $\chi_0|G_0 \approx 1$ and $\widetilde{\chi}_0 = \varkappa_0$ (see Theorem 2.2.21 and the remark after it). We define an internal function $\varphi_{i\chi_0} \colon G \longrightarrow {}^*\mathbb{C}$ by setting

$$\varphi_{i\chi_0}(y) = \begin{cases} 0 & \text{if } y \notin x_i + K, \\ \chi_0(y - x_i) & \text{if } y \in x_i + K \end{cases}$$

for any $y \in G$.

Since $K + G_0 = K$, it follows easily that $\forall y \in G_f\ ({}^\circ\varphi_{i\chi_0}(y) = f_{i\varkappa_0}(y^\#))$, that is, $\varphi_{i\chi_0}$ is a lifting of the function $f_{i\varkappa_0}$.

It follows from the boundedness of $\varphi_{i\chi_0}$ and the fact that its support lies in the internal set $x_i + K \subseteq G_f$ that $\varphi_{i\chi_0} \in \mathcal{S}_{2,\Delta}(G)$. In view of Lemma 1.3.2 it suffices to show that $\Phi^G_\Delta(\varphi_{i\chi_0})$ is an $\mathcal{S}_{2,\widehat{\Delta}}$-integrable lifting of the function $\mathcal{F}^{G^\#}_\Delta(f_{i\varkappa_0})$. A direct computation yields the following equalities $\forall \varkappa \in G^{\#\frown}\ \forall \chi \in G^\frown$:

$$\mathcal{F}^{G^\#}_\Delta(f_{i\varkappa_0})(\varkappa) = \begin{cases} \overline{\varkappa(\xi_i)} \cdot \mu_\Delta(K^\#) & \text{if } \varkappa|K^\# = \varkappa_0, \\ 0 & \text{if } \varkappa|K^\# \neq \varkappa_0; \end{cases}$$

$$\Phi^G_\Delta(\varphi_{i\chi_0})(\chi) = \begin{cases} \overline{\chi(x_i)} \cdot \Delta|K| & \text{if } \chi|K = \chi_0, \\ 0 & \text{if } \chi|K \neq \chi_0. \end{cases}$$

Since $j^{-1}(K^\#) = K$, we have $\mu_\Delta(K^\#) = {}^\circ(\Delta \cdot |K|)$. If $\varkappa = \widetilde{\chi}$, it obviously follows that $\varkappa|K^\# = \varkappa_0 \iff \chi|K = \chi_0$ (Lemma 2.2.20). It is now clear that ${}^\circ\Phi^G_\Delta(\varphi_{i\chi_0}) = \mathcal{F}^{G^\#}_\Delta(f_{i\varkappa_0})(\widetilde{\varkappa})$, and $\Phi^G_\Delta(\varphi_{i\chi})$ is a lifting of $\mathcal{F}^{G^\#}_\Delta(f_{i\varkappa})$ because $\widetilde{\chi}$ coincides with $\chi^\#$. The internal function $\Phi^G_\Delta(\varphi_{i\chi_0})$ is bounded, and its support lies in the internal set $\{\chi \in G^\frown \mid \chi|K = \chi_0\} \subseteq H_f$, that is, $\Phi^G_\Delta(\varphi_{i\chi_0}) \in \mathcal{S}_{2,\widehat{\Delta}}(H_f)$. □

§3. The case of compact groups

This section is devoted to a proof of Theorem 2.2.21 for the case when the group $G^\#$ is compact. By Theorem 2.2.4, $G^\#$ is compact if and only if G_f is an internal subgroup of G. Therefore, we can assume without loss of generality that $G_f = G$. In view of the remark after Theorem 2.2.21 in the preceding section it suffices to prove that every character $\varkappa \in G^{\#\frown}$ has the form $\widetilde{\chi}$ for some $\chi \in G^\frown$ such that $\chi|G_0 \approx 1$ (Lemma 2.2.14). The last assertion follows easily from the completeness of the system of characters of the form $\widetilde{\chi}$. This completeness is proved here with the help of a certain modification of the proof in **[26]** of the Peter–Weyl theorem on completeness of the system of characters of irreducible representations of a compact group. We shall actually prove a more general assertion than Theorem 2.2.21 relating to the case when the group G is noncommutative. The proof of Theorem 2 in the Introduction is based on it.

All the arguments in the preceding section except those relating to the character group carry over unchanged to the case when the internal hyperfinite group G is noncommutative, and the external subgroups $G_0 \subseteq G_f$ satisfying conditions (A) and (B) of §2, 2° are normal subgroups of G.

The notation + is preserved below for the group operation.

1°. Here G is an internal hyperfinite group, and G_0 is a subgroup of G that is the intersection of a countable family of internal sets and has the property that for any internal F with $G_0 \subseteq F \subseteq G$ there exists a standardly finite $B \subseteq G$ such that $F + B = G$. In this case $G^\#$ is a compact group by Theorem 2.2.4.

An internal function $\varphi\colon G \longrightarrow {}^*\mathbb{C}$ is said to be S-continuous if $\forall g_1, g_2 \in G$ $((g_1 - g_2) \in G_0 \Longrightarrow \varphi(g_1) \approx \varphi(g_2))$.

According to Lemma 2.2.14, if $\varphi\colon G \longrightarrow {}^*\mathbb{C}$ is S-continuous and finite, that is, $\forall g \in G\ ({}^\circ|\varphi(g)| < +\infty)$, then the function $\widetilde{\varphi}\colon G^\# \longrightarrow \mathbb{C}$ defined by $\forall g \in G$ $(\widetilde{\varphi}(g^\#) = {}^\circ\varphi(g))$ is continuous.

THEOREM 2.3.1. *Every continuous function $f\colon G^\# \longrightarrow \mathbb{C}$ has the form $f = \widetilde{\varphi}$ for some S-continuous finite function $\varphi\colon G \longrightarrow {}^*\mathbb{C}$.*

PROOF. Denote by $CS(G)$ the set of all S-continuous finite internal functions $\varphi\colon G \longrightarrow {}^*\mathbb{C}$. It is clear that $CS(G)$ is an external subalgebra of the internal algebra ${}^*\mathbb{C}^G$. To prove the theorem it suffices to show that the algebra $\mathfrak{G} = \{\widetilde{\varphi} \mid \varphi \in CS(G)\}$ is uniformly closed and separates the points of $G^\#$.

LEMMA 2.3.2. *If $x, y \in G$ and $x - y \notin G_0$, then there exists a $\varphi \in CS(G)$ such that $\varphi(x) = 0$ and $\varphi(y) = 1$.*

PROOF OF LEMMA 2.3.2. It is clear that this lemma is a 'discrete' version of Uryson's lemma, and therefore it can be proved by modifying the proof of the latter. It follows from the ω^+-saturation of the nonstandard universe that $y \notin x + C_k$ for some $k \in \mathbb{Z}$ (see Lemma 2.2.1). Let $V_0 = C_k$ and $V_n = C_{k-n}$ for any $n \in \mathbb{Z}$. Then $\{V_n \mid n \in \mathbb{N}\}$ is a sequence of symmetric internal sets such that $V_{n+1} + V_{n+1} \subseteq V_n$. The ω^+-saturation implies the existence of an internal sequence $\{B_n \mid n \in {}^*\mathbb{N}\}$ of symmetric subsets of G such that $\forall n \in {}^*\mathbb{N}\ (B_{n+1} + B_{n+1} \subseteq B_n)$ and $B_k = V_k$ for any standard $k \in \mathbb{N}$. Let us fix an infinite $N \in {}^*\mathbb{N}$. Suppose that $0 \leq m < n \leq N$, and let $V_{m,n} = B_{m+1} + \cdots + B_n$. By induction on n we see immediately from the inclusion $B_{n+1} + B_{n+1} \subseteq B_n$ that $V_{m,n} + B_n \subseteq B_m$. We now consider the rational numbers of the form

$$r = \frac{a_1}{2} + \frac{a_2}{2^2} + \cdots + \frac{a_n}{2^n}, \qquad a_i \in \{0, 1\}. \tag{2.3.1}$$

For $a \in \{0, 1\}$ let

$$B^a = \begin{cases} B & \text{if } a = 1, \\ \{0\} & \text{if } a = 0. \end{cases}$$

Let $W_r = B_1^{a_1} + B_2^{a_2} + \cdots + B_n^{a_n}$. Then $W_r \subseteq B_1 + B_2 + \cdots + B_n = V_{0N} \subseteq B_0 = V_0 = C_k$, hence $y \notin x + B_0$, and thus $y \notin x + W_r$. It is easy to verify that $W_r \subseteq W_{r'}$ for $r < r'$. We define an internal function $\varphi\colon G \longrightarrow {}^*\mathbb{R}$ by setting $\varphi(g) = \min\{r \mid g \in x + W_r\}$. If for any r of the form (2.3.1) $g \notin W_r + x$, then $\varphi(g) = 1$, in particular, $\varphi(y) = 1$. Since $x \in x + \{0\}$, it follows that $\varphi(x) = 0$. We show now that, for any $k \in [0, N]$, $|f(u) - f(v)| \leq \frac{1}{2^{k-1}}$ if $u - v \in B_k$. This implies that φ is S-continuous, because if $u - v \in G_0$, then $u - v \in B_k$ for any standard k. Since B_k is symmetric, it can be assumed that $\varphi(u) < \varphi(v)$, and it can be proved that $\varphi(v) - \varphi(u) \leq \frac{1}{2^{k-1}}$. We note that $\varphi(u) < 1$ because $\max \varphi = 1$. Let q be the number among $1, 2, \ldots, 2^k$ $(k \leq N)$ with $\frac{q-1}{2^k} \leq \varphi(u) < \frac{q}{2^k}$. If $q = 2^k$ or $q = 2^k - 1$, then $1 - \varphi(u) \leq \frac{1}{2^{k-1}}$, and $\varphi(v) - \varphi(u) \leq \frac{1}{2^{k-1}}$ because $\varphi(v) \leq 1$.

Suppose now that $q < 2^{k-1} - 1$ and $r = \frac{q}{2^k}$. Then $\varphi(u) < r$, that is, $u \in W_r + x$, and since $v - u \in B_k$, it follows that $v \in W_r + B_k + x$. In our case $r = \frac{a_1}{2} + \cdots + \frac{a_k}{2^k}$, $a_i \in \{0, 1\}$, and $\exists i\ (a_i = 0)$ because $q < 2^k - 1$. We choose the largest such i. Then

$$W_r + B_k = B_1^{a_1} + \cdots + B_{i+1}^{a_{i+1}} + \cdots + B_k^{a_k} + B_k \subseteq B_1^{a_1} + \cdots + B_{i-1}^{a_{i-1}} + B_i$$

(it was used that $B_s + B_s \subseteq B_{s-1}$). On the other hand, $r' = r + \frac{1}{2^k} = \frac{a_1}{2} + \cdots + \frac{a_{i-1}}{2^{i-1}} + \frac{1}{2^k}$. Thus, $W_r + B_k = W_{r'}$, that is, $v \in W_{r'} + x$ and $\varphi(v) \leq r + \frac{1}{2^k}$. Finally, $r - \frac{1}{2^k} \leq \varphi(u) < \varphi(v) \leq r + \frac{1}{2^k}$, and hence $\varphi(v) - \varphi(u) \leq \frac{1}{2^{k-1}}$. □

Assume now that $\xi, \eta \in G^\#$ and $\xi \neq \eta$. If $\xi = x^\#$ and $\eta = y^\#$, then $x - y \notin G_0$, and by the lemma just proved there exists a $\varphi \in CS(G)$ such that $\varphi(x) = 0$ and $\varphi(y) = 1$. But then $\widetilde{\varphi}(\xi) = 0$, while $\widetilde{\varphi}(\eta) = 1$. Thus, the subalgebra $\mathfrak{G} \subseteq C(G^\#)$ separates the points of $G^\#$.

To conclude the proof of Theorem 2.3.1 it remains to prove the following lemma.

LEMMA 2.3.3. *The subalgebra* $\mathfrak{G}$ *is uniformly closed in* $C(G^{\#})$.

PROOF. Obviously, $\sup\{\widetilde{\psi}(\xi) \mid \xi \in G^{\#}\} = {}^{\circ}\max\{\psi(g) \mid g \in G\}$ for any $\psi \in CS(G)$. Thus, if a sequence $\{\widetilde{\varphi}_n\}$ of functions converges in $C(G^{\#})$ to some function f, then ${}^{\circ}\max\{|\varphi_n(g) - \varphi_m(g)| \mid g \in G\} \to 0$ as $n, m \to \infty$. Therefore, there exists a standard function $N(m)$ such that $\forall n_1, n_2 > N(m)$

$$\max\{|\varphi_{n_1}(g) - \varphi_{n_2}(g)| \mid g \in G\} < \frac{1}{m}. \tag{2.3.2}$$

We consider the family of internal sets

$$\{\{\varphi \mid \max\{|\varphi_n(g) - \varphi(g)| \mid g \in G\} < 1/m\} \mid n > N(m)\}.$$

By (2.3.2), this family has the finite intersection property. The ω^+-saturation of the nonstandard universe gives us the existence of an internal function $\varphi\colon G \longrightarrow {}^*\mathbb{C}$ such that

$${}^{\circ}\max\{|\varphi_n(g) - \varphi(g)| \mid g \in G\} \underset{n\to\infty}{\longrightarrow} 0.$$

The finiteness and S-continuity of φ are trivial to prove, and hence $\widetilde{\varphi}_n \to \widetilde{\varphi}$ as $n \to \infty$. □

We present some auxiliary propositions.

PROPOSITION 2.3.4. *If the triples* $\langle G', G'_0, G'_f\rangle$ *and* $\langle G'', G''_0, G''_f\rangle$ *of groups satisfy the conditions of Theorem* 2.3.1, *then so does the triple* $\langle G' \times G'', G'_0 \times G''_0, G'_f \times G''_f\rangle$, *and the group* $(G' \times G'')^{\#}$ *is topologically isomorphic to* $G'^{\#} \times G''^{\#}$.

The trivial proof of this is omitted. The proposition implies that in our case, when $G_f = G$, the group $(G\times G)^{\#} = (G\times G)/(G_0\times G_0)$ is topologically isomorphic to $G^{\#} \times G^{\#} = (G/G_0) \times (G/G_0)$.

LEMMA 2.3.5. *Let* $K\colon G^{\#2} \longrightarrow \mathbb{C}$ *be a continuous function, and let* $K = \widetilde{k}$, *where the internal function* $k\colon G \times G \longrightarrow {}^*\mathbb{C}$ *is* S*-continuous. For any* $g \in G$ *let* $K_{g^{\#}}\colon G^{\#} \longrightarrow \mathbb{C}$ *be the function defined by* $K_{g^{\#}}(\cdot) = K(g^{\#}, \cdot)$, *and let* $k_g\colon G \longrightarrow {}^*\mathbb{C}$ *be the internal function defined by* $k_g(\cdot) = k(g, \cdot)$. *Then* k_g *is* S*-continuous, and* $K_{g^{\#}} = \widetilde{k}_g$.

PROOF. Obvious. □

LEMMA 2.3.6. *If* $f\colon G^{\#} \longrightarrow \mathbb{C}$ *is a continuous even function, then there exists an internal* S*-continuous even function* φ *such that* $f = \widetilde{\varphi}$. *If, furthermore,* $K\colon G^{\#2} \longrightarrow \mathbb{C}$ *and* $k\colon G^2 \longrightarrow {}^*\mathbb{C}$ *are defined by* $K(\xi, \eta) = f(\xi - \eta)$ *and* $k(g_1, g_2) = \varphi(g_1 - g_2)$, *respectively, then* $K = \widetilde{k}$.

PROOF. If $f = \widetilde{\psi}$, where $\psi\colon G \longrightarrow {}^*\mathbb{C}$ is an internal S-continuous function, then we set $\varphi(g) = \frac{1}{2}[\psi(g) + \psi(-g)]$. □

2°. We study the connection between integral equations on the group $G^{\#}$ and the corresponding systems of linear equations on the group G.

First of all we note that since $G^{\#}$ is a compact group, Haar measure is finite and it can be assumed that $\mu(G^{\#}) = 1$. This measure corresponds to the Loeb measure on G determined by the multiplier $\Delta = |G|^{-1}$. It follows from the definition of a lifting of a measurable function $f\colon G^{\#} \longrightarrow \mathbb{C}$ (see Theorem 2.2.10 and its Corollary 2.2.11) that if $f = \widetilde{\varphi}$ for some finite S-continuous function φ, then φ is a lifting of f. Further, $\varphi \in \mathcal{S}_p(G)$ for any $p \in [1, \infty)$ (condition 2 of Definition 1.2.1 follows from the finiteness of φ; condition 3 holds because $X = M$ in this case). Of course, the last remark applies also to functions defined on $G^{\#2}$.

LEMMA 2.3.7. *Suppose that $k\colon G^2 \longrightarrow {}^*\mathbb{C}$ is an internal finite S-continuous function, $\varphi\colon G \longrightarrow {}^*\mathbb{C}$ is an internal finite function, and the internal function $\psi\colon G \longrightarrow {}^*\mathbb{C}$ is defined by*

$$\psi(g) = |G|^{-1} \sum_{h \in G} k(g, h)\varphi(h) \qquad \forall g \in G.$$

Then ψ is finite and S-continuous. Further, if φ is S-continuous, then

$$\widetilde{\psi}(\xi) = \int_{G^{\#}} \varphi(\eta)\widetilde{k}(\xi, \eta)\, d\mu(\eta) \qquad \forall \xi \in G^{\#}.$$

PROOF. The finiteness of ψ follows in an obvious way from that of k and φ. By Lemma 2.3.5 and the S-continuity of k, $k(g_1, h) - k(g_2, h) \approx 0$ for any $h \in G$ if $g_1 - g_2 \in G_0$, that is, there exists an $\alpha \approx 0$ such that $\forall h \in G$ $(|k(g_1, h) - k(g_2, h)| \leq \alpha)$. Let $C > 0$ be a standard number such that $\forall h \in G$ $(|\varphi(h)| \leq C)$. Then $|\psi(g_1) - \psi(g_2)| \leq C\alpha \approx 0$, which proves that ψ is S-continuous. The second assertion is an immediate consequence of Theorem 1.3.6. □

In our case, when $\Delta = |G|^{-1}$, we simply write $\mathcal{L}_2(G)$ instead of $\mathcal{L}_{2,\Delta}(G)$. We fix in $\mathcal{L}_2(G)$ the canonical orthonormal basis $\{e_h \mid h \in G\}$ with $e_h(g) = |G|^{1/2}\delta_{hg}$.

We now study equations of the form

$$\varphi(g) = \lambda \cdot |G|^{-1} \sum_{h \in G} k(g, h)\varphi(h) \tag{2.3.3}$$

with finite symmetric internal function k. Those λ for which (2.3.3) has nonzero solutions will be called eigenvalues of the equation (2.3.3), and solutions corresponding to them will be called eigenfunctions of (2.3.3) belonging to the eigenvalue λ. Thus, the eigenvalues of (2.3.3) are the reciprocals of the nonzero eigenvalues of the operator A given in the canonical orthonormal basis by the matrix $(a_{gh})_{g,h \in G}$ with $a_{gh} = |G|^{-1}k(g, h)$. It follows from the finiteness of k that ${}^\circ\sum_{g,h \in G} |a_{gh}|^2 < +\infty$, that is, A satisfies the conditions of Proposition 1.3.18, which immediately gives us

LEMMA 2.3.8. *The equation* (2.3.3) *does not have infinitesimal eigenvalues. If λ is a finite eigenvalue of this equation, and R_λ is the subspace of eigenfunctions belonging to λ, then* $\dim R_\lambda \in \mathbb{N}$, *that is, is a standard number.*

Below we assume that the finite symmetric internal function k in (2.3.3) is S-continuous.

LEMMA 2.3.9. *If λ is finite, then for every eigenfunction φ of* (2.3.3) *belonging to R_λ there exists an S-continuous finite function that is a multiple of it.*

PROOF. If $\varphi \in R_\lambda$ and $\varphi \neq 0$, then $\varphi_1 = \varphi / \max\{|\varphi(g)| \mid g \in G\}$ is bounded, and $\varphi_1 \in R_\lambda$. We now apply Lemma 2.3.7. $\square$

If $\varphi \in \mathfrak{L}_2(G)$, then $\|\varphi\|^2 = |G|^{-1} \sum_{g \in G} |\varphi(g)|^2$ and $\|\varphi\|_\infty = \max\{|\varphi(g)| \mid g \in G\}$. Thus, $\|\varphi\| \leq \|\varphi\|_\infty$. Nevertheless, we have

LEMMA 2.3.10. *If φ is an eigenfunction of (2.3.3) belonging to the finite eigenvalue λ and $\|\varphi\| = 1$, then φ is S-continuous.*

PROOF. Suppose that $\varphi_1 = C\varphi$ is an S-continuous eigenfunction of (2.3.3) with $\|\varphi_1\|_\infty = 1$ (Lemma 2.3.9). Then $\|\widetilde{\varphi}\|_\infty = 1$, hence $\int_{G^\#} |\widetilde{\varphi}_1|^2 \, d\mu > 0$. But φ_1 is a lifting of $\widetilde{\varphi}_1$, and $^\circ\|\varphi_1\|^2 = \int_{G^\#} |\widetilde{\varphi}_1|^2 \, d\mu$ in view of the finiteness of these functions and Corollary 2.2.11, and thus $0 < {}^\circ\|\varphi_1\| < +\infty$. Then $\varphi_2 = \varphi_1 / \|\varphi_1\|$ is S-continuous and finite, and $\|\varphi_2\| = 1$. Since $\varphi_2 = C_1\varphi$, $\|\varphi\| = 1$, and $C_1 > 0$, it follows that $C_1 = 1$, that is, $\varphi_2 = \varphi$. $\square$

The assertion of the lemma is valid for any eigenfunction φ with $0 < {}^\circ\|\varphi\| < +\infty$, of course.

LEMMA 2.3.11. *Let $f \colon G^\# \longrightarrow \mathbb{C}$ be a continuous function that is a solution of the equation*

$$f(\xi) = \gamma \int_{G^\#} \widetilde{k}(\xi, \eta) f(\eta) \, d\mu(\eta), \tag{2.3.4}$$

where γ is a standard number, $\gamma \neq 0$. Then there exist a standard natural number n, eigenvalues $\lambda_1, \dots, \lambda_n$ of (2.3.3), and S-continuous finite eigenfunctions $\varphi_1, \dots, \varphi_n$ such that $\lambda_i \approx \gamma$ and $\varphi_i \in R_{\lambda_i}$ for any $i \in [1, n]$, and f is a linear combination of $\widetilde{\varphi}_1, \dots, \widetilde{\varphi}_n$.

PROOF. We use Proposition 1.3.20. According to Theorem 1.3.6 (see also Lemma 2.3.7), the operator A described above with matrix $a_{gh} = |G|^{-1/2} k(g, h)$ in the canonical orthonormal basis of $\mathfrak{L}_2(G)$ is a hyperfinite-dimensional approximation of the integral operator $\mathcal{A}$ with kernel $\widetilde{k} \colon G^{\#2} \longrightarrow \mathbb{C}$. This means that the diagram

$$\begin{array}{ccc} L_2(G^\#) & \xrightarrow{\mathcal{A}} & L_2(G^\#) \\ {\scriptstyle j_2}\big\downarrow & & \big\downarrow{\scriptstyle j_2} \\ \mathfrak{L}_2(G)^\# & \xrightarrow{A} & \mathfrak{L}_2(G)^\# \end{array}$$

is commutative. (We recall that the mapping j_2 assigns to each function $f \in L_2(G^\#)$ the class of an $\mathcal{S}_2(G)$-lifting of it, that is, an $\mathcal{S}_2(G)$-lifting of the function $f \circ j$, where $j \colon G \longrightarrow G^\#$ is the natural projection). It is clear from the diagram that $j_2(f)$ is an eigenvector of $A^\#$ belonging to the eigenvalue γ^{-1}. According to Proposition 1.3.5, there exists a $\lambda^{-1} \approx \gamma^{-1}$ that is an eigenvalue of A. By Lemma 2.3.10, every normalized eigenfunction φ of A belonging to λ^{-1} is S-continuous $(0 < {}^\circ|\lambda^{-1}| < \infty)$, and $\widetilde{\varphi}$ is an eigenfunction of $\mathcal{A}$ belonging to γ^{-1} in view of Lemma 2.3.7. Since the eigenvalue γ^{-1} of $\mathcal{A}$ has finite multiplicity ($\mathcal{A}$ is a compact operator), there exists only a standardly finite number of eigenvalues of A that are infinitely close to γ^{-1}, and each of them has standardly finite multiplicity. This means that the conditions of Proposition 1.3.20 hold, and according to that proposition, for some standard n there exist $\lambda_1, \dots, \lambda_n \approx \gamma^{-1}$ and orthonormal

$\varphi_1, \dots, \varphi_n \in \mathcal{L}_2(G)$ such that $\varphi_i \in R_{\lambda_i^{-1}}$ $\forall i \le n$ and $j_2(f) = \sum_{i=1}^n C_i \varphi_i^\#$. Since φ_i is S-continuous, φ_i is an $\mathcal{S}_2(G)$-lifting of $\widetilde{\varphi}_i$, hence $\varphi_i^\# = j_2(\widetilde{\varphi}_i)$, and thus $f = \sum_{i=1}^n C_i \widetilde{\varphi}_i$. □

3°. Let $T\colon G \longrightarrow GL(V)$ be a hyperfinite-dimensional unitary representation of G in an internal unitary space V, that is, $\dim V = n \in {}^*\mathbb{N}$. We say that the representation T is S-continuous if $\forall g \in G_0$ $(\|T(g) - I\| \approx 0)$, where I is the identity operator.

LEMMA 2.3.12. *Every S-continuous irreducible unitary representation of G has a standard dimension.*

PROOF. This is analogous to the proof of Theorem 22.13 in the book [**29**], which asserts that every irreducible representation of a compact group is finite-dimensional, but it is somewhat simpler, since here we have to treat hyperfinite groups, which can be dealt with just like finite groups.

We fix a vector $\xi \in V$ and consider the sesquilinear form $\varphi_\xi(\eta, \zeta) = |G|^{-1} \cdot \sum_{g \in G} (T(g)\xi, \eta) \cdot \overline{(T(g), \xi, \zeta)}$.

Let $B_\xi\colon V \longrightarrow V$ be a linear operator such that $\varphi_\xi(\eta, \zeta) = (B_\xi \eta, \zeta)$. A trivial computation shows that B_ξ commutes with all the operators $T(g)$, and hence by Shur's lemma $B_\xi = \alpha(\xi) \cdot I$, where $\alpha(\xi) \in {}^*\mathbb{C}$. Thus, $\varphi_\xi(\eta, \zeta) = \alpha(\xi)(\eta, \zeta)$. Setting $\eta = \zeta$, we get that $\varphi_\xi(\zeta, \zeta) = \alpha(\xi) \cdot \|\zeta\|^2 = \alpha(\zeta) \cdot \|\xi\|^2$. It follows from the last equality that there is a $D \in {}^*\mathbb{R}$ such that $\forall \xi \in V$ $(\alpha(\xi) = D \cdot \|\xi\|^2)$. Let the vector ξ be such that $\|\xi\| = 1$. Then $\varphi_\xi(\xi, \xi) = \alpha(\xi) = D$. Thus, for any $\xi \in V$ with $\|\xi\| = 1$

$$D = |G|^{-1} \cdot \sum_{g \in G} |(T(g)\xi, \xi)|^2. \tag{2.3.5}$$

We show that ${}^\circ D > 0$. Let us consider the internal function $\psi\colon G \longrightarrow {}^*\mathbb{R}$ given by $\psi(g) = |(T(g)\xi, \xi)|^2$. It is trivial to verify that ψ is S-continuous. Consequently, $\|\widetilde{\psi}\|^2 = {}^\circ\|\psi\|^2 = D$, where the left-hand side is the norm in $L_2(G^\#)$. It is clear from the definition of ψ that $\psi(e) = 1$, where e is the identity of G. Consequently, $\widetilde{\psi}(e^\#) = 1$, and since $\widetilde{\psi}$ is a continuous function, it follows that $\|\psi\| > 0$, as required to prove.

Now let $\theta_1, \dots, \theta_n \in V$ be an arbitrary orthonormal basis. Then

$$|G|^{-1} \cdot \sum_{g \in G} |(T(g)\theta_k, \theta_1)|^2 = \varphi_{\theta_k}(\theta_1, \theta_1) = \alpha(\theta_k) \cdot \|\theta_1\|^2 = D.$$

Since $T(g)$ is unitary, the system $\{T(g)\theta_k \mid k = 1, \dots, n\}$ forms an orthonormal basis, that is, $\sum_{k=1}^n |(T(g)\theta_k, \theta_1)|^2 = \|\theta_1\|^2 = 1$. Summing the last equality over g and multiplying by $|G|^{-1}$, we get by the previous equality that $n \cdot D = 1$. The standardness of n now follows from the fact that ${}^\circ D > 1$. □

Suppose now that $T\colon G \longrightarrow {}^*\mathbb{U}_n$ is an irreducible S-continuous unitary representation, where $\mathbb{U}_n$ is the group of unitary matrices of standard order n. It determines a continuous representation $\widetilde{T}$ of the group $G^\#$ by the formula $\widetilde{T}(g^\#) = {}^\circ T(g)$, which is clearly also unitary. Moreover, it is clear that the character $\varkappa$ of the representation $\widetilde{T}$ is equal to $\widetilde{\chi}$, where χ is the character of T. Thus, $\|\varkappa\| = {}^\circ\|\chi\| = 1$, because T is an irreducible representation. This implies that $\widetilde{T}$ is an irreducible representation. Theorem 2.2.21 is the special case of the next proposition relating to commutative groups.

PROPOSITION 2.3.13. *Every irreducible unitary representation of the group* $G^{\#}$ *has the form* $\widetilde{T}$ *for some* S*-continuous irreducible representation* T *of the group* G.

According to the well-known criterion for the completeness of a system of irreducible representations of a compact group (see, for example, [**81**]), this proposition follows immediately from the next lemma.

LEMMA 2.3.14. *The set of linear combinations of functions of the form* $\widetilde{\psi}$, *where* $\psi(g)$ *is a matrix element of some* S*-continuous irreducible unitary representation of* G, *is dense in* $C(G^{\#})$.

PROOF. In the proof of Theorem 32 of [**26**] it is established that the space of linear combinations of eigenfunctions of all the integral equations with kernels of the form $f(x-y)$, where $f\colon G^{\#} \longrightarrow \mathbb{C}$ is a continuous even function, is dense in $C(G^{\#})$. Using Lemmas 2.3.6 and 2.3.11, we conclude from this that the space of linear combinations of functions of the form $\widetilde{\varphi}$, where φ is an S-continuous eigenfunction of an equation of the form (2.3.3) with $0 < {}^{\circ}|\lambda| < +\infty$ and $k(g,h) = f(g-h)$ for some S-continuous even function $f\colon G \longrightarrow {}^{*}\mathbb{C}$, is dense in $C(G^{\#})$.

In essence, the rest of the proof copies the proof of Theorem 32 in [**26**] and is given here only for completeness. It is assumed below that $k(g,h) = f(g-h)$ in (2.3.3) for a function f of the indicated form. According to Lemma 2.3.8, the dimension of R_λ is standardly finite. Let $\varphi_1, \dots, \varphi_n$ be a complete orthonormal system of eigenfunctions of (2.3.3) which, as already mentioned more than once, are S-continuous. Obviously, if $\varphi(g) \in R_\lambda$, then $\varphi(a+g) \in R_\lambda$ $\forall a \in G$. Thus, $\varphi_1(a+g), \dots, \varphi_n(a+g)$ is also a complete orthonormal system of eigenfunctions of (2.3.3), and hence there exists a unitary matrix $U(a) = (u_{ij}(a))_{i,j=1}^{n}$ such that

$$\varphi_i(a+g) = \sum_{j=1}^{n} u_{ij}(a)\varphi_j(g). \tag{2.3.6}$$

We show that $\{U(a) \mid a \in G\}$ is a representation of the group G. Indeed,

$$\begin{aligned}\varphi_i(a+b+g) &= \sum_{j=1}^{n} u_{ij}(a+b)\varphi_j(g)\\ &= \sum_{k=1}^{n}\Bigl(\sum_{j=1}^{n} u_{ij}(a)u_{jk}(b)\Bigr)\varphi_k(g),\end{aligned}$$

therefore, $u_{ij}(a+b) = \sum_{k=1}^{n} u_{ik}(a)u_{kj}(b)$.

It follows from the orthonormality of the system $\{\varphi_i \mid i = 1, \dots, n\}$ that

$$u_{ij}(a) = |G|^{-1} \sum_{g \in G} \varphi_i(a+g)\varphi_j(g). \tag{2.3.7}$$

It follows easily from (2.3.7) and the finiteness and S-continuity of the functions φ_i that $u_{ij}(a)$ is S-continuous. Since $U(a)$ is a unitary representation of G, there exists a unitary matrix V such that $\forall a \in G\ (U(a) = VX(a)V^{-1})$, where

$$X(a) = \begin{pmatrix} T_1(a) & & 0 \\ & \ddots & \\ 0 & & T_n(a) \end{pmatrix},$$

and the T_i are irreducible unitary representations of G for $i = 1, \dots, n$. Since $X(a) = V^{-1}U(a)V$, all the matrix elements of the representations T_i are standardly finite linear combinations of S-continuous functions, that is, the representations T_i are themselves S-continuous. Similarly, the u_{ij} are standardly finite linear combinations of the matrix elements of the T_i with finite coefficients. Setting $g = 0$ in (2.3.6), we get that each φ_i is a standardly finite linear combination of the u_{ij} with finite coefficients, and hence also of some matrix elements ψ_k of the representations T_i. It is clear that if $\varphi_i = \sum_{j=1}^n C_j \cdot \psi_j$, then $\widetilde{\varphi}_i = \sum_{j=1}^n {}^\circ(C_j) \cdot \widetilde{\psi}_j$. This concludes the proof of the lemma, and with it of Proposition 2.3.13. □

§4. Hyperfinite approximations of LCA groups

In this section we introduce and study the basic concept of this chapter: the concept of a hyperfinite approximation of a topological group. All the main results relate to the case of LCA groups. In essence, only Theorems 1 and 2 of the Introduction and their nonstandard reformulations are known for noncommutative groups. Therefore, as in the preceding section, commutativity of the group is not assumed here in the determination of an approximation, though the group operation is denoted by the symbol $+$. In the definition of a hyperfinite approximation it is always assumed that commutative topological groups are approximated by commutative hyperfinite groups.

1°. Recall that if $\mathfrak{G}$ is a topological Abelian group, then $\mu_{\mathfrak{G}}(0) = \cap\{{}^*U \mid 0 \in U, U \subseteq \mathfrak{G}, U \text{ is open}\}$, $\xi_1 \overset{\mathfrak{G}}{\approx} \xi_2 \iff \xi_1 - \xi_2 \in \mu_{\mathfrak{G}}(0)$, and $\mathrm{Ns}({}^*\mathfrak{G}) = \{\xi \in {}^*\mathfrak{G} \mid \exists \eta \in \mathfrak{G}\ (\xi \approx \eta)\}$. The mapping $\mathrm{st}\colon \mathrm{Ns}({}^*\mathfrak{G}) \longrightarrow \mathfrak{G}$ with $\mathrm{st}(\xi) \approx \xi\ \forall \xi \in \mathrm{Ns}({}^*\mathfrak{G})$ is an epimorphism with kernel $\mu_{\mathfrak{G}}(0)$, that is, $\mathfrak{G} \cong \mathrm{Ns}({}^*\mathfrak{G})/\mu_{\mathfrak{G}}(0)$. We simply write $\mu(0)$ and $\xi_1 \approx \xi_2$ instead of $\mu_{\mathfrak{G}}(0)$ and $\xi_1 \overset{\mathfrak{G}}{\approx} \xi_2$ when this does not lead to confusion.

DEFINITION 2.4.1. Let $\mathfrak{G}$ be a standard topological group, G an internal hyperfinite group, and $j\colon G \longrightarrow {}^*\mathfrak{G}$ an internal mapping. The pair $\langle G, j\rangle$ is called a hyperfinite approximation (HA) of the group $\mathfrak{G}$ if the following conditions hold:

1) $\forall \xi \in \mathrm{Ns}({}^*\mathfrak{G})\ \exists g \in G\ (j(g) \approx \xi)$;
2) $g_1, g_2 \in j^{-1}(\mathrm{Ns}({}^*\mathfrak{G})) \Longrightarrow j(g_1 + g_2) \approx j(g_1) + j(g_2)$;
3) $g \in j^{-1}(\mathrm{Ns}({}^*\mathfrak{G})) \Longrightarrow j(-g) \approx -j(g)$;
4) $j(0) = 0$.

We could have postulated that $j(0) \approx 0$ in the fourth condition, of course, but if the mapping is changed so that exact equality holds, then the properties 1)–3) are preserved.

Let $G_f = j^{-1}(\mathrm{Ns}({}^*\mathfrak{G}))$, $G_0 = j^{-1}(\mu(0))$, and $\widetilde{j} = \mathrm{st} \circ j|G_f$. Then conditions 1)–4) are equivalent to requiring that $\widetilde{j}\colon G_f \longrightarrow \mathfrak{G}$ be an epimorphism with kernel $\ker \widetilde{j} = G_0$. The isomorphism $\widetilde{j}$ induces between $G^\# = G_f/G_0$ and $\mathfrak{G}$ will be denoted by $\bar{j}$. The natural projection of G_f onto $G^\#$ will be denoted simply by $\#$.

PROPOSITION 2.4.2. *If $\mathfrak{G}$ is a separable LC group, and $\langle G, j\rangle$ is a hyperfinite approximation of $\mathfrak{G}$, then the triple $\langle G, G_0, G_f\rangle$ satisfies the conditions of Theorem 2.2.4, and $\overline{j}\colon G^{\#} \longrightarrow \mathfrak{G}$ is a topological isomorphism.*

PROOF. In view of local compactness and separability it can be assumed that $\mathfrak{G} = \bigcup_{n=1}^{\infty} U_n$, where each U_n is an open relatively compact set. Then it is easy to see that $G_f = \bigcup_{n=1}^{\infty} j^{-1}({}^*U_n)$. In exactly the same way, if $\{V_n \mid n \in \omega\}$ is a countable base consisting of relatively compact neighborhoods of zero in $\mathfrak{G}$, then $G_0 = \bigcap_{n\in\omega} j^{-1}({}^*V_n)$. Thus, G_f is a countable union and G_0 a countable intersection of internal sets, that is, the canonical topology is defined on $G^{\#}$ (see Theorem 2.2.3). It remains to show that $\overline{j}$ and $\overline{j}^{-1}$ are continuous at zero. The fact that $G^{\#}$ is locally compact implies the conditions of Theorem 2.2.4 in this case.

Fixing an arbitrary neighborhood V of zero in $\mathfrak{G}$, we choose a relatively compact neighborhood V^1 of zero such that $V^1 \subseteq V$, and consider the internal set $F = j^{-1}({}^*V^1)$. It is obvious that $G_0 \subseteq F$, that is, $\overset{\circ}{F}{}^{\#}$ is a neighborhood of zero in $G^{\#}$. It is easy to verify that $\overline{j}(\overset{\circ}{F}{}^{\#}) \subseteq V^1 \subseteq V$, that is, $\overline{j}$ is continuous at zero.

We now fix a neighborhood of zero in $G^{\#}$ of the form $\overset{\circ}{F}{}^{\#}$, where F is an internal subset of G_f with $F \supseteq G_0$. Since $G_0 = \cap\{j^{-1}({}^*U) \mid U$ a neighborhood of zero in $\mathfrak{G}\}$, the ω^+-saturation of the nonstandard universe implies the existence of a relatively compact neighborhood U of zero in $\mathfrak{G}$ such that $j^{-1}({}^*U) \subseteq F$. Let $V \subseteq \mathfrak{G}$ be a neighborhood of zero with $V+V+V \subseteq U$. Using the obvious inclusion $V \subseteq V+V$, we get that $j^{-1}({}^*V)+j^{-1}({}^*V) \subseteq j^{-1}({}^*U)$. This inclusion gives us that if $\xi \in V$ and $\overline{j}^{-1}(\xi) = g^{\#}$ (that is, $j(g) \approx \xi$), then $g \in \overset{\circ}{F}$, that is, $\overline{j}^{-1}(V) \subseteq \overset{\circ}{F}{}^{\#}$. $\square$

REMARKS. 1. If the group $\mathfrak{G}$ is compact, then $G_f = G$, and every standardly finite-dimensional S-continuous unitary representation of G determines (as mentioned in the preceding section) a unitary representation $\widetilde{T}$ of $G^{\#}$ of the same dimension. From this representation we can construct the representation $\widetilde{T} \circ \overline{j}$ of $\mathfrak{G}$ equivalent to it. Proposition 2.3.13 shows that if $\mathfrak{G}$ has the HA $\langle G, j\rangle$, then every irreducible unitary representation of it has such a form for some S-continuous irreducible unitary representation T of G.

2. All the remaining results in this chapter relate to separable LCA groups. For most of them the assumption about separability can be omitted if instead of ω^+-saturation of the nonstandard universe we require that it be λ^+-saturated, where λ is the weight of $\mathfrak{G}$ (the minimal cardinality of a base for the topology).

DEFINITION 2.4.3. We say that $\langle G, j\rangle$ is a good hyperfinite approximation of a separable LCA group $\mathfrak{G}$ if the corresponding triple $\langle G, G_0, G_f\rangle$ is admissible (see Definition 2.2.24).

EXAMPLE 2.4.4. Let $N = 2L+1$ be an infinitely large hypernatural number, let $\Delta \approx 0$ be such that $N\Delta \sim +\infty$, and let $G = \{-L, \dots, L\}$ be the additive group of the ring ${}^*\mathbb{Z}/N{}^*\mathbb{Z}$ (see §2, 1°). We define a mapping $j\colon G \longrightarrow {}^*\mathbb{R}$ by setting $j(k) = k\Delta\ \forall k \in G$. Obviously, $\langle G, j\rangle$ is a hyperfinite approximation of the additive group of the field $\mathbb{R}$. The equivalence (2.2.1) and Theorem 2.1.1 show that this HA is good.

THEOREM 2.4.5. *Any separable LCA group $\mathfrak{G}$ with a compact open subgroup U admits a good hyperfinite approximation.*

PROOF. Theorems 2.2.23 and 2.2.25 show that in this case any HA is good, that is, it suffices to prove only the existence of some HA.

We denote by $\mathcal{D}$ the factor group $\mathfrak{G}/U$ and consider the short exact sequence $U \subseteq \mathfrak{G} \xrightarrow{\pi} \mathcal{D}$, where π is the natural projection. The ω^+-saturation of the nonstandard universe and the countability of $\mathcal{D}$ imply the existence of a hyperfinite set $T \subseteq {}^*\mathcal{D}$ such that $\mathcal{D} \subseteq T$. Denote by ${}^*\mathcal{D}(T)$ the internal subgroup of ${}^*\mathcal{D}$ generated by T and let $\mathcal{H} = {}^*\pi^{-1}({}^*\mathcal{D}(T))$. Thus, we have the following commutative diagram whose bottom row is a short exact sequence:

$$\begin{array}{ccccc} U & \xrightarrow{\subseteq} & \mathfrak{G} & \xrightarrow{\pi} & \mathcal{D} \\ {\scriptstyle \mathrm{id}}\downarrow & & {\scriptstyle \mathrm{id}}\downarrow & & {\scriptstyle \mathrm{id}}\downarrow \\ U & \xrightarrow{\subseteq} & \mathcal{H} & \xrightarrow{\varepsilon} & {}^*\mathcal{D}(T), \end{array} \tag{2.4.1}$$

where $\varepsilon = {}^*\pi|{}^*\mathcal{H}$ is an internal mapping.

Using the basic theorem on finitely generated Abelian groups and the transfer principle, we get that ${}^*\mathcal{D}(T) = \mathcal{D}_1 \oplus \mathcal{D}_2$, where $\mathcal{D}_1$ is a hyperfinite Abelian group and $\mathcal{D}_2$ is a free (in the nonstandard universe) Abelian group with a hyperfinite set of generators. Letting $\mathcal{H}_i = \varepsilon^{-1}(\mathcal{D}_i)$ $(i = 1, 2)$, we get that, first, $\mathcal{H} = \mathcal{H}_1 + \mathcal{H}_2$ and $\mathcal{H}_1 \cap \mathcal{H}_2 = {}^*U$, and second, the following sequences are exact:

$$ {}^*U \subseteq \mathcal{H}_1 \xrightarrow{\varepsilon_1} \mathcal{D}_1, \tag{2.4.2}$$

$$ {}^*U \subseteq \mathcal{H}_2 \xrightarrow{\varepsilon_2} \mathcal{D}_2. \tag{2.4.3}$$

Here $\varepsilon_i = \varepsilon|\mathcal{H}_i$ $(i = 1, 2)$. Let us study the sequence (2.4.2) in greater detail. Using a theorem of van Kampen ([**29**], Chapter 2, Theorem 9.5) and the transfer principle, and taking an infinitely small neighborhood V of zero contained in ${}^*U$ (that is, $V \subseteq \mu(0)$), we find a hypernatural number k, a hyperfinite group R, and a continuous epimorphism $\varphi\colon {}^*U \longrightarrow {}^*S^k \oplus R$ such that $\ker\varphi \subseteq V$. Here S is the unit circle.

According to the theory of group extensions, there exist an internal group L and an internal homomorphism $\gamma\colon \mathcal{H}_1 \longrightarrow L$ such that the following diagram commutes:

$$\begin{array}{ccccc} {}^*U & \xrightarrow{\subseteq} & \mathcal{H}_1 & \xrightarrow{\varepsilon_1} & \mathcal{D}_1 \\ {\scriptstyle \varphi}\downarrow & & \downarrow{\scriptstyle \gamma} & & \downarrow{\scriptstyle \mathrm{id}} \\ {}^*S^k \oplus R & \xrightarrow{\varkappa} & L & \xrightarrow{\delta} & \mathcal{D}_1, \end{array} \tag{2.4.4}$$

and its bottom row is a short exact sequence. Since φ is an epimorphism, it follows from the diagram (2.4.4) that γ is an epimorphism. The equality $\mathrm{Ext}(\mathcal{D}_1, S^k \oplus R) = \mathrm{Ext}(\mathcal{D}_1, S^k) \oplus \mathrm{Ext}(\mathcal{D}_1, R)$ implies the existence of the short exact sequences

$$ {}^*S^k \xrightarrow{\varkappa_1} L_1 \xrightarrow{\delta_1} \mathcal{D}_1, \tag{2.4.5}$$

$$ R \xrightarrow{\varkappa_2} L_2 \xrightarrow{\delta_2} \mathcal{D}_1, \tag{2.4.6}$$

from which the bottom row of (2.4.4) is obtained (up to an isomorphism) as follows: $L = \{\langle l_1, l_2 \rangle \in L_1 \oplus L_2 \mid \delta_1(l_1) = \delta_2(l_2)\}$, $\varkappa = \langle \varkappa_1, \varkappa_2 \rangle$, and $\delta(l_1, l_2) = \delta_1(l_1) = \delta_2(l_2)$. Since ${}^*S^k$ is a divisible group, the sequence (2.4.5) splits, that is, there exists a monomorphism $\chi\colon \mathcal{D}_1 \longrightarrow L_1$ that is a right inverse of δ_1. The group L_2

is hyperfinite in view of the hyperfiniteness of the groups R and $\mathcal{D}_1$. Note that φ is open, being a continuous epimorphism of compact groups, and thus $\varphi(V)$ is a neighborhood of zero in ${}^*S^k \oplus R$, so that $\varphi(V) \cap {}^*S^k$ is a neighborhood of zero in ${}^*S^k$. Since the unit circle contains arbitrarily dense finite subgroups, there exists a hyperfinite subgroup $F \subseteq {}^*S^k$ such that $F + (\varphi(V) \cap {}^*S^k) = {}^*S^k$. We now consider the hyperfinite subgroup $M \subseteq L$ given by $M = \{\langle \varkappa_1(f) + \chi(d), l\rangle \mid f \in F,\, d \in \mathcal{D}_1,\, l \in L_2,\, \delta_2(l) = d\}$. Since γ is an epimorphism, it follows that $\forall m \in M\ (\gamma^{-1}(m) \neq \emptyset)$. We choose one element g_m from each set $\gamma^{-1}(m)$ in an internal way, and we set $G_1 = \{g_m \mid m \in M\}$. The operation $+_1$ is defined in G_1 by setting $g_{m_1} +_1 g_{m_2} = g_{m_1+m_2}$. Thus, $\gamma(g_{m_1} +_1 g_{m_2}) = \gamma(g_{m_1}) + \gamma(g_{m_2}) = m_1 + m_2$. Obviously, $\langle G_1, +_1\rangle$ is a hyperfinite Abelian group. Let us study some of its properties.

LEMMA 2.4.6. 1. $\forall m_1, m_2 \in M\ (g_{m_1} +_1 g_{m_2} \approx g_{m_1} + g_{m_2})$.
2. $\forall m \in M\ ((-_1 g_m) \approx -g_m)$.

PROOF. Since $\gamma(g_{m_1} +_1 g_{m_2}) = \gamma(g_{m_1}) + \gamma(g_{m_2})$, it follows from the commutativity of the diagram (2.4.4) that $\varepsilon_1(g_{m_1} +_1 g_{m_2} - g_{m_1} - g_{m_2}) = 0$, that is, $g_{m_1} +_1 g_{m_2} - g_{m_1} - g_{m_2} \in {}^*U$. Using the first square of this diagram and the fact that $\varkappa$ is a monomorphism, we get that $\varphi(g_{m_1} +_1 g_{m_2} - g_{m_1} - g_{m_2}) = 0$, that is, $g_{m_1} +_1 g_{m_2} - g_{m_1} - g_{m_2} \in V \subseteq \mu(0)$, which proves assertion 1. The proof of assertion 2 is similar. □

LEMMA 2.4.7. $\forall h \in \mathcal{H}_1\ \exists g \in G_1\ (g \approx h)$.

PROOF. Let $h \in \mathcal{H}_1$. Using the structure of the group L and the fact that the sequence (2.4.5) splits, we get that $\gamma(h) = \langle \varkappa_1(s) + \chi(d), l\rangle$, where $l \in L_2$ and $\delta_2(l) = d$. Since $F + (\varphi(V) \cap {}^*S^k) = {}^*S^k$, there is an $f \in F$ such that $s - f \in \varphi(V) \cap {}^*S^k$. Then $m = \langle \varkappa_1(f) + \chi(d), l\rangle \in M$. We show that $g_m \approx h$. First of all, $\varepsilon_1(h) = \varepsilon_1(g_m) = d$, and thus $l - g_m \in {}^*U$. Moreover, $\gamma(l) - \gamma(g_m) = \langle \varkappa_1(f-s), 0\rangle = \varkappa(\langle f-s, 0\rangle) = \varkappa\varphi(l - g_m)$. Thus, $\varphi(l - g_m) = \langle f - s, 0\rangle \in \varphi(V) \cap {}^*S^k \subseteq \varphi(V)$. Then $l - g_m \in \varphi^{-1}(\varphi(V)) = V + \ker\varphi \subseteq V + V \subseteq \mu(0)$. □

Let $G_U = {}^*U \cap G_1$. Using the lemmas just proved and the fact that U is a compact-open subgroup of $\mathfrak{G}$, we immediately get the next lemma.

LEMMA 2.4.8. *G_U is a subgroup of G_1. The pair $\langle G_U, \hookrightarrow\rangle$ is a hyperfinite approximation of the group U.*

The restriction of the operation $+_1$ to G_U will be denoted by $+_U$.

We pass to a study of the sequence (2.4.3). Let $\nu_i\colon {}^*\mathcal{D}(T) \longrightarrow \mathcal{D}_i$ be the natural projection. We choose a hypernatural number m such that $(\nu_2(T) - \nu_2(T)) \cap m\mathcal{D}_2 = \{0\}$. To prove the existence of such an m we must apply the transfer principle to the following obvious proposition: "If P is a finite subset of a free finitely generated Abelian group H, then there exists a natural number m such that $P \cap mH \subseteq \{0\}$." Let $Q = \mathcal{D}_2/m\mathcal{D}_2$. Then Q is a hyperfinite Abelian group. The natural projection of $\mathcal{D}_2$ on Q is denoted by λ. By construction, $\nu_2(T)$ is injective. We choose in an internal way one element d_q from each set $\lambda^{-1}(q)$ for $q \in Q$ in such a manner that if $\lambda^{-1}(q) \cap \nu_2(T) \neq \emptyset$, then $d_q \in \nu_2(T)$ (here such a d_q is unique because λ is injective on $\nu_2(T)$). Let $G_3 = \{d_q \mid q \in Q\}$. We define an operation $+_3$ on G_3 by $d_{q_1} +_3 d_{q_2} = d_{q_1+q_2}$. Thus, $\lambda(d_{q_1} +_3 d_{q_2}) = \lambda(d_{q_1}) + \lambda(d_{q_2})$.

LEMMA 2.4.9. 1) $\forall d \in \mathcal{D}\ \exists q \in Q\ (\nu_2(d) = d_q)$.
2) *If* $d_{q_1}, d_{q_2} \in \nu_2(\mathcal{D})$, *then* $d_{q_1} +_3 d_{q_2} = d_{q_1} + d_{q_2}$.

The right-hand side of the last equality contains the addition operation in $\mathcal{D}_2$. Note that $\nu_2(\mathcal{D})$ is in general an external subgroup of $\mathcal{D}_2$.

PROOF. The first assertion of the lemma follows from the definition of d_q and the fact that $\nu_2(\mathcal{D}) \subseteq \nu_2(T)$. If $d_{q_1}, d_{q_2} \in \nu_2(\mathcal{D})$, then $d_{q_1} + d_{q_2} \in \nu_2(\mathcal{D})$. Let $d_{q_1} + d_{q_2} = d_{q_3}$. Since λ is a homomorphism, it follows that $q_3 = \lambda(d_{q_1} + d_{q_2}) = \lambda(d_{q_1}) + \lambda(d_{q_2}) = q_1 + q_2$. On the other hand, $\lambda(d_{q_1} +_3 d_{q_2}) = \lambda(d_{q_1}) + \lambda(d_{q_2}) = q_1 + q_2 = q_3$. Thus, $\lambda^{-1}(q_3) \cap \nu_2(T) = \{d_{q_1} + d_{q_2}\}$, that is, $d_{q_3} = d_{q_1} + d_{q_2} = d_{q_1} +_3 d_{q_2}$.□

Since $\mathcal{D}_2$ is a free Abelian group, the sequence (2.4.3) splits, that is, there exists a monomorphism $\mu_2 \colon \mathcal{D}_2 \longrightarrow \mathcal{H}_2$ that is a right inverse of ε_2. We consider the set $G_2 = \{g + \mu(d_q) \mid g \in G_U,\ q \in Q\}$ and define an operation $+_2$ on G_2 by setting $(g_1 + \mu(d_{q_1})) +_2 (g_2 + \mu(d_{q_2})) = g_1 +_U g_2 + m(d_{q_1} +_3 d_{q_2})$. By using the fact that $\mathcal{H}_2 = {}^*U \oplus \mu_2(\mathcal{D}_2)$ and the fact that $G_U \subseteq {}^*U$ it is easy to get that $\langle G_2, +_2\rangle$ is a hyperfinite Abelian group, and $G_2 \cap {}^*U = G_U$.

Suppose now that $G = G_1 \times G_2$. We define $j \colon G \longrightarrow {}^*\mathfrak{G}$ by setting $j(\langle g_1, g_2\rangle) = g_1 + g_2$, and we show that $\langle G, j\rangle$ is a hyperfinite approximation of $\mathfrak{G}$.

Let $\xi \in \mathfrak{G}$. Since $\mathfrak{G} \subseteq \mathcal{H} = \mathcal{H}_1 + \mathcal{H}_2$, it follows that $\xi = h_1 + h_2$, where $h_1 \in \mathcal{H}_1$ and $h_2 \in \mathcal{H}_2$. Then $d = \pi(\xi) = \varepsilon_1(h_1) + \varepsilon_2(h_2)$ (see the diagram (2.4.1)). Thus, $\varepsilon_2(h_2) = \nu_2(d) \in \nu_2(\mathcal{D})$, because $d \in \mathcal{D}$. By Lemma 2.4.9, $\nu_2(d) = d_q$. Then there exists a $g \in {}^*U$ such that $h_2 = g + \mu(d_q)$. According to Lemma 2.4.8, G_U approximates ${}^*U$, and since U is compact, there exists a $g_0 \in G_U$ such that $g_0 \approx g$. Then $g_2 = g_0 + \mu(d_q) \approx g + \mu(d_q) = h_2$ and $g_2 \in G_2$. In view of Lemma 2.4.7 there is a $g_1 \in G_1$ such that $g_1 \approx h_1$, Consequently, $g_1 + g_2 \approx h_1 + h_2 = \xi$. This verifies the first condition in Definition 2.4.1 (the fourth condition is obvious). It remains to verify the second and third conditions.

Suppose that $g_1 + g_2 \approx \xi \in \mathfrak{G}$ and $g_1' + g_2' \approx \xi' \in \mathfrak{G}$, where $g_1, g_1' \in G_1$, and $g_2, g_2' \in G_2$. It is required to show that $g_1 +_1 g_1' + g_2 +_2 g_2' \approx \xi + \xi'$. By Lemma 2.4.6, $g_1 +_1 g_1' \approx g_1 + g_1'$. Consequently, it suffices to show that $g_2 +_2 g_2' \approx g_2 + g_2'$. Let $\pi(\xi) = d$. Since $g_1 + g_2 \approx \xi$, it follows that $g_1 + g_2 - \xi \in {}^*U$, and hence $\varepsilon(g_1 + g_2) = d$ (see the diagram (2.4.1)). The fact that $g_i \in \mathcal{H}_i$ implies that $\varepsilon_2(g_2) = \nu_2(d) \in \nu_2(\mathcal{D})$. Similarly, $\varepsilon_2(g_2') = \nu_2(d_2') \in \nu_2(\mathcal{D})$, where $d' = \pi(\xi')$. From this we get immediately that $g_2 = \widetilde{g} + \mu(\nu_2(d))$ and $g_2' = \widetilde{g}' + \mu(\nu_2(d'))$. Then by Lemma 2.4.9, $g_2 +_2 g_2' = \widetilde{g} +_U \widetilde{g}' + \mu(\nu_2(d) + \nu_2(d')) = \widetilde{g} +_U \widetilde{g}' + \mu(\nu_2(d)) + \mu(\nu_2(d'))$. According to Lemma 2.4.8, $\widetilde{g} +_U \widetilde{g}' \approx \widetilde{g} + \widetilde{g}'$, which yields what was required. The third condition is verified similarly.

LEMMA 2.4.10. *If separable LCA groups* $\mathfrak{G}_1$ *and* $\mathfrak{G}_2$ *admit good HA's, then* $\mathfrak{G}_1 \times \mathfrak{G}_2$ *also admits a good HA.*

PROOF. Obvious. □

It is well known that every LCA group can be represented as a direct product of $\mathbb{R}^m$ (for some $m \geq 0$) and a group having a compact open subgroup (see, for example, Theorem 1 in §10.3 of Chapter 2 in [**14**]). Comparing this fact with Theorem 2.4.5, Lemma 2.4.10, and Example 2.4.4, we get the next theorem.

THEOREM 2.4.11. *Any separable LCA group admits a good hyperfinite approximation.*

2°. Let $\langle G, j\rangle$ be an HA of the group $\mathfrak{G}$. If $U \subseteq \mathfrak{G}$ is a neighborhood of zero with compact closure, then $G_0 \subseteq j^{-1}({}^*U) \subseteq G_f$, that is, $\Delta = |j^{-1}({}^*U)|^{-1}$ is a normalizing multiplier of the triple $\langle G, G_0, G_f\rangle$ (see Definition 2.2.8 and the remarks after it). The Loeb measure ν_Δ on G induces the Haar measure μ_Δ on $G^\#$. The topological isomorphism $\overline{j}$ determines from μ_Δ a Haar measure $\overline{\mu}_\Delta$ on $\mathfrak{G}$. Obviously, any Haar measure on $\mathfrak{G}$ can be obtained in this way.

If $f\colon \mathfrak{G} \longrightarrow \mathbb{R}$ is a measurable function, a lifting of the function $f \circ \overline{j}$ will be called a lifting of f (see the remarks after the proof of Theorem 2.2.10).

PROPOSITION 2.4.12. *Let $\langle G, j\rangle$ be an HA of a separable LCA group $\mathfrak{G}$ and Δ an NM of the triple $\langle G, G_0, G_f\rangle$, let $p \in [1, \infty)$ be standard, and let $f\colon \mathfrak{G} \longrightarrow \mathbb{C}$. Then $f \in L_p(\mathfrak{G})$ if and only if f has an $\mathcal{S}_{p,\Delta}$-integrable lifting. Further, if $p = 1$ and $\varphi\colon G \longrightarrow {}^*\mathbb{C}$ is an $\mathcal{S}_{1,\Delta}$-integrable lifting of f, then*

$$\int f\, d\overline{\mu}_\Delta = {}^\circ\Big(\Delta \sum_{g\in G} \varphi(g)\Big).$$

This proposition is simply a reformulation of Corollary 2.2.11.

It now follows directly from Definition 1.2.17 that the triple $\langle G, j, \Delta\rangle$ is a hyperfinite realization of the space $\langle \mathfrak{G}, \mu_\Delta\rangle$, and we can reformulate Proposition 1.2.18.

PROPOSITION 2.4.13. *If under the conditions of Proposition 2.4.12 the function $f \in L_1(\mathfrak{G})$ is bounded and continuous almost everywhere with respect to Haar measure and satisfies the condition*

$$\forall B \in {}^*\mathcal{P}(G)\ \Big(B \subseteq G\setminus G_f \Longrightarrow \Delta \sum_{g\in B} |{}^*f(j(g))| \approx 0\Big), \tag{2.4.7}$$

then the function $\varphi = {}^*f \circ j$ is an $\mathcal{S}_{1,\Delta}$-integrable lifting of f, and hence

$$\int f\, d\mu_\Delta = {}^\circ\Big(\Delta \sum_{g\in G} {}^*f(j(g))\Big). \ \square$$

If $\langle G, j\rangle$ is an HA of $\mathfrak{G}$, then it is natural to approximate the group $\widehat{\mathfrak{G}}$ with the help of the group $\widehat{G}$. If $\langle \widehat{G}, i\rangle$ is an HA of $\widehat{\mathfrak{G}}$, then $\widehat{G}_f = i^{-1}(\mathrm{Ns}({}^*\widehat{\mathfrak{G}}))$ and $\widehat{G}_0 = i^{-1}\mu_{\widehat{\mathfrak{G}}}(\mathbf{1})$. Here $\mathbf{1}$ is the neutral element of $\widehat{\mathfrak{G}}$, that is, the character identically equal to 1.

LEMMA 2.4.14. *Let $\chi \in {}^*\widehat{\mathfrak{G}}$ and $\varkappa \in \widehat{\mathfrak{G}}$. Then $\chi \overset{\widehat{\mathfrak{G}}}{\approx} \varkappa$ if and only if $\forall \xi \in \mathrm{Ns}({}^*\mathfrak{G})$ $(\chi(\xi) \approx \varkappa(\xi))$.*

PROOF. Let $\chi \approx {}^*\varkappa$. If $\xi \in \mathrm{Ns}({}^*\mathfrak{G})$, then $\xi \approx \eta \in \mathfrak{G}$. If u is a relatively compact neighborhood of η in $\mathfrak{G}$, then $\xi \in \overline{u}$. By assumption, $\forall^{\mathrm{st}} k$ $(\chi - {}^*\varkappa \in {}^*\mathcal{W}(\overline{u}, \Lambda_k))$ (see the proof of Theorem 2.2.15), but this means that $\chi(\xi) \approx {}^*\varkappa(\xi)$.

Conversely, suppose that $\forall \xi \in \mathrm{Ns}({}^*\mathfrak{G})$ $(\chi(\xi) \approx {}^*\varkappa(\xi))$. In this case if F is a compact subset of $\mathfrak{G}$, then ${}^*F \subseteq \mathrm{Ns}({}^*\mathfrak{G})$, and hence $\forall \xi \in {}^*F$ $(\chi(\xi) \approx {}^*\varkappa(\xi))$, so that $\chi - {}^*\varkappa \in {}^*\mathcal{W}(F, \Lambda_k)$ for any standard k. $\square$

The external subgroups G_0 and G_f determined by the hyperfinite approximation $\langle G, j\rangle$ of the group $\mathfrak{G}$ determine in turn external subgroups $H_0, H_f \subseteq \widehat{G}$ by means of the relation (2.2.5).

DEFINITION 2.4.15. Let $\langle G, j\rangle$ and $\langle \widehat{G}, i\rangle$ be HA's of the separable LCA groups $\mathfrak{G}$ and $\widehat{\mathfrak{G}}$, respectively. We say that $\langle \widehat{G}, i\rangle$ is dual to $\langle G, j\rangle$ if the following two conditions hold: 1) $H_0 \subseteq \widehat{G}_0$; 2) $\forall h \in \widehat{G}_f\ \forall g \in G_f\ (i(h)(j(g)) \approx h(g))$.

REMARKS. 1. If the group $\mathfrak{G}$ is compact, then $G_f = G$, and the first condition holds automatically in view of Lemma 2.2.20.

2. Dual approximations of the character groups of the circle group and of discrete groups can be constructed directly, and thus a careful analysis of the proof of Theorem 2.4.5 shows that any separable LCA group containing a compact open subgroup has a pair of dual hyperfinite approximations. This is therefore true also for arbitrary separable LCA groups, because for $\mathbb{R}$ a pair of dual HA's can also be constructed directly (see §1 of this chapter).

THEOREM 2.4.16. *Let $\langle G, j\rangle$ be a good HA of a separable LCA group $\mathfrak{G}$, let $\langle \widehat{G}, i\rangle$ be an HA of $\widehat{\mathfrak{G}}$ dual to it, and let Δ be an NM of the triple $\langle G, G_0, G_f\rangle$ constructed from $\langle G, j\rangle$. Then:*

1) *$(|G|\cdot\Delta)^{-1}$ is an NM of the triple $\langle \widehat{G}, \widehat{G}_0, \widehat{G}_f\rangle$ constructed from $\langle \widehat{G}, i\rangle$;*

2) *if $\mathcal{F}\colon L_2(\mathfrak{G}, \overline{\mu}_\Delta) \longrightarrow L_2(\widehat{\mathfrak{G}}, \overline{\mu}_{\widehat{\Delta}})$ is the Fourier transformation, then $\mathcal{F}$ preserves the inner product;*

3) *the discrete Fourier transformation $\Phi^G_\Delta\colon \mathcal{L}_{2,\Delta}(G) \longrightarrow \mathcal{L}_{2,\widehat{\Delta}}(\widehat{G})$ is a hyperfinite-dimensional approximation of $\mathcal{F}$.*

REMARK. If $|f|^2$ satisfies the conditions of Proposition 2.4.13, and $|\mathcal{F}(f)|^2$ satisfies the same conditions with $\mathfrak{G}, G, \Delta, G_f$ replaced by $\widehat{\mathfrak{G}}, \widehat{G}, \widehat{\Delta}, \widehat{G}_f$, respectively, then the third assertion of the theorem is equivalent to the relation

$$(\Delta|G|)^{-1}\sum_{h\in\widehat{G}} |\mathcal{F}(f)(i(h)) - \Phi_\Delta({}^*f\circ j)(h)|^2 \approx 0, \tag{2.4.8}$$

which can be written in greater detail as follows:

$$(\Delta|G|)^{-1}\sum_{h\in\widehat{G}}\left|\int_{{}^*\mathfrak{G}} {}^*f(\xi)\cdot\overline{i(h)(\xi)}\,d\mu_\Delta(\xi) - \Delta\sum_{g\in G} {}^*f(j(g))\cdot\overline{h(g)}\right|^2 \approx 0.$$

PROOF OF THEOREM 2.4.16. We prove first that $H_0 = \widehat{G}_0$ and $H_f = \widehat{G}_f$. If $h \in \widehat{G}_0$, then $i(h) \approx \mathbf{1}$, that is, $\forall \xi \in \mathrm{Ns}({}^*\mathfrak{G})\ (i(h)(\xi) \approx 1)$ (Lemma 2.4.14). Consequently, $\forall g \in G_f\ (i(h)(j(g)) \approx 1)$, so that $h(g) \approx 1$ in view of condition 2 in Definition 2.4.15, and hence $h \in H_0$. Thus, $\widehat{G}_0 \subseteq H_0$. The reverse inclusion is condition 1 of Definition 2.4.15. The inclusion $\widehat{G}_f \subseteq H_f$ is a trivial consequence of condition 2 of Definition 2.4.15 and Lemma 2.4.14. The reverse inclusion has a somewhat more complicated proof. Let $h \in H_f$. Then $\widetilde{h} \in G^{\#\wedge}$ (recall that $\widetilde{h}(g^\#) = {}^\circ h(g)$). If the mapping $\overline{i}$ is defined from i as $\overline{j}$ is defined from j, then $\overline{i}\colon \widehat{G}^\# \longrightarrow \widehat{\mathfrak{G}}$ is a topological isomorphism in view of Proposition 2.4.2. Let $\varkappa = \widetilde{h}\circ\overline{j}^{-1} \in \widehat{\mathfrak{G}}$. Since $\widehat{G}^\# = \widehat{G}_f/\widehat{G}_0$, there is an $h_1 \in \widehat{G}_f$ such that $\overline{i}(h_1^\#) = \widetilde{h}\circ\overline{j}^{-1}$. If $g \in G_f$, then $\overline{i}(h_1^\#)(\overline{j}(g^\#)) = \mathrm{st}\, i(h_1)(\mathrm{st}\, j(g)) \approx i(h_1)j(g) \approx h_1(g)$ by virtue of condition 2 in Definition 2.4.15. On the other hand, $\overline{i}(h_1^\#)(\overline{j}(g^\#)) = \widetilde{h}\overline{j}^{-1}(\overline{j}(g^\#)) = \widetilde{h}(g^\#) \approx h(g)$. Thus, $h(g) \approx h_1(g)$ for any $g \in G_f$. This means that $h\cdot\overline{h}_1 \in H_0 = \widehat{G}_0 \subseteq \widehat{G}_f$. Since $h_1 \in \widehat{G}_f$, it follows that $h \in \widehat{G}_f$, which proves the second of the necessary equalities. The first assertion of the theorem now follows from the fact that $\langle G, j\rangle$ is a good HA.

Denote by $\gamma\colon L_2(G^{\#},\mu) \longrightarrow \mathcal{L}_{2,\Delta}(G)^{\#}$ the inclusion induced by the natural projection $\#\colon G_f \longrightarrow G^{\#}$,[1] and by $\widetilde{\gamma}\colon L_2(G^{\#\frown},\mu_{\widehat{\Delta}}) \longrightarrow \mathcal{L}_{2,\widehat{\Delta}}(\widehat{G})^{\#}$ the imbedding induced by the natural projection $\widehat{\#}\colon \widehat{G}_f \longrightarrow \widehat{G}^{\#}$. Since $\langle G, j\rangle$ is a good HA, the equalities just proved give us that $\widehat{G}^{\#}$ is canonically isomorphic to $G^{\#\frown}$ (the isomorphism is established by associating the character $\widetilde{h}$ with an element $h^{\#} \in G^{\#\frown}$, where $h \in \widehat{G}_f = H_f$) and the diagram

$$\begin{array}{ccc} L_2(G^{\#},\mu_\Delta) & \xrightarrow{\mathcal{F}_\Delta^{G^{\#}}} & L_2(G^{\#\frown},\mu_{\widehat{\Delta}}^{\#}) \\ {\scriptstyle\gamma}\big\downarrow & & \big\downarrow{\scriptstyle\widehat{\gamma}} \\ \mathcal{L}_{2,\Delta}(G)^{\#} & \xrightarrow{(\Phi_\Delta^G)^{\#}} & \mathcal{L}_{2,\widehat{\Delta}}(\widehat{G})^{\#} \end{array} \tag{2.4.9}$$

is commutative (see Definition 2.2.24). Since the topological isomorphisms $\overline{j}$ and $\overline{i}$ carry the measures μ_Δ on $G^{\#}$ and $\mu_{\widehat{\Delta}}$ on $G^{\#\frown}$ into the measures $\overline{\mu}_\Delta$ on $\mathfrak{G}$ and $\overline{\mu}_{\widehat{\Delta}}$ on $\widehat{\mathfrak{G}}$, isomorphisms $j_*\colon L_2(\mathfrak{G},\overline{\mu}_\Delta) \longrightarrow L_2(G^{\#},\mu_\Delta)$ and $i_*\colon L_2(\widehat{\mathfrak{G}},\overline{\mu}_{\widehat{\Delta}}) \longrightarrow L_2(\widehat{G}^{\#},\mu_{\widehat{\Delta}})$ are defined such that $j_*(f) = f\circ j$ and $i_*(\varphi) = \varphi\circ i$. Furthermore, the diagram

$$\begin{array}{ccc} L_2(\mathfrak{G},\overline{\mu}_\Delta) & \xrightarrow{\mathcal{F}} & L_2(\widehat{\mathfrak{G}},\overline{\mu}_{\widehat{\Delta}}) \\ {\scriptstyle j_*}\big\downarrow & & \big\downarrow{\scriptstyle i_*} \\ L_2(G^{\#},\mu_\Delta) & \xrightarrow{\mathcal{F}_\Delta^{G^{\#}}} & L_2(G^{\#\frown},\mu_{\widehat{\Delta}}) \end{array} \tag{2.4.10}$$

is commutative.

It follows immediately from the definitions that $\gamma\circ j_*(f)$ is the class of an $\mathcal{S}_{2,\Delta}$-lifting of f, that is, $\gamma\circ j_*(f) = j_{2,\Delta}(f)$. Similarly, $\widehat{\gamma}\circ i_* = j_{2,\widehat{\Delta}}$, and $j_{2,\Delta}$ and $j_{2,\widehat{\Delta}}$ are induced by j and i, respectively (see the diagram (1.3.1)). Comparing the diagrams (2.4.9) and (2.4.10), we get that the diagram

$$\begin{array}{ccc} L_2(\mathfrak{G},\overline{\mu}_\Delta) & \xrightarrow{\mathcal{F}} & L_2(\widehat{\mathfrak{G}},\overline{\mu}_{\widehat{\Delta}}) \\ {\scriptstyle j_{2,\Delta}}\big\downarrow & & \big\downarrow{\scriptstyle j_{2,\widehat{\Delta}}} \\ \mathcal{L}_{2,\Delta}(G^{\#},\mu_\Delta) & \xrightarrow{(\Phi_\Delta^G)^{\#}} & \mathcal{L}_{2,\widehat{\Delta}}(G^{\#\frown}) \end{array}$$

is commutative, and this proves the third assertion of the theorem. The second assertion follows immediately from the third. □

§5. Examples of hyperfinite approximations

1°. We return to Example 2.4.4, in which a good HA of the additive group of the field $\mathbb{R}$ was constructed. In this example $G = \{-L,\dots,L\}$ is the additive group of the ring ${}^*\mathbb{Z}/N{}^*\mathbb{Z}$, $N = 2L+1$, $j\colon G \longrightarrow {}^*\mathbb{R}$ is defined by $j(k) = k\Delta$,

[1] Thus, γ assigns to a function $f \in L_2(G^{\#},\mu_\Delta)$ the class of an $\mathcal{L}_{2,\Delta}(G_f)$-lifting of f (that is, of the function $f\circ\#$) in $\mathcal{L}_{2,\Delta}(G)^{\#}$.

$\Delta \approx 0$, and $N\Delta \sim +\infty$. In this case the dual group $\widehat{G}$ is $\cong G$. The isomorphism assigns to each $n \in G$ the character χ_n with $\chi_n(m) = \exp(2\pi inm/N)$. The group $\widehat{\mathbb{R}}$ is isomorphic to $\mathbb{R}$, the isomorphism assigning to each $t \in \mathbb{R}$ the character $\varkappa_t$ with $\varkappa_t(x) = \exp(2\pi itx)$. A dual hyperfinite approximation $\langle \widehat{G}, \iota \rangle$ is defined by $\iota(n) = \frac{n}{N\Delta}$, or, more precisely, $\iota(\chi_n)(x) = \exp(\frac{2\pi in}{N\Delta}x)$. The relations (2.2.1) show that the first of the conditions in Definition 2.4.15 holds. It is trivial to verify the second condition. Indeed, $j(m) = m\Delta \approx x$, and $\iota(m) = \frac{m}{N\Delta} \approx t$, so that $\exp(2\pi itx) = \exp 2\pi i\iota(m)(j(n)) \approx \exp(2\pi imn/N)$. The corresponding hyperfinite-dimensional approximation of the Fourier transformation was investigated at length in §1 of this chapter.

2°. Let us construct a hyperfinite approximation of the unit circle S, which it is convenient to represent, as above, in the form of the interval $[-1/2, 1/2)$. The group operation $+_S$ is addition modulo 1.

The dual group $\widehat{S}$ is isomorphic to the additive group $\mathbb{Z}$. The isomorphism assigns to each $n \in \mathbb{Z}$ the character $\varkappa_n(x) = \exp(2\pi inx)$. We take G to be the same group $\{-L, \dots, L\}$ as in the preceding subsection, where $2L+1 = N \sim +\infty$, and we define the mapping $j\colon G \longrightarrow {}^*S$ by $j(m) = m/N$ $\forall m \in G$. The mapping $\iota\colon \widehat{G} \longrightarrow {}^*\mathbb{Z}$ is defined on the dual HA $\langle \widehat{G}, \iota \rangle$ by $\iota(n) = n$ (more precisely, $\iota(\chi_n) = \varkappa_n$, where the character χ_n is defined in 1°). Since the group $\mathbb{Z}$ is discrete, $\widehat{G}_0 = \{0\}$, and $\widehat{G}_f = \mathbb{Z}$. Since S is compact, it suffices to verify only the second condition of Definition 2.4.15, which is just as trivial as in the preceding subsection.

Applying Theorem 2.4.16 to this case, we get

PROPOSITION 2.5.1. *Let $f\colon [-1/2, 1/2) \longrightarrow \mathbb{C}$ be a Riemann-integrable function. Then*

$$\sum_{n=-L}^{L} \left| \int_{-1/2}^{1/2} f(x) \exp(-2\pi inx)\, dx - \frac{1}{N} \sum_{m=-L}^{L} f(m/N) \exp(-2\pi imn/N) \right|^2 \approx 0$$

if $N = 2L+1$ is an infinite natural number.

Of course, by a suitable change of variables we can get from Proposition 2.5.1 that

$$\sum_{n=-L}^{L} \left| \int_{-l}^{l} f(x) \exp(\pi inx/l)\, dx - \Delta \sum_{m=-L}^{L} f(m\Delta) \exp(-2\pi imn/N) \right|^2 \approx 0 \tag{2.5.1}$$

for any Riemann-integrable function f on $[-l, l]$ and for any N and Δ such that $N = 2L+1 \sim +\infty$ and $^\circ(N\Delta) = 2l$.

In essence, the hyperfinite approximation considered in this subsection was studied thoroughly in the paper [**45**], although the concept of an HA itself was not introduced there. The relation (2.5.1) was obtained in [**45**] for continuous functions, and there are many other interesting applications of nonstandard analysis to the theory of Fourier series, but its possibilities were limited by the fact that the theory of Loeb measures was lacking at the time, and the property of saturation of models had not yet been used in nonstandard analysis.

3°. In this subsection we construct hyperfinite approximations of profinite Abelian groups.

We consider a standard sequence $\{\langle K_n, \varphi_n \rangle \mid n \in \mathbb{N}\}$, where K_n is a finite Abelian group for any $n \in \mathbb{N}$, and $\varphi_n\colon K_{n+1} \longrightarrow K_n$ is a surjective homomorphism.

Let $\langle K, \psi\rangle = \varprojlim\langle K_n, \varphi_n\rangle$, that is, $\psi = \{\psi_n \mid n \in \mathbb{N}\}$, where $\psi_n\colon K \longrightarrow K_n$ is a surjective homomorphism such that $\varphi_n \circ \psi_{n+1} = \psi_n$ for any $n \in \mathbb{N}$. We fix an $N \in {}^*\mathbb{N} \setminus \mathbb{N}$ and let $G = {}^*K_N$. Then ${}^*\psi_N\colon K \longrightarrow {}^*K_N = G$ is a surjective homomorphism.

Let $j\colon G \longrightarrow {}^*K$ be an internal mapping (not a homomorphism in general) that is a right inverse of ${}^*\psi_N$ with $j(0) = 0$.

PROPOSITION 2.5.2. *The pair $\langle G, j\rangle$ is a hyperfinite approximation of K.*

PROOF. Since the topology in K is induced by the product topology in $\prod_n K_n$, it follows that

$$\forall \alpha, \beta \in {}^*K\ (\alpha \overset{K}{\approx} \beta \Longleftrightarrow \forall^{\mathrm{st}} n\ ({}^*\psi_n(\alpha) = {}^*\psi_n(\beta))). \tag{2.5.2}$$

The group K is compact, so it suffices to show that $\forall a, b \in G\ (j(a+b) \approx j(a) + j(b) \wedge j(-a) \approx -j(a))$.

For $n > m$ we define a homomorphism $\varphi_{nm}\colon K_n \longrightarrow K_m$ by

$$\varphi_{nm} = \varphi_{n-1} \circ \cdots \circ \varphi_m. \tag{2.5.3}$$

Then $\varphi_{n,n-1} = \varphi_{n-1}$ and $\varphi_{nm} \circ \psi_n = \psi_m$, and now $\forall^{\mathrm{st}} n\ ({}^*\psi_n(j(a+b)) = {}^*\varphi_{Nn} \circ {}^*\psi_N(j(a+b)) = {}^*\varphi_{Nn}(a+b) = {}^*\varphi_{Nn}(a) + {}^*\varphi_{Nn}(b))$. Similarly, ${}^*\psi_n(j(a) + j(b)) = {}^*\psi_n(j(a)) + {}^*\psi_n(j(b)) = {}^*\varphi_{Nn}(a) + {}^*\varphi_{Nn}(b)$, and this proves the first of the required relations by virtue of (2.5.2). The second can be proved analogously. □

COROLLARY 2.5.3. *Under the conditions of Proposition* 2.5.2

$$G_0 = \{a \in G \mid \forall^{\mathrm{st}} n\ ({}^*\varphi_{Nn}(a) = 0)\}, \tag{2.5.4}$$

where φ_{nm} is defined by the formula (2.5.3), *and $K \cong G/G_0 = G^\#$.*

REMARK 2.5.4. Proposition 2.5.2 and its corollary remain true also when K_n is a ring. In the last case K and G are also rings, and $G \longrightarrow {}^*K$ is a 'near-homomorphism' of rings, that is, the relation $j(ab) \approx j(a)j(b)$ holds in addition to the relations already proved. Further, the object G_0 defined by (2.5.4) is a two-sided ideal.

In this case the dual group $\widehat{K}$ is $\varinjlim\langle \widehat{K}_n, \widehat{\varphi}_n\rangle$, where the imbedding $\widehat{\varphi}_n\colon \widehat{K}_n \longrightarrow \widehat{K}_{n+1}$ is defined by $\widehat{\varphi}_n(\chi) = \chi \circ \varphi_n\ \forall \chi \in \widehat{K}_n$. The imbeddings $\widehat{\psi}_n\colon \widehat{K}_n \longrightarrow \widehat{K}$ are defined similarly. For $n > m$ the imbeddings $\widehat{\varphi}_{nm}\colon \widehat{K}_m \longrightarrow \widehat{K}_n$ are defined such that $\widehat{\varphi}_{nm} = \widehat{\varphi}_m \circ \cdots \circ \widehat{\varphi}_{n-1}$ (see (2.5.3)), and the relation $\widehat{\psi}_m = \widehat{\psi}_n \circ \widehat{\varphi}_{nm}$ holds.

PROPOSITION 2.5.5. *If the HA $\langle G, j\rangle$ of the group K is defined as in Proposition* 2.5.2, *that is, $G = {}^*K_N$ and $j\colon G \longrightarrow {}^*K$ is the right inverse of ${}^*\psi_N$, then $\langle \widehat{K}, \widehat{\psi}_N\rangle$ is an HA of $\widehat{K}$ dual to $\langle G, j\rangle$.*

PROOF. First of all we note that conditions 2)–4) of Definition 2.4.1 hold automatically because $\widehat{\psi}_N$ is a homomorphism. The fact that $\widehat{K}$ is the inductive limit of the $\widehat{K}_n$ implies that $\widehat{K} = \bigcup_{n\in\mathbb{N}} A_n$, where $A_n = \{\chi \circ \psi_n \mid \chi \in \widehat{K}_n\}$. By the transfer principle, ${}^*\widehat{K} = \bigcup_{n\in{}^*\mathbb{N}} A_n$. Since $\widehat{K}_n$ is a standard finite set, it follows that ${}^*A_n = \{\chi \circ {}^*\psi_n \mid \chi \in \widehat{K}_n\}$. Thus, every standard element $\varkappa \in {}^*\widehat{K}$ has the form $\varkappa = \chi \circ {}^*\psi_n$ for some standard n and $\chi \in \widehat{K}_N$. Then ${}^*\widehat{\varphi}_{nM}(\chi) \in \widehat{K}_n$ and

$\widehat{\psi}_N({}^*\widehat{\varphi}_{nN}(\chi)) = \varkappa$. The first condition of Definition 2.4.1 is thus satisfied. The first condition of Definition 2.4.15 holds automatically because K is compact. If $\chi \in \widehat{K}_n$, then ${}^*\widehat{\psi}_N(\chi)(j(a)) = \chi({}^*\psi_N(j(a))) \doteq \chi(a)$, that is, the second condition in Definition 2.4.15 also holds.

Applying Theorem 2.4.16 and the relation (2.4.8) to this case, we obtain

PROPOSITION 2.5.6. *If under the conditions of Propositions 2.5.2 and 2.5.5 the function $f\colon K \longrightarrow \mathbb{C}$ is bounded and continuous almost everywhere with respect to Haar measure, then*

$$\sum_{\chi\in\widehat{K}_N}\left|\int_{{}^*K} f(\alpha)\overline{\chi(\psi_N(\alpha))}\,d^*\mu_K(\alpha) - |K_N|^{-1}\sum_{a\in K_N} f(j(a))\overline{\chi(a)}\right|^2 \approx 0, \tag{2.5.5}$$

where μ_K is the Haar measure on K with $\mu_K(K) = 1$.

4°. Here the considerations of the preceding subsection are applied to the construction of a hyperfinite approximation of the ring Δ_τ of τ-adic integers (see [**29**], [**14**]).

Let $\tau = \langle a_n \mid n \in \mathbb{N}\rangle$ be a standard sequence of natural numbers such that $a_n > 1$ and $a_n \mid a_{n+1}$. Denote by A_n the ring $\mathbb{Z}/a_n\mathbb{Z}$, which in our case is conveniently regarded as the ring of smallest positive residues modulo a_n, that is, $A_n = \{0, 1, \dots, a_n - 1\}$. Let $\varphi_n\colon A_{n+1} \longrightarrow A_n$ be the surjective homomorphism with $\varphi_n(a) = \operatorname{rem}(a, a_n)$ $\forall a \in A_{n+1}$. The ring $\Delta_\tau = \varprojlim\langle A_{n+1}, \varphi_n\rangle$ is called the ring of τ-adic integers. We define an imbedding $\nu\colon \mathbb{Z} \longrightarrow \Delta_\tau$ by setting $\nu(a)_n = \operatorname{rem}(a, a_n)$ $\forall a \in \mathbb{Z}$, $\forall n \in \mathbb{N}$. Then the sequence $\nu(a)$ is in $\prod_{n\in\mathbb{N}} A_n$, and $\varphi_n(\nu(a)_{n+1}) = \nu(a)_n$, that is, $\nu(a) \in \Delta_\tau$. It is easy to verify that $\nu(\mathbb{Z})$ is dense in Δ_τ. Let $j_n = \nu|A_n\colon A_n \longrightarrow \Delta_\tau$. Then j_n is a right inverse of the mapping $\psi_n\colon \Delta_\tau \longrightarrow A_n$. Indeed, if $\xi = \langle \xi_n \mid n \in \mathbb{N}\rangle \in \Delta_\tau$, then $\psi_n(\xi) = \xi_n$. Considering that $\operatorname{rem}(a, a_n) = a$ for $a \in A_n$, we then get that $\psi_n(\nu(a)) = a$.

Proposition 2.5.2 and its Corollary 2.5.3 now yield

PROPOSITION 2.5.7. *If $N \in {}^*\mathbb{N}\setminus\mathbb{N}$, then the pair $\langle {}^*A_N, {}^*j_N\rangle$ is an HA of the ring Δ_τ. Further, Δ_τ is topologically isomorphic to ${}^*A_N/G_0$, where*

$$G_0 = \{a \in {}^*A_N \mid \forall^{\mathrm{st}} n\ (a_n \mid a)\}. \tag{2.5.6}$$

We now describe the group $\widehat{\Delta}_\tau$ (see [**29**]). Let $\mathbb{Q}^{(\tau)} = \{\frac{m}{a_n} \mid m \in \mathbb{Z},\ n \in \mathbb{N}\}$. Since $a_n \mid a_m$ for $n \le m$, it follows that $\mathbb{Q}^{(\tau)}$ is a subgroup of the additive group $\mathbb{Q}$. Obviously, $\mathbb{Z} \subseteq \mathbb{Q}^{(\tau)}$. Let $\mathbb{Z}^{(\tau)} = \mathbb{Q}^{(\tau)}/\mathbb{Z}$. It is known that $\widehat{\Delta}_\tau \cong \mathbb{Z}^{(\tau)}$. To describe the isomorphism we introduce the following notation. If $\xi = \langle \xi_n \mid n \in \mathbb{N}\rangle \in \Delta_\tau$, then we write that $\xi_n = \operatorname{rem}(\xi, a_n)$. This notation agrees with the case when $\xi = \nu(a)$ for $a \in \mathbb{Z}$: $\operatorname{rem}(a, a_n) = \operatorname{rem}(\nu(a), a_n)$. In what follows, $\nu(a)$ is identified with a, that is, it is assumed that $\mathbb{Z} \subseteq \Delta_\tau$. Then the equality $\xi = \eta a_n + \operatorname{rem}(\xi, a_n)$ holds in Δ_τ for some $\eta \in \Delta_\tau$. Let $\{\xi/a_n\} = (\operatorname{rem}(\xi, a_n))/a_n \in \mathbb{Q}^{(\tau)}$. Then it is easy to verify the formulas $\{C\xi/a_n\} \equiv C\{\xi/a_n\} \pmod{\mathbb{Z}}$ and $\{\xi/a_n + \eta\} = \{\xi/a_n\}$, where $C \in \mathbb{Z}$ and $\eta \in \Delta_\tau$. If now (C/a_n) is the class of the element $C/a_n \in \mathbb{Q}^{(\tau)}$ in $\mathbb{Z}^{(\tau)}$, then $\chi_{(C/a_n)} \in \widehat{\Delta}_\tau$ is given by the formula

$$\chi_{(C/a_n)}(\xi) = \exp(2\pi i\{\xi/a_n\}), \qquad \xi \in \Delta_\tau.$$

It is now easy to describe the imbedding $\widehat{\psi}_n \colon \widehat{A}_n \longrightarrow \widehat{\Delta}_\tau$. In our case $\widehat{A}_n$ is isomorphic to A_n. The isomorphism assigns to each m the character $\chi_m \in \widehat{A}_n$ given by $\chi_m(a) = \exp(2\pi i m a/a_n)$, $a \in A_n$. Hence, $\widehat{\psi}_n(\chi_m)(\xi) = \chi_m(\psi_n(\xi)) = \chi_n(\mathrm{rem}(\xi, a_n)) = \exp(2\pi i m \,\mathrm{rem}(\xi, a_n)/a_n) = \exp(2\pi i\{m\xi/a_n\})$. Thus, after the obvious identifications we can assume that $\widehat{\psi}_n \colon A_n \longrightarrow \mathbb{Z}^{(\tau)}$ is given by the formula $\widehat{\psi}_n(m) = (m/a_n)$. It now follows from Theorem 2.4.16 that for $N \in {}^*\mathbb{N} \setminus \mathbb{N}$ the pair $\langle {}^*A_N, {}^*\widehat{\psi}_N\rangle$ is an HA of $\widehat{\Delta}_\tau$ dual to $\langle {}^*A_N, {}^*j_N\rangle$, and the next result follows from the relations (2.4.8).

PROPOSITION 2.5.8. *If $N \in {}^*\mathbb{N} \setminus \mathbb{N}$ and $f \colon \Delta_\tau \longrightarrow \mathbb{C}$ is a standard bounded function that is continuous almost everywhere with respect to the Haar measure μ_τ ($\mu_\tau(\Delta_\tau) = 1$), then*

$$\sum_{k=0}^{{}^*a_N - 1} \left| \int_{{}^*\Delta_\tau} {}^*f(\xi) \exp(-2\pi i\{k\xi/{}^*a_N\})\, d\mu_\tau(\xi) - a_N^{-1} \sum_{n=0}^{{}^*a_N - 1} {}^*f(n) \exp(-2\pi i k n/{}^*a_N) \right|^2 \approx 0. \tag{2.5.7}$$

REMARK 2.5.9. Proposition 2.4.13 shows that for a function f satisfying the conditions of Proposition 2.5.8

$$\int_{\Delta_\tau} f(\xi)\, d\mu_\tau(\xi) = {}^\circ\left(\frac{1}{a_N} \sum_{n=0}^{{}^*a_N - 1} {}^*f(n)\right),$$

which is equivalent to the standard equality

$$\int_{\Delta_\tau} f(\xi)\, d\mu_\tau(\xi) = \lim_{N\to\infty} \frac{1}{a_N} \sum_{n=0}^{a_N - 1} f(n).$$

For continuous functions we have the more general equality

$$\int_{\Delta_\tau} f(\xi)\, d\mu_\tau(\xi) = \lim_{N\to\infty} \frac{1}{N} \sum_{n=0}^{N-1} f(n). \tag{2.5.8}$$

This follows from the strict ergodicity of the shift by 1 in Δ_τ, which follows in turn from the fact that $\chi(1) \neq 1$ for any nontrivial character $\chi \in \widehat{\Delta}_\tau$ ([**21**], Chapter 4, §1, Theorem 1). For the case of the sequence $\tau = \{(n+1)! \mid n \in \mathbb{N}\}$ the equality (2.5.8) was obtained earlier in works on analytic number theory for a certain broader class of functions that includes all the functions satisfying Proposition 2.5.8 (see, for example, [**82**]). This result shows that the condition of boundedness and continuity almost everywhere of f is not necessary for the function ${}^*f \circ j$ to be a lifting of f (here f is defined on a compact Abelian group $\mathfrak{G}$, and $\langle G, j\rangle$ is an HA of $\mathfrak{G}$).

If τ is taken to be the sequence $\{p^{n+1} \mid n \in \mathbb{N}\}$, then Δ_τ is the ring $\mathbb{Z}_p$ of p-adic integers. Thus (Proposition 2.5.7),

$$\mathbb{Z}_p \cong K_p/K_{p0}, \tag{2.5.9}$$

where $K_p = {}^*\mathbb{Z}/p^N\,{}^*\mathbb{Z}$, $K_{p0} = \{a \in K_p \mid \forall^{\mathrm{st}} n\ (p^n \mid a)\}$, and $N \sim +\infty$.

An important role in number theory is played by the ring $\Delta_{\{(n+1)!|n\in\mathbb{N}\}}$, which we denote simply by Δ. It is sometimes called the ring of polyadic integers. Proposition 2.5.7 shows that

$$\Delta \cong K/K_0, \tag{2.5.10}$$

where $K = {}^*\mathbb{Z}/N!{}^*\mathbb{Z}$, $K_0 = \{a \in K \mid \forall^{\mathrm{st}} n\ (n \mid a)\}$, and $N \sim +\infty$.

To illustrate an application of the HA we give a simple proof that

$$\Delta \cong \prod_{p\in P} \mathbb{Z}_p,$$

where P is the set of all prime numbers.

Let $\Delta' = \prod_{p\in P} \mathbb{Z}_p$. Then by the transfer principle, ${}^*\Delta' = \prod_{p\in P} {}^*\mathbb{Z}_p$. If $\xi = \langle \xi_p \mid p \in {}^*P\rangle \in {}^*\Delta'$, then it follows from the nonstandard properties of the product topology (see, for example, [**15**]) that

$$\xi \overset{\Delta'}{\approx} 0 \Longleftrightarrow \forall p \in P\ \bigl(\xi_p \overset{\mathbb{Z}_p}{\approx} 0\bigr). \tag{2.5.11}$$

Let K be defined as in (2.5.10), and let $P_N = \{p \in {}^*P \mid p \leq N\}$, so that $P \subseteq P_N \subseteq {}^*P$. For any $p \in P_N$ let $N_p = [N/p] + [N/p^2] + \cdots$. Then $N! = \prod_{p\in P_N} p^{N_p}$, and N_p is infinite for standard p. We define a mapping $j\colon K \longrightarrow {}^*\Delta'$ by

$$j(k) = \begin{cases} \mathrm{rem}(k, p^{N_p}) & \text{for } p \in P_N, \\ 0 & \text{for } p \notin P_N, \end{cases}$$

$\forall k \in K$ (it can be assumed that $\mathrm{rem}(k, p^{N_p}) \in {}^*\mathbb{Z} \subseteq {}^*\mathbb{Z}_p$). In view of (2.5.11) it is trivial to verify that $j(k_1 + k_2) \overset{\Delta'}{\approx} j(k_1) + j(k_2)$, $j(k_1k_2) \overset{\Delta'}{\approx} j(k_1)j(k_2)$, and $j(-k) \overset{\Delta'}{\approx} -j(k)$. We show that $\forall \eta \in \Delta'\ \exists k \in K\ (j(k) \approx \eta)$. Let $\eta = \langle \eta_p \mid p \in P\rangle$. For any $p \in P_N$ we set $a_p = \mathrm{rem}({}^*\eta_p, p^{N_p})$. By the Chinese remainder theorem, $\exists! k \in K\ \forall p \in P_N\ (k \equiv a_p \pmod{p^{N_p}})$. If $j(a) = \xi = \langle \xi_p \mid p \in {}^*P\rangle$, then $\forall^{\mathrm{st}} p$ $(\xi_p = \mathrm{rem}(k, p^{N_p}) = a_p = \mathrm{rem}({}^*\eta_p, p^{N_p}))$, that is, $\xi_p \approx {}^*\eta_p$, and $\xi \approx {}^*\eta$ by virtue of (2.5.11). Thus, $\langle K, j\rangle$ is an HA of Δ'. We show that $j^{-1}(\mu_{\Delta'}(0)) = K_0$. Obviously, $j(k) \overset{\Delta'}{\approx} 0 \Longleftrightarrow \forall^{\mathrm{st}} p\ \bigl(\mathrm{rem}(k, p^{N_p}) \overset{\mathbb{Z}_p}{\approx} 0\bigr) \Longleftrightarrow \forall^{\mathrm{st}} p\ \forall^{\mathrm{st}} m\ (p^m \mid \mathrm{rem}(k, p^{N_p})) \Longleftrightarrow \forall^{\mathrm{st}} p\ \forall^{\mathrm{st}} m\ (p^m \mid k)$ (recall that $N_p \sim +\infty$) $\Longleftrightarrow \forall^{\mathrm{st}} n\ (n \mid k) \Longleftrightarrow k \in K_0$.

Using (2.5.10) and Proposition 2.5.7, we get what is required.

5°. Here we study hyperfinite approximation of the τ-adic solenoid. Recall ([**29**], [**14**]) that the τ-adic solenoid Σ_τ can be represented as $[0,1) \times \Delta_\tau$ with the group operation $+_\tau$ defined by

$$\langle x, \xi\rangle +_\tau \langle y, \eta\rangle = \langle \{x + y\}, \xi + \eta + [x + y]\rangle.$$

The topology in Σ_τ is given by the system $\{V_n \mid n \in \mathbb{N}\}$ of neighborhoods of zero, where

$$\begin{aligned} V_n = \{\langle x, \xi\rangle \mid 0 \leq x < 1/a_n,\ \forall k \leq n\ (\mathrm{rem}(x, a_k) = 0)\} \\ \cup \{\langle x, \xi\rangle \mid 1 - 1/a_n < x < 1,\ \forall k \leq n\ (\mathrm{rem}(x+1, a_k) = 0)\} \end{aligned}$$

(recall that $\tau = \{a_n \mid n \in N\}$).

Thus, if $\langle x, \xi\rangle \in {}^*\Sigma_\tau$, then

$$\langle x, \xi\rangle \overset{\Sigma_\tau}{\approx} 0 \Longleftrightarrow x \approx 0 \wedge \xi \overset{\Delta_\tau}{\approx} 0 \vee x \approx 1 \wedge \xi + 1 \overset{\Delta_\tau}{\approx} 0. \tag{2.5.12}$$

To construct an HA of Σ_τ we fix an $N \in {}^*\mathbb{N} \setminus \mathbb{N}$, let $G = {}^*\mathbb{Z}/{}^*a_N^2{}^*\mathbb{Z} = \{0, 1, \ldots, {}^*a_N^2 - 1\}$, and define $j\colon G \longrightarrow \Sigma_\tau$ by setting

$$j(a) = \langle \{a/a_N\}, [a/a_N] \rangle,$$

$\forall a \in G$.

As in the preceding subsection, it is assumed that $\mathbb{Z} \subseteq \Delta_\tau$. Moreover, note that $[a/a_N] < a_N$. We show that $\langle G, j\rangle$ is an HA of Σ_τ.

Since $a +_G b \equiv a + b \pmod{a_N^2}$, we have $a +_G b \equiv a + b \pmod{a_N}$, so

$$\begin{aligned}\{(a +_G b)/a_N\} &= (\operatorname{rem}(a +_G b, a_N))/a_N \\ &= (\operatorname{rem}(a + b, a_N))/a_N = \{\{a/a_N\} + \{b/a_N\}\}.\end{aligned}$$

We show that $[(a+_G b)/a_N] \overset{\Delta_\tau}{\approx} [(a+b)/a_N] = [a/a_N]+[b/a_N]+[\{a/a_N\}+\{b/a_N\}]$. In view of (2.5.12) this implies that $j(a+_G b) \overset{\Sigma_\tau}{\approx} j(a)+j(b)$. Let $a = q(a)a_N + r(a)$ and $b = q(b)a_N + r(b)$, that is, $q(a) = [a/a_N]$ and $q(b) = [b/a_N]$, and let $a + b = q(a+b)a_N + r(a+b)$. If $q(a+b) = sa_N + r$, then $a+b = sa_N^2 + ra_N + r(a+b)$, and $ra_N + r(a+b) \le (a_N - 1)a_N + a_N - 1 = a_N^2 - 1$. Thus, $a +_G b = \operatorname{rem}((a+b), a_N^2) = ra_N + r(a+b)$, and hence $q(a +_G b) = r$, therefore, $q(a+b) \equiv q(a +_G b) \pmod{a_N}$. Since $a_n \mid a_N$ $\forall^{\mathrm{st}} n$, it follows that $q(a+b) \approx q(a +_G b)$, as required.

To prove the relation $j(-_G a) \approx j(a)$ just represent a in the form $a = qa_N + r$, use the fact that $-_G a = a_N^2 - a$, and treat the two cases $r = 0$ and $r \ne 0$.

To show that $\langle G, j\rangle$ is an HA of the group Σ_τ it remains to prove (according to Definition 2.4.1) that $\forall \langle x, \xi\rangle \in \Sigma_\tau\ \exists a \in G\ (j(a) \approx \langle x, \xi\rangle)$. We choose an $r < a_N$ with $r/a_N \le x < (r+1)/a_N$ and set $q = \operatorname{rem}({}^*x, a_N)$. Then $q < a_N$, and we get what is required by taking $a = qa_N + r$. We have thereby constructed a hyperfinite approximation of Σ_τ. According to Proposition 2.5.7 and the equivalence (2.5.12), $\Sigma_\tau \cong G/G_0$, where

$$G_0 = \{a \in G \mid \{a/a_N\} \approx 0\ \wedge\ [a/a_N] \overset{\Delta_\tau}{\approx} 0\ \vee\ \{a/a_N\} \approx 1\ \wedge\ [a/a_N] + 1 \overset{\Delta_\tau}{\approx} 0\}.$$

We can also give a different, more concise, description of G_0.

LEMMA 2.5.10. $G_0 = \{a \in G \mid \forall^{\mathrm{st}} n\ (\exp(2\pi i a/(a_N a_n)) \approx 1)\}$.

PROOF. Let $\exp(2\pi i a/(a_N a_n)) \approx 1$ $\forall^{\mathrm{st}} n$. Then $\forall^{\mathrm{st}} n\ \exists k \in \mathbb{Z}\ (a/(a_N a_n) \approx k)$. If $a = qa_N + r$, then $a/(a_N a_n) = (q + r/a_N)/a_n \approx k \in {}^*\mathbb{Z}$. Since a_n is standard, it follows that $q + r/a_N \approx ka_n \in {}^*\mathbb{Z}$, where $q \in {}^*\mathbb{Z}$ and $0 \le r/a_N < 1$, so that the two cases $r/a_N \approx 0$ and $r/a_N \approx 1$ are possible. In the first case $q \approx ka_n$, and $q = ka_n$ because both these numbers are integers. Therefore, $q \overset{\Delta_\tau}{\approx} 0$. Thus, $r/a_N \approx 0$ and $q \overset{\Delta_\tau}{\approx} 0$, hence $a \in G_0$. In the second case (that is, when $r/a_N \approx 1$) if $q \not\equiv -1 \pmod{a_n}$, then $q + 1 = ta_n + s$, where $0 < s < a_n$, and therefore $q + r/a_N = s - 1 + ta_n + r/a_N \approx s + ta_n \approx ka_n$. Then $s/a_n + t \approx k$, which is impossible if $s \ne 0$. Thus, $q \equiv -1 \pmod{a_n}$, and again $a \in G_0$.

Suppose, conversely, that $a \in G_0$, that is, $a = qa_N + r$. Two cases should be considered: 1) $r/a_N \approx 0$, $q \overset{\Delta_\tau}{\approx} 0$; 2) $r/a_N \approx 1$, $q \overset{\Delta_\tau}{\approx} 0$. In the first case $\exp(2\pi i a/(a_N a_n)) = \exp(2\pi i(a/a_n + r/(a_N a_n))) = \exp(2\pi i a/(a_N a_n)) \approx 1$, because $q/a_n \in {}^*\mathbb{Z}$. The second case is treated similarly.

Finally, we get the following result.

PROPOSITION 2.5.11. *If τ is a standard sequence $\{a_n \mid n \in \mathbb{N}\}$ of natural numbers with $a_n > 1$ and $a_n \mid a_{n+1}$, $G = \{0, 1, \ldots, a_N^2 - 1\}$ is the additive group of the ring ${}^*\mathbb{Z}/a_N^2\,{}^*\mathbb{Z}$, and $G_0 = \{a \in G \mid \forall^{\mathrm{st}} n \exp(2\pi i a/(a_N a_n)) \approx 1\}$, then the τ-adic solenoid Σ_τ is topologically isomorphic to $G^\# = G/G_0$.*

We now construct a dual HA of the group $\widehat{\Sigma}_\tau$.

It is known that $\widehat{\Sigma}_\tau \cong \mathbb{Q}^{(\tau)} = \{m/a_n \mid m \in \mathbb{Z},\, n \in \mathbb{N}\}$.

The isomorphism associates with each $\alpha = m/a_n \in \mathbb{Q}^{(\tau)}$ the character $\varkappa_\alpha \in \widehat{\Sigma}_\tau$ such that $\forall x \in [0,1]\ \forall \xi \in \Delta_\tau$

$$\varkappa_\alpha(x, \xi) = \exp(2\pi i \alpha(x + \mathrm{rem}(\xi, a_n)))$$

(see [29], [14]).

As in the case of an arbitrary finite group, $\widehat{G} \cong G$ in our case, and the isomorphism associates with each $b \in G$ the character χ_b with $\chi_b(a) = \exp(2\pi i ab/a_N^2)\ \forall a \in G$. As a dual to the HA of Σ_τ constructed above we consider the pair $\langle G, \iota\rangle$ such that $\iota\colon G \longrightarrow {}^*\mathbb{Q}^{(\tau)}$ is defined by $\iota(b) = b/a_N$ (more precisely, $\iota\colon \widehat{G} \longrightarrow \widehat{\Sigma}_\tau$ associates with each $\chi_b \in \widehat{G}$ the character $\varkappa_{b/a_n} \in {}^*\widehat{\Sigma}_\tau$), except that here G is represented as the group of absolutely smallest residues, that is, $G = \{-\frac{1}{2}a_N^2, \ldots, \frac{1}{2}a_N^2 - 1\}$.

It is a trivial matter to verify the conditions of Definition 2.4.1 in this case. In Definition 2.4.15 it is again necessary to check only the second condition, except that here we even have exact equality. Indeed, if $a \in G$ and $a = qa_N + r$, then $[a/a_N] = q < a_N$, that is, $\mathrm{rem}([a/a_N], a_N) = [a/a_N] = q$. Now $\varkappa_{\iota(b)}(j(a)) = \exp(2\pi i (b/a_N)(r/a_N + q)) = \exp(2\pi i ab/a_N^2) = \chi_b(a)$.

Using Theorem 2.4.16 and the relation (2.4.8), we now get

PROPOSITION 2.5.12. *Let $f\colon \Sigma_\tau \longrightarrow \mathbb{C}$ be a bounded function that is continuous almost everywhere with respect to the Haar measure μ_{Σ_τ} ($\mu_{\Sigma_\tau}(\Sigma_\tau) = 1$), and let $N \in {}^*\mathbb{N} \setminus \mathbb{N}$. Then:*

$$1)\quad \int_{\Sigma_\tau} f\, d\mu_\tau = {}^\circ\!\left(\frac{1}{a_N^2} \sum_{k=0}^{a_N^2 - 1} f(\{k/a_N\}, [k/a_N])\right);$$

$$2)\quad \sum_{m=-a_N^2/2}^{a_N^2/2-1} \left| \int_{\Sigma_\tau} f(x, \xi) \exp(-2\pi i (m/a_N)(x + \mathrm{rem}(\xi, a_N)))\, d\mu_{\Sigma_\tau} - \frac{1}{a_N^2} \sum_{k=0}^{a_N^2-1} f(\{k/a_N\}, [k/a_N]) \exp(-2\pi i k m/a_N^2) \right|^2 \approx 0.$$

6°. We construct a hyperfinite approximation of the additive group of the field $\mathbb{Q}_p$ of p-adic numbers (p a standard prime number). Let M and N be fixed, with $M, N, N - M \in {}^*\mathbb{N} \setminus \mathbb{N}$. As a hyperfinite Abelian group G we consider the additive group of the ring ${}^*\mathbb{Z}/p^N\,{}^*\mathbb{Z}$, which is conveniently represented here as the group of smallest positive residues, that is, $G = \{0, 1, \ldots, p^N - 1\}$. We also define a mapping $j\colon G \longrightarrow {}^*\mathbb{Q}_p$ by setting $j(n) = n/p^M \in {}^*\mathbb{Q} \subseteq \mathbb{Q}_p\ \forall n \in G$.

LEMMA 2.5.13. *If $n \in G$, then*

$$j(n) \in \mathrm{Ns}({}^*\mathbb{Q}_p) \iff \exists^{\mathrm{st}} k \in \mathbb{N}\ (p^{M-k} \mid n).$$

In this case if

$$n = a_{-k}p^{M-k} + a_{-k+1}p^{M-k+1} + \cdots + a_{N-M-1}p^{N-1}, \tag{2.5.13}$$

where $0 \le a_i < p$, *then*[2]

$$\widetilde{j}(n) = \mathrm{st}(j(n)) = \sum_{i=-k}^{\infty} a_i p^i. \tag{2.5.14}$$

PROOF. If $p^{M-k} \mid n$, then n has the form (2.5.13), and the equality (2.5.14) follows from the fact that $N - M - 1$ is an infinite number. Conversely, suppose that $np^{-M} \approx \xi \in \mathbb{Q}_p$ and $|\xi|_p = p^k$, where $k \in \mathbb{Z}$. Since $|np^{-M} - \xi|_p \approx 0$, there exists an infinite $b \in {}^*\mathbb{N}$ such that $np^{-M} - \xi = p^b\varepsilon_1$, where ε_1 is a unit of the ring ${}^*\mathbb{Z}_p$. By assumption, $\xi = p^{-k}\varepsilon_2$, where ε_2 is a unit of the ring $\mathbb{Z}_p$. This implies that $n = p^{M-k}\varepsilon_2 + p^{M+b}\varepsilon_1$. Since k is standard, it follows that $M - k < M + b$, that is, $p^{M-k} \mid n$ in ${}^*\mathbb{Z}_p$, and hence also in $\mathbb{Z}$. □

PROPOSITION 2.5.14. *The pair* $\langle G, j\rangle$ *is an HA of the additive group* $\mathbb{Q}_p^+$ *of the field* $\mathbb{Q}_p$. *Further,*

$$\begin{aligned} G_f &= \{n \in G \mid \exists^{\mathrm{st}} k\ (p^{M-k} \mid n)\}, \\ G_0 &= \{n \in G \mid \forall^{\mathrm{st}} k\ (p^{M+k} \mid n)\}. \end{aligned} \tag{2.5.15}$$

PROOF. The equalities (2.5.15) follow immediately from Lemma 2.5.13. To verify the conditions of Definition 2.4.1 it suffices to show that $\widetilde{j}\colon G_f \longrightarrow \mathbb{Q}_p$ is a surjective homomorphism. We denote by $\oplus$ the addition operation in G. Let $n = n_1 \oplus n_2$, that is, let $n_1 + n_2 = n + tp^M$. Then $|n_1p^{-M} + n_2p^{-M} - np^{-M}|_p \le p^{N-M}$, and hence $\widetilde{j}(n_1 + n_2) = \widetilde{j}(n_1 \oplus n_2)$ because $N - M$ is infinite. Using the fact that $\mathrm{st}\colon \mathrm{Ns}({}^*\mathbb{Q}_p) \longrightarrow \mathbb{Q}_p$ is a homomorphism, we get that $\widetilde{j}(n_1 + n_2) = \widetilde{j}(n_1) + \widetilde{j}(n_2)$. This proves that $\widetilde{j}$ is a homomorphism. To verify that it is surjective we observe that if ξ has the form (2.5.14), then, defining n by the formula (2.5.13), we get that $j(n) \approx \xi$. □

COROLLARY 2.5.15. $\mathbb{Q}_p^+ \cong G_f/G_0 = G^{\#}$.

Let $G^{(0)} = \{n \in G \mid p^M | n\}$. Then $G^{(0)}$ is an internal subgroup of G such that $G_0 \subseteq G^{(0)} \subseteq G_f$. Since $|G^{(0)}| = p^{N-M}$, the number $\Delta = p^{M-N}$ can be taken as a normalizing multiplier of the triple $\langle G, G_0, G_f\rangle$.

COROLLARY 2.5.16. $\widetilde{j}(G^{(0)}) = \mathbb{Z}_p$. *Further, the normalizing multiplier* $\Delta = p^{M-N}$ *induces on* $\mathbb{Q}_p$ *the Haar measure* $\overline{\mu}_\Delta$ *with* $\overline{\mu}_\Delta(\mathbb{Z}_p) = 1$ (*this measure is denoted by* μ_p *below*).

We now write out the standard equivalent of condition (2.4.7) for this case.

[2] The notation $\widetilde{j}$ was introduced after Definition 2.4.1.

LEMMA 2.5.17. *The function* $f\colon \mathbb{Q}_p \longrightarrow \mathbb{C}$ *satisfies the condition* (2.4.7) *if and only if*

$$\lim_{m,n\to\infty} \frac{1}{p^n} \sum_{\substack{0\le k<p^{m+n+l} \\ p^l \nmid k}} |f(k/p^{m+l})| = 0 \tag{2.5.16}$$

uniformly with respect to l.

PROOF. If $B \subseteq G\setminus G_f$, then $\forall n \in B\ \forall^{\mathrm{st}} k\ (p^{M-k} \nmid L)$. This condition is equivalent to the condition that $p^{M-L} \nmid n$ for some infinite L. Thus, $B \subseteq G\setminus B_f$ if and only if $B \subseteq \{n \mid p^{M-L} \nmid n\} = B_L$, but B_L is an internal set for any infinite L. Hence, condition (2.4.7) can be reformulated as follows: for any infinite $L < M < N$ such that $N - M$ is infinite,

$$\frac{1}{p^{N-M}} \sum_{\substack{0\le k<p^N \\ p^{M-L} \nmid k}} |{}^*f(k/p^M)| \approx 0.$$

Denoting $N - M$ by n, L by m, and $M - L$ by l, we get that for any infinite m and n and for any l

$$\frac{1}{p^n} \sum_{\substack{0\le k<p^{m+n+l} \\ p^l \nmid k}} |{}^*f(k/p^{m+l})| \approx 0.$$

This condition is precisely equivalent to (2.5.16).

COROLLARY 2.5.18. *If the function* $f\colon \mathbb{Q}_p \longrightarrow \mathbb{C}$ *is bounded and continuous almost everywhere with respect to Haar measure and if it satisfies* (2.5.16), *then it is integrable, and*

$$\int_{\mathbb{Q}_p} f\,d\mu_p = {}^\circ\!\left(\frac{1}{p^{N-M}} \sum_{0\le k<p^N} {}^*f(k/p^M)\right),$$

or, in standard terms,

$$\int_{\mathbb{Q}_p} f\,d\mu_p = \lim_{m,n\to\infty} \frac{1}{p^n} \sum_{0\le k<p^{m+n}} {}^*f(k/p^m).$$

We construct a dual HA of the group $\widehat{\mathbb{Q}}_p^+$. Recall that $\widehat{\mathbb{Q}}_p^+ \cong \mathbb{Q}_p^+$. The isomorphism associates with each $\xi \in \mathbb{Q}_p^+$ the character $\varkappa_\xi(\eta) = \exp(2\pi i\{\xi\eta\})$, where $\{\zeta\}$ is the fractional part of the p-adic number ζ. Identifying $\widehat{\mathbb{Q}}_p^+$ and $\mathbb{Q}_p^+$ in this way, we construct an HA $\langle G, \widehat{j}\rangle$ of the group $\mathbb{Q}_p$ such that $\widehat{j}(n) = n/p^{N-M}$ $\forall n \in G$. The fact that this really is an HA was established in Proposition 2.5.14. To verify the fact that $\langle G, \widehat{j}\rangle$ is dual to $\langle G, j\rangle$ we must check both the conditions in Definition 2.4.15. Checking the second condition is quite trivial here: $\varkappa_{j(m)}(\widehat{j}(n)) = \exp(2\pi i\{\widehat{j}(n)j(m)\}) = \exp(2\pi i\{nm/p^N\}) = \exp(2\pi inm/p^N) = \chi_m(n)$. To verify the first condition we must show that $\forall m\ (\exists^{\mathrm{st}} k\ (p^{M-k} \mid m) \longrightarrow \exp(2\pi imn/p^N) \approx 1) \Longrightarrow \forall^{\mathrm{st}} k\ (p^{N-M+k} \mid n)$. If this is not so, then there is a k such that $n = qp^{N-M+k} + r$ and $0 < r < p^{N-M+k}$. Two cases are possible:

1) ${}^\circ(r/p^{N-M+k}) = 0$. Let $a = [p^{N-M+k}/(2r)]$ and $m = ap^{M-k}$ (obviously, $m < p^N$). Then $\exp(2\pi imn/p^N) = \exp(2\pi i[p^{N-M+k}/(2r)](r/p^{N-M+k})) \approx \exp \pi i = -1$, which is impossible.

2) ${}^\circ(r/p^{N-M+k}) = \alpha$, $0 < \alpha \leq 1$. Setting $m = p^{M-k-1}$, we get that $\exp(2\pi imn/p^N) \approx \exp(2\pi i\alpha/p) \neq 1$, and again we arrive at a contradiction. Thus, $\langle G, \widehat{j}\rangle$ is an HA dual to $\langle G, j\rangle$, and we get the next result from Theorem 2.4.16 and the relation (2.4.8).

PROPOSITION 2.5.19. *For the FT* $\mathcal{F}\colon L_2(\mathbb{Q}_p) \longrightarrow L_2(\mathbb{Q}_p)$ *with* $\mathcal{F}(f)(\xi) = \int_{\mathbb{Q}_p} f(\eta)\exp(-2\pi i\{\xi\eta\})\, d\mu_p(\eta)$ *let* $f \in L_2(\mathbb{Q}_p)$ *be such that* $|f|^2$ *and* $|\mathcal{F}(f)|^2$ *are bounded and continuous almost everywhere, and suppose that they satisfy* (2.5.16). *Then*

$$\frac{1}{p^M}\sum_{k=0}^{p^N-1}\Bigg|\int_{\mathbb{Q}_p} f(\eta)\exp(-2\pi i\{\eta k/p^{N-M}\})\, d\mu_p(\eta) - \frac{1}{p^{N-M}}\sum_{n=0}^{p^N-1} f(n/p^M)\exp(-2\pi ikn/p^N)\Bigg|^2 \approx 0$$

if $N, M, N-M \in {}^*\mathbb{N}\setminus\mathbb{N}$.

REMARK 2.5.20. It is easy to see that in contrast to the case of $\mathbb{Z}_p$ the mapping constructed here does not approximate multiplication in $\mathbb{Q}_p$. Indeed, let $m, n \in G$ be $m = p^{M-1}$ and $n = p^{M+1}$. Then $j(m) = p^{-1}$ and $j(n) = p$, that is, $j(m)j(n) = 1$. Two cases are possible here.

1) $2M \leq N$. In this case $j(mn) = p^M \approx 0$.

2) $2M > N$. Then since we are dealing with the multiplication in $\mathbb{Z}/p^N\mathbb{Z}$, it follows that $mn = p^{M-(N-M)}$, and hence $mn \notin G_f$ because $N - M$ is infinite.

It can be shown similarly that, whatever Δ, the HA in 1° of the additive group of the field $\mathbb{R}$ is not an approximation of the multiplication in $\mathbb{R}$.

7°. The field of p-adic numbers is generalized by the ring $\mathbb{Q}_\alpha$ of α-adic numbers, where $\alpha = \{a_n \mid n \in \mathbb{Z}\}$ is a doubly infinite sequence of natural numbers such that $a_n \mid a_{n+1}$ for $n \geq 0$ and $a_{n+1} \mid a_n$ for $n < 0$. This ring is described in detail in §3.7 of Chapter 2 in the book [**14**]. It is shown in [**29**] that the group $\widehat{\mathbb{Q}_\alpha^+}$ is isomorphic to $\mathbb{Q}_{\widehat{\alpha}}^+$, where $\widehat{\alpha}(n) = \alpha(-n)$.

PROPOSITION 2.5.21. *Let* $M, N \in {}^*\mathbb{N}\setminus\mathbb{N}$, *let* $G = \{0, 1, \dots, {}^*a_M{}^*a_N - 1\}$ *be the additive group of the ring* ${}^*\mathbb{Z}/{}^*(a_M a_N){}^*\mathbb{Z}$, *and let* $j\colon G \longrightarrow \mathbb{Q}_\alpha^+$ *and* $\widehat{j}\colon G \longrightarrow \mathbb{Q}_{\widehat{\alpha}}^+$ *be the mappings* $j(a) = aa_{-M}^{-1}$ *and* $\widehat{j}(a) = aa_N^{-1}$ $\forall a \in G$. *Then the pair* $\langle G, j\rangle$ *is an HA of the group* $\mathbb{Q}_\alpha^+$, *and the pair* $\langle \widehat{G}, \widehat{j}\rangle$ *is an HA of* $\mathbb{Q}_{\widehat{\alpha}}^+$ *dual to it. Further,* $G_f = \{a \in G \mid \exists^{\mathrm{st}} k \in \mathbb{Z}\ (a_{-M}a_k^{\operatorname{sgn} k} \mid a)\}$, $G_0 = \{a \in G \mid \forall^{\mathrm{st}} k \in \mathbb{Z}\ (a_{-M}a_k^{\operatorname{sgn} k} \mid a)\}$, *and* $\Delta = a_N^{-1}$ *is a normalizing multiplier of the triple* $\langle G, G_0, G_f\rangle$. *Similar definitions apply to* $\widehat{G}_f$, $\widehat{G}_0$, *and the normalizing multiplier* $\widehat{\Delta} = a_{-M}^{-1}$ *of the triple* $\langle G, \widehat{G}_0, \widehat{G}_f\rangle$.

The proof of this proposition is analogous to that of the corresponding assertions in the preceding subsection. It is also easy to get assertions analogous to Lemma 2.5.17, Corollary 2.5.18, and Proposition 2.5.19. It is easy to prove (as it was proved above that Δ and $\prod_p \mathbb{Z}_p$ are isomorphic) that if $\alpha(n) = (|n|!)^{(-1)^n}$, then $\mathbb{Q}_\alpha^+$ is isomorphic to the disconnected component of the additive group of the adele ring.

8°. In the conclusion of this chapter we consider briefly the standard versions of the results obtained in it. For those results relating to hyperfinite-dimensional approximations of the Fourier transformation in concrete cases (Theorem 2.1.1 and Propositions 2.5.6, 2.5.8, 2.5.12 2), and 2.5.19) the corresponding standard versions follow immediately from Definition 1.3.15, Proposition 1.3.16, and the remarks after Proposition 1.3.17, therefore, we do not dwell on them but concentrate on the standard version of the definition of a hyperfinite approximation of a topological group.

First of all we note that since the pair $\langle G, j\rangle$ in Definition 2.4.1 is a nonstandard object, Nelson's algorithm cannot be applied to the proposition "$\langle G, j\rangle$ is an HA of the group $\mathfrak{G}$", because it is applicable solely to sentences containing only standard parameters.

Here it is natural to proceed as follows.

DEFINITION 2.5.22. *A standard sequence* $\{\langle G_n, j_n\rangle \mid n \in \mathbb{N}\}$ *with* G_n *a finite Abelian group and* $j_n \colon G_n \longrightarrow \mathfrak{G}$ $\forall n$ *is called an approximating sequence of the separable LCA group* $\mathfrak{G}$ *if* $\langle G_N, j_N\rangle$ *is an HA of* $\mathfrak{G}$ $\forall N \in {}^*\mathbb{N} \setminus \mathbb{N}$.

Nelson's algorithm can now be applied to this definition. We should point out that far from every HA can be obtained in this way, especially if the nonstandard universe ${}^*V(\mathbb{R})$ is not an ultrapower of $V(\mathbb{R})$ with respect to an ultrafilter on $\mathbb{N}$. However, a careful analysis of the proof of Theorem 2.4.5 shows that for any separable LCA group G there is an approximating sequence.

PROPOSITION 2.5.23. *Let* $\{\langle G_n, j_n\rangle \mid n \in \mathbb{N}\}$ *be a sequence of finite Abelian groups* G_n *and mappings* $j_n \colon G_n \longrightarrow \mathfrak{G}$*, where* $\mathfrak{G}$ *is a separable LCA group, and let* $\mathbb{T}_0$ *be a base of relatively compact neighborhoods of zero in* $\mathfrak{G}$*. This sequence is approximating for* $\mathfrak{G}$ *if and only if the following conditions hold:*

1) $\forall \xi \in \mathfrak{G}\ \forall U \in \mathbb{T}_0\ \exists f \in \prod_{n\in\mathbb{N}} G_n\ \exists n_0 \in \mathbb{N}\ \forall n > n_0\ (\xi - j_n(f_n) \in U)$;

2) $\forall$ *compact* $K \subseteq \mathfrak{G}\ \forall U \in \mathbb{T}_0\ \exists m \in \mathbb{N}\ \forall n > m\ \forall g, h \in G_n\ (j_n(g), j_n(h) \in K \longrightarrow (j_n(g+h) - j_n(g) - j_n(h) \in U) \wedge (j_n(g) + j_n(-g) \in U))$.

This proposition can be proved by a simple application of Nelson's algorithm and use of the fact that $\mathrm{Ns}({}^*\mathfrak{G}) = \cup\{{}^*K \mid K \subseteq \mathfrak{G} \text{ is compact}\}$.

DEFINITION 2.5.24. A function $f \colon \mathfrak{G} \longrightarrow \mathbb{C}$ is said to be rapidly decreasing with respect to the approximating sequence $\{\langle G_n, j_n\rangle \mid n \in \mathbb{N}\}$ of the group $\mathfrak{G}$ if for any relatively compact neighborhood $U \subseteq \mathfrak{G}$ of zero and any infinite $N \in {}^*\mathbb{N}$ the condition (2.4.7) holds with $\Delta = |j_N^{-1}({}^*U)| \cdot |G_n|^{-1}$.

In the subsequent propositions $\mathcal{K}$ ($\widehat{\mathcal{K}}$) is the family of all compact subsets of $\mathfrak{G}$ ($\widehat{\mathfrak{G}}$), and $\mathbb{T}_0$ and $\widehat{\mathbb{T}}_0$ are bases of relatively compact neighborhoods of the neutral elements in $\mathfrak{G}$ and $\widehat{\mathfrak{G}}$, respectively.

PROPOSITION 2.5.25. *Let* $\{\langle G_n, j_n\rangle \mid n \in \mathbb{N}\}$ *be an approximating sequence of the separable LCA group* $\mathfrak{G}$*. Then:*

1) *the function* $f \colon \mathfrak{G} \longrightarrow \mathbb{C}$ *is rapidly decreasing with respect to this sequence if and only if*

$$\forall U \in \mathbb{T}_0\ \forall \varepsilon > 0\ \exists n_0 \in \mathbb{N}\ \exists K \in \mathcal{K}\ \forall n > n_0\ \forall B \subseteq j_n^{-1}(\mathfrak{G} \setminus K)$$
$$\frac{1}{|j_n^{-1}(U)|} \sum_{g\in B} |f(j_n(g))| < \varepsilon;$$

2) *for any Haar measure* μ *on* $\mathfrak{G}$ *there is a* $U \in \mathbb{T}_0$ *such that the equality*

$$\int f\,d\mu = \lim_{n\to\infty} \frac{1}{|j_n^{-1}(U)|} \sum_{g\in G_n} |f(j_n(g))|$$

holds for any bounded function $f\colon \mathfrak{G} \longrightarrow \mathbb{C}$ *that is continuous almost everywhere with respect to* μ *and rapidly decreasing with respect to the given sequence.*

The first assertion is obtained by applying Nelson's algorithm to the condition (2.4.7), and the second by applying it to Proposition 2.4.13.

DEFINITION 2.5.26. Suppose that under the conditions of the preceding proposition $\{\langle \widehat{G}_n, \widehat{j}_n\rangle\}$ is an approximating sequence of $\widehat{\mathfrak{G}}$. It is said to be dual to $\{\langle G_n, j_n\rangle\}$ if for any $N \in {}^*\mathbb{N}\setminus\mathbb{N}$ the HA $\langle \widehat{G}_N, \widehat{j}_N\rangle$ is dual to the HA $\langle G_N, j_N\rangle$.

PROPOSITION 2.5.27. *Under the conditions of Definition* 2.5.26 *the sequence* $\{\langle \widehat{G}_n, \widehat{j}_n\rangle\}$ *is dual to* $\{\langle G_n, j_n\rangle\}$ *if and only if the following two conditions hold*:

1) $\forall V \in \widehat{\mathbb{T}}_0\ \exists n_0 \in \mathbb{N}\ \exists K \in \mathcal{K}\ \exists \varepsilon > 0\ \forall n > n_0\ \forall \chi \in \widehat{G}_n$
$$(\forall g \in j_n^{-1}(K)\ (|\chi(g) - 1| < \varepsilon) \longrightarrow \widehat{j}_n(\chi) \in V);$$
2) $\forall K \in \mathcal{K}\ \forall L \in \widehat{\mathcal{K}}\ \forall \varepsilon > 0\ \exists n_0 \in \mathbb{N}\ \forall n > n_0\ \forall g \in j_n^{-1}(K)\ \forall \chi \in \widehat{j}_n^{-1}(L)$
$$(|\widehat{j}_n(\chi)(j_n(g)) - \chi(g)| < \varepsilon).$$

This can be proved by applying Nelson's algorithm to Definition 2.4.15.

PROPOSITION 2.5.28. *Suppose that* $\{\langle G_n, j_n\rangle\}$ *is an approximating sequence of the separable LCA group* $\mathfrak{G}$, $\{\langle \widehat{G}_n, \widehat{j}_n\rangle\}$ *is an approximating sequence of* $\widehat{\mathfrak{G}}$ *dual to it,* μ *is Haar measure on* $\mathfrak{G}$, $U \in \mathbb{T}_0$ *corresponds to* μ *according to Proposition* 2.5.25, 2), *and* $\mathcal{F}\colon L_2(\mathfrak{G}) \longrightarrow L_2(\widehat{\mathfrak{G}})$ *is the FT. In this case if* f *and* $|\mathcal{F}(f)|$ *are bounded and continuous almost everywhere with respect to Haar measure, and* $|f|^2$ *and* $|\mathcal{F}(f)|^2$ *are rapidly decreasing with respect to the given approximating sequences, then*

$$\lim_{n\to\infty} \frac{|j_n^{-1}(U)|}{|G_n|} \sum_{\chi\in\widehat{G}_n} \Bigg| \int_{\mathfrak{G}} f(\xi)\widehat{j}_n(\chi)(\xi)\,d\mu(\xi) - \frac{1}{|j_n^{-1}(U)|} \sum_{g\in G_n} f(j_n(g))\chi(g)\Bigg|^2 = 0.$$

To prove this, apply Nelson's algorithm to Theorem 2.4.16. □

It is not hard to see that the standard definitions obtained here of an approximating sequence and one dual to it are equivalent to those given in the Introduction. By analogous arguments it is easy to show that the equicontinuity of a sequence φ_n of unitary representations of the groups G_n is equivalent to the S-continuity of the unitary representation φ_N of the group G_N for any infinite N. Proposition 1 of the Introduction now follows from Lemma 2.3.12, and Theorem 2 from Proposition 2.3.13 and Remark 2 after Proposition 2.4.2.

Bibliography

1. S. Albeverio, J. Fenstad, R. Høegh-Krohn, and T. Lindstrøm, *Nonstandard methods in stochastic analysis and mathematical physics*, Academic Press, NY, 1986.
2. D. A. Vladimirov, *Boolean algebras*, "Nauk", Moscow, 1969; German transl., Akademie-Verlag, Berlin, 1972.
3. E. I. Gordon, *On a representation of generalized functions by nonstandard finite sequences*, Izv. Vyssh. Uchebn. Zaved. Mat. **1986**, no. 6, 57–60; English transl. in Soviet Math. (Iz. VUZ) **30** (1986).
4. ———, *Nonstandard finite-dimensional analogues of operators in* $L_2(\mathbb{R}^n)$, Sibirsk. Mat. Zh. **29** (1988), no. 2, 45–59; English transl. in Siberian Math. J. **29** (1988).
5. ———, *Relatively standard elements in Nelson's internal set theory*, Sibirsk. Mat. Zh. **30** (1989), no. 1, 89–95; English transl. in Siberian Math. J. **30** (1989).
6. ———, *On the Fourier transformation in nonstandard analysis*, Izv. Vyssh. Uchebn. Zaved. Mat. **1989**, no. 2, 17–25; English transl. in Soviet Math. (Iz. VUZ) **33** (1989).
7. ———, *Hyperfinite approximation of locally compact Abelian groups*, Dokl. Akad. Nauk SSSR **314** (1990), 1044–1047; English transl. in Soviet Math. Dokl. **42** (1991).
8. ———, *Nonstandard analysis and compact Abelian groups*, Sibirsk. Mat. Zh. **32** (1991), no. 2, 26–40; English transl. in Siberian Math. J. **32** (1991).
9. ———, *On Loeb measures*, Izv. Vyssh. Uchebn. Zaved. Mat. **1991**, no. 2, 25–33; English transl. in Soviet Math. (Iz. VUZ) **35** (1991).
10. ———, *The concept of relative standardness and Benninghofen–Richter π-monads*, Optimizatsiya **1990**, no. 48 (65), 58–62. (Russian)
11. ———, *Construction of the axiomatics of nonstandard analysis on the basic of the concept of class*, Proc. Eleventh Interrepublic Conference on Mathematical Logic (Kazan, 6–8 October), Izdat. Kazan. Gos. Univ., Kazan, 1992, p. 43. (Russian)
12. E. I. Gordon and I. A. Korchagina, *Hyperfinite approximation of compact groups and their representations*, Sibirsk Mat. Zh. **36** (1995), no. 2, 308–327; English transl. in Siberian Math. J. **36** (1995).
13. E. I. Gordon, *On approximations of topological groups and their approximations*, Ross. Akad. Nauk Dokl. (1996); English transl. in Russian Acad. Sci. Dokl. Math. (to appear).
14. V. P. Gurariĭ, *Group methods of commutative harmonic analysis*, Itogi Nauki i Tekhniki: Sovremennye Problemy Mat.: Fundamental. Napravleniya, vol. 25, VINITI, Moscow, 1988, pp. 5–311; English transl. in *Encyclopaedia of Math. Sci., vol.* 25 [*Commutative Harmonic Analysis* II], Springer–Verlag, Berlin (to appear).
15. M. Davis, *Applied nonstandard analysis*, Wiley, NY, 1977; English transl. of V. A. Uspenskiĭ's introduction to the Russian transl. ("Mir", 1980) in Selecta Math. Soviet. **5** (1986), 357–369.
16. A. K. Zvonkin and M. A. Shubin, *Nonstandard analysis and singular perturbations of ordinary differential equations*, Uspekhi Mat. Nauk **39** (1984), no. 2, 77–127; English transl. in Russian Math. Surveys **39** (1984).
17. L. V. Kantorovich and G. P. Akilov, *Functional analysis*, 2nd rev. ed., "Nauka", Moscow, 1977; English transl., Pergamon Press, Oxford, 1982.
18. V. G. Kanoveĭ, *Undecidable hypotheses in Nelson's internal set theory*, Uspekhi Mat. Nauk **46** (1991), no. 6 (282), 3–50; English transl. in Russian Math. Surveys **46** (1991).
19. C. C. Chang and H. J. Keisler, *Model theory*, North-Holland, Amsterdam, American Elsevier, NY, 1973.
20. A. G. Kusraev and S. S. Kutateladze, *Nonstandard methods of analysis*, "Nauka", Novosibirsk, 1990; English transl., Kluwer, Dordrecht, 1994.

21. I. P. Kornfel′d [Cornfeld], Ya. G. Sinaĭ, and S. V. Fomin, *Ergodic theory*, "Nauka", Moscow, 1980; English transl., Springer-Verlag, Berlin, 1982.
22. V. È. Lyantse and T. S. Kudrik, *On functions of a discrete variable. I. Discrete differentiation and integration*, Manuscript No. 274-Uk.87, deposited at NIINTI (01/08/87), Lvov, 1987. (Russian)
23. ———, *On functions of a discrete variable. II. Cauchy problems and boundary value problems*, Manuscript No. 275-Uk.87, deposited at NIINTI (01/08/87), Lvov, 1987. (Russian)
24. V. A. Molchanov, *On the use of iterated nonstandard extensions in topology*, Sibirsk. Mat. Zh. **30** (1989), no. 3, 64–71; English transl. in Siberian Math. J. **30** (1989).
25. Y. Péraire, *A general theory of infinitesimals*, Sibirsk. Mat. Zh. **31** (1990), no. 3, 103–124; English transl. in Siberian Math. J. **31** (1990).
26. L. S. Pontryagin, *Topological groups*, 4th ed., "Nauka", Moscow, 1984; English transl., *Selected works, Vol. 2: Topological groups*, Gordon & Breach, NY, 1986.
27. M. Reed and B. Simon, *Methods of modern mathematical physics*, Vols. 1, 2, Academic Press, NY, 1972, 1975.
28. V. A. Uspenskiĭ, *What is nonstandard analysis?*, "Nauka", Moscow, 1987. (Russian)
29. Edwin Hewitt and Kenneth A. Ross, *Abstract harmonic analysis*, Vol. 1, Springer-Verlag, Berlin, 1963.
30. R. M. Anderson, *Star-finite representations of measure spaces*, Trans. Amer. Math. Soc. **271** (1982), 667–687.
31. B. Bennighofen and M. Richter, *A general theory of superinfinitesimals*, Fund. Math. **128** (1987), no. 3, 199–215.
32. A. R. Bernstein and A. Robinson, *Solution of the invariant subspace problem of K. T. Smith and P. R. Halmos*, Pacific J. Math. **16** (1966), 421–431.
33. A. R. Bernstein and F. Wattenberg, *Nonstandard measure theory*, Applications of Model Theory to Algebra, Analysis, and Probability Theory (Proc. Internat. Symp. on Nonstandard Analysis), Holt, Rinehart, and Winston, NY, 1969, pp. 171–185.
34. E. I. Gordon, *Nonstandard analysis and locally compact Abelian groups*, Acta Appl. Math. **25** (1991), 221–239.
35. N. Cutland, *Nonstandard measure theory and its applications*, Bull. London Math. Soc. **15** (1983), 530–589.
36. M. D. Farukh, *Applications of nonstandard analysis to quantum mechanics*, J. Math. Phys. **16** (1975), no. 2, 177–200.
37. C. W. Henson, *On the nonstandard representation of measures*, Trans. Amer. Math. Soc. **172** (1972), 437–446.
38. ———, *Unbounded Loeb measures*, Proc. Amer. Math. Soc. **74** (1979), 143–150.
39. C. W. Henson and L. C. Moore, *Nonstandard analysis and the theory of Banach spaces*, Nonstandard Analysis: Recent Developments, Lecture Notes in Math., Vol. 983, Springer-Verlag, Berlin, 1983, pp. 27–112.
40. S. Heinrich, *Ultraproducts in Banach space theory*, J. Reine Angew. Math. **313** (1980), 72–104.
41. H. J. Keisler, *An infinitesimal approach to stochastic analysis*, Mem. Amer. Math. Soc. **48** (1984).
42. P. Loeb, *A nonstandard representation of measurable spaces and L_∞*, Bull. Amer. Math. Soc. **77** (1977), 540–544.
43. ———, *Conversion from nonstandard to standard measure spaces and applications in probability*, Trans. Amer. Math. Soc. **211** (1975), 113–122.
44. W. A. J. Luxemburg, *A general theory of monads*, Applications of Model Theory to Algebra, Analysis, and Probability (W. A. J. Luxemburg, ed.), Holt, Rinehart, and Winston, NY, 1969, pp. 18–86.
45. ———, *A nonstandard approach to Fourier analysis*, Contributions to Nonstandard Analysis, North-Holland, Amsterdam, 1972, pp. 16–39.
46. L. C. Moore, *Hyperfinite extensions of bounded operators on a separable Hilbert space*, Trans. Amer. Math. Soc. **218** (1976), 285–295.
47. S. Moore, *Nonstandard analysis and generalized functions*, Rev. Colombiana Mat. **14** (1980), 73–94.
48. E. Nelson, *Internal set theory. A new approach to nonstandard analysis*, Bull. Amer. Math. Soc. **83** (1977), 1165–1198.

49. Y. Péraire, *Une nouvelle théorie des infinitesimaux*, C. R. Acad. Sci. Paris Sér. I **301** (1985), no. 5, 157–159.
50. ———, *Théorie relative des ensemble internes*, Osaka J. Math. **29** (1992), 267–297.
51. K. D. Stroyan, *Superinfinitesimals and inductive limits*, Nonstandard Analysis and its Applications (N. Cutland, ed.), Cambridge Univ. Press, NY, 1988, pp. 298–320.
52. K. D. Stroyan, B. Benninghofen, and M. Richter, *Superinfinitesimals in topology and functional analysis*, Proc. London Math. Soc. **59** (1989), no. 3, 153–181.
53. F. Wattenberg, *Nonstandard measure theory. Hausdorff measure*, Proc. Amer. Math. Soc. **65** (1977), 326–331.
54. ———, *Nonstandard measure theory: Avoiding pathological sets*, Trans. Amer. Math. Soc. **250** (1979), 357–368.
55. A. Tarski, *A decision method of elementary algebra and geometry*, Univ. Calif. Press, Berkeley, 1951.
56. Paul J. Cohen, *Set theory and the continuum hypothesis*, Benjamin, NY, 1966.
57. E. Engeler, *Metamathematik der Elementarmathematik*, Springer-Verlag, Berlin, 1983.
58. S. Feferman, *The number systems. Foundations of algebra and analysis*, Addison-Wesley, Reading, MA, 1963.
59. Yu. I. Manin, *The provable and the unprovable*, "Sov. Radio", Moscow, 1979; English transl. in A Course in Mathematical Logic, Springer-Verlag, Berlin, 1977.
60. I. M. Glazman and Yu. I. Lyubich, *Finite-dimensional linear analysis: A systematic presentation in problem form*, "Nauka", Moscow, 1969; English transl., M.I.T. Press, Cambridge, MA, 1974.
61. A. Robinson, *Nonstandard analysis*, North-Holland, Amsterdam, 1966.
62. K. Hrbáček, *Axiomatic foundations for nonstandard analysis*, Fund. Math. **98** (1978), no. 1, 1–19.
63. D. Ballard and K. Hrbáček, *Standard foundations for nonstandard analysis*, J. Symbolic Logic **57** (1992), 741–748.
64. T. Kawai, *Axiom systems of nonstandard set theory*, Logic Symposia (Hakone, 1979, 1980), Lecture Notes in Math., Vol. 891, Springer-Verlag, Berlin, 1981, pp. 57–65.
65. V. Kanovei, *External power in a theory of internal sets* (to appear).
66. R. Solovay, *A model of set theory in which every set of reals is Lebesgue measurable*, Ann. of Math (2) **92** (1970), 1–56.
67. P. Martin-Löf, *The definition of a random sequence*, Inform. and Control **9** (1966), 602–619.
68. Thomas J. Jech, *Lectures in set theory, with particular emphasis on the method of forcing*, Lecture Notes in Math., Vol. 217, Springer-Verlag, Berlin, 1971.
69. H. Furstenberg, Y. Katznelson, and D. Ornstein, *The ergodic-theoretic proof of Szemerédi's theorem*, Bull. Amer. Math. Soc. **7** (1982), 527–552.
70. V. Bergelson, *Ergodic Ramsey theory*, Comtemporary Math. **57** (1987), 63–87.
71. Frederich P. Greenleaf, *Invariant means on topological groups and their applications*, Van Nostrand Reinholt, NY, 1969.
72. Elliott Mendelson, *Introduction to mathematical logic*, Van Nostrand, Princeton, NJ, 1964.
73. A. A. Bushuev and E. I. Gordon, *Nonstandard theory of classes*, in print.
74. A. Sochor, *Metamathematics of the alternative set theory*. I, II, III, Comment. Math. Univ. Carolin. **20** (1979), 697–722; **23** (1982), 55–79; **24** (1983), 137–154.
75. Ronald L. Graham, *Rudiments of Ramsey theory*, Amer. Math. Soc., Providence, RI, 1981.
76. A. I. Mal′tsev, *Algebraic systems*, "Nauka", Moscow, 1970; English transl., Springer-Verlag, Berlin, 1973.
77. D. McDuff, *On the structure of* II_1 *factors*, Uspekhi Mat. Nauk **25** (1970), no. 6 (156), 29–51; English transl. in Russian Math. Surveys **25** (1970).
78. Petr Vopěnka, *Mathematics in the alternative set theory*, Teubner, Leipzig, 1979.
79. A. M. Vershik and E. I. Gordon, *Groups that are locally imbedded in the class of finite groups*, Algebra i Analiz **8** (1997), no. 1, 70–85; English transl. in St. Petersburg Math. J. **8** (1997).
80. G. M. Fikhtengol′ts, *A course of differential and integral calculus*, Vol. II, 7th ed., "Nauka", Moscow, 1969; German transl. of 4th ed., VEB Deutscher Verlag Wiss., Berlin, 1966.
81. M. A. Naĭmark, *Theory of group representations*, "Nauka", Moscow, 1976; English transl., Springer-Verlag, Berlin, 1981.
82. A. G. Postnikov, *Introduction to analytic number theory*, "Nauka", Moscow, 1971; English transl., Amer. Math. Soc., Providence, RI, 1988.

83. J. Guričan and P. Zlatoš, *Biequivalences and topology in the alternative set theory*, Comment. Math. Univ. Carolin. **26** (1985), 525–552.
84. M. Šmíd and P. Zlatoš, *Biequivalence vector spaces in the alternative set theory*, Comment. Math. Univ. Carolin. **32** (1991), 517–544.
85. J. Nater, P. Pulmann, and P. Zlatoš, *Dimensional compactness in biequivalence vector spaces*, Comment. Math. Univ. Carolin. **33** (1992), 681–688.
86. C. W. Henson and P. Zlatoš, *Indiscernibles and dimensional compactness*, Comment. Math. Univ. Carolin. **37** (1996), 199–203.
87. C. W. Henson, N. J. Kalton, N. T. Peck, and P. Zlatoš, *Some Ramsey type theorems for normed and quasinormed spaces* (to appear).

Selected Titles in This Series

(Continued from the front of this publication)

126 **M. L. Agranovskiĭ,** Invariant function spaces on homogeneous manifolds of Lie groups and applications, 1993

125 **Masayoshi Nagata,** Theory of commutative fields, 1993

124 **Masahisa Adachi,** Embeddings and immersions, 1993

123 **M. A. Akivis and B. A. Rosenfeld,** Élie Cartan (1869–1951), 1993

122 **Zhang Guan-Hou,** Theory of entire and meromorphic functions: deficient and asymptotic values and singular directions, 1993

121 **I. B. Fesenko and S. V. Vostokov,** Local fields and their extensions: A constructive approach, 1993

120 **Takeyuki Hida and Masuyuki Hitsuda,** Gaussian processes, 1993

119 **M. V. Karasev and V. P. Maslov,** Nonlinear Poisson brackets. Geometry and quantization, 1993

118 **Kenkichi Iwasawa,** Algebraic functions, 1993

117 **Boris Zilber,** Uncountably categorical theories, 1993

116 **G. M. Fel′dman,** Arithmetic of probability distributions, and characterization problems on abelian groups, 1993

115 **Nikolai V. Ivanov,** Subgroups of Teichmüller modular groups, 1992

114 **Seizô Itô,** Diffusion equations, 1992

113 **Michail Zhitomirskiĭ,** Typical singularities of differential 1-forms and Pfaffian equations, 1992

112 **S. A. Lomov,** Introduction to the general theory of singular perturbations, 1992

111 **Simon Gindikin,** Tube domains and the Cauchy problem, 1992

110 **B. V. Shabat,** Introduction to complex analysis Part II. Functions of several variables, 1992

109 **Isao Miyadera,** Nonlinear semigroups, 1992

108 **Takeo Yokonuma,** Tensor spaces and exterior algebra, 1992

107 **B. M. Makarov, M. G. Goluzina, A. A. Lodkin, and A. N. Podkorytov,** Selected problems in real analysis, 1992

106 **G.-C. Wen,** Conformal mappings and boundary value problems, 1992

105 **D. R. Yafaev,** Mathematical scattering theory: General theory, 1992

104 **R. L. Dobrushin, R. Kotecký, and S. Shlosman,** Wulff construction: A global shape from local interaction, 1992

103 **A. K. Tsikh,** Multidimensional residues and their applications, 1992

102 **A. M. Il′in,** Matching of asymptotic expansions of solutions of boundary value problems, 1992

101 **Zhang Zhi-fen, Ding Tong-ren, Huang Wen-zao, and Dong Zhen-xi,** Qualitative theory of differential equations, 1992

100 **V. L. Popov,** Groups, generators, syzygies, and orbits in invariant theory, 1992

99 **Norio Shimakura,** Partial differential operators of elliptic type, 1992

98 **V. A. Vassiliev,** Complements of discriminants of smooth maps: Topology and applications, 1992 (revised edition, 1994)

97 **Itiro Tamura,** Topology of foliations: An introduction, 1992

96 **A. I. Markushevich,** Introduction to the classical theory of Abelian functions, 1992

95 **Guangchang Dong,** Nonlinear partial differential equations of second order, 1991

94 **Yu. S. Il′yashenko,** Finiteness theorems for limit cycles, 1991

93 **A. T. Fomenko and A. A. Tuzhilin,** Elements of the geometry and topology of minimal surfaces in three-dimensional space, 1991

92 **E. M. Nikishin and V. N. Sorokin,** Rational approximations and orthogonality, 1991

91 **Mamoru Mimura and Hirosi Toda,** Topology of Lie groups, I and II, 1991

90 **S. L. Sobolev,** Some applications of functional analysis in mathematical physics, third edition, 1991

(See the AMS catalog for earlier titles)